OXFORD SERIES ON MATERIALS MODELLING

Series Editors

Adrian P. Sutton, FRS
Department of Physics, Imperial College London

Robert E. Rudd
Lawrence Livermore National Laboratory

OXFORD SERIES ON MATERIALS MODELLING

Materials modelling is one of the fastest growing areas in the science and engineering of materials, both in academe and in industry. It is a very wide field covering materials phenomena and processes that span ten orders of magnitude in length and more than twenty in time. A broad range of models and computational techniques has been developed to model separately atomistic, microstructural and continuum processes. A new field of multi-scale modelling has also emerged in which two or more length scales are modelled sequentially or concurrently. The aim of this series is to provide a pedagogical set of texts spanning the atomistic and microstructural scales of materials modelling, written by acknowledged experts. Each book will assume at most a rudimentary knowledge of the field it covers and it will bring the reader to the frontiers of current research. It is hoped that the series will be useful for teaching materials modelling at the postgraduate level.

APS, London
RER, Livermore, California

1. M. W. Finnis: *Interatomic forces in condensed matter*
2. K. Bhattacharya: *Microstructure of martensite—Why it forms and how it gives rise to the shape-memory effects*
3. V. V. Bulatov, W. Cai: *Computer simulations of dislocations*

Forthcoming:
A. S. Argon: *Strengthening mechanisms in crystalline solids*
L. P. Kubin, B. Devincre: *Multi-dislocation dynamics and interactions*
T. N. Todorov, M. Di Ventra: *Electrical conduction in nanoscale systems*

Computer Simulations of Dislocations

Vasily V. Bulatov

Lawrence Livermore National Laboratory, University of California

Wei Cai

Department of Mechanical Engineering, Stanford University

OXFORD

UNIVERSITY PRESS

*This book has been printed digitally and produced in a standard specification
in order to ensure its continuing availability*

OXFORD
UNIVERSITY PRESS

Great Clarendon Street, Oxford OX2 6DP

Oxford University Press is a department of the University of Oxford.
It furthers the University's objective of excellence in research, scholarship,
and education by publishing worldwide in

Oxford New York

Auckland   Cape Town Dar es Salaam   Hong Kong   Karachi
Kuala Lumpur Madrid Melbourne Mexico City   Nairobi
New Delhi Shanghai Taipei Toronto
With offices in
Argentina Austria Brazil Chile Czech Republic France Greece
Guatemala Hungary Italy Japan South Korea Poland Portugal
Singapore Switzerland Thailand Turkey Ukraine Vietnam

Oxford is a registered trade mark of Oxford University Press
in the UK and in certain other countries

Published in the United States
by Oxford University Press Inc., New York

ISBN 978-0-19-852614-8

Printed and bound by CPI Antony Rowe, Eastbourne

To Marina and Bei

PREFACE

This book is intended mainly for the students interested in learning the methods of computer simulations applicable in a wide context of materials physics. The other group that can benefit from this text are researchers who are already active in the field of materials simulations who wish to expand their repertoire of computational methods.

The book presents a multitude of methods, some mature and some in their infancy, for computer simulation of crystal dislocations. Unlike most other texts on the methods of computer simulations, here we focus on a single physical object, the dislocation, and use it as a common basis for laying out the ideas behind diverse computational methods, whose applicability ranges well beyond dislocations *per se*. Given the wide range of physical length and time scales straddled by dislocations, we bring in different computational methods to repeatedly come back to the same aspects of dislocation behavior. Our hope is that, by looking at the same problem from different angles, the reader will become more aware of the strengths and weaknesses of various computational approaches. Thus, this book regards crystal dislocations as a microcosm of computational materials sciences.

Even though crystal dislocations are complex and fascinating objects in their own right, we have had to resist the temptation to write more about the physics of dislocation behavior. Our deliberate choice for this book was to focus on the computational details rather than on the physics. The latter topic is arguably more interesting (even to us), but is also much better covered by numerous texts and monographs published over the last 40–50 years, including such classic texts as *Theory of Dislocations* by J. P. Hirth and J. Lothe (1982) and two books still to appear in this series, *Strengthening Mechanisms in Crystal Plasticity* by A. S. Argon and *Mesoscale Simulations of Dislocations and Plasticity* by L. P. Kubin and B. Devincre (These titles are referenced in the main text.). At the same time, a great many methods for computer simulation of dislocations have been developed, but their descriptions are scattered all over the literature. This book attempts to gather some of the most important methods into one text.

The practical aspect of this book is that it is a *how to* text written in the style of numerical recipes. Almost every one of its 46 sections offers a set of problems and many of the sections contain case studies describing what goes into a particular simulation and walking the reader through its steps. The case studies and the problem sets are the main tools we provide to help the reader to develop practical hands-on experience in materials simulation. Since our case studies and problems are computational, we supply a suite of computer programs and utilities that can be used for solving the problems. Thus, every case study and almost every exercise problem has its own sub-directory on the book's web site at http://micro.stanford.edu. For example, Problem 8.3.2 given in the end of Section 3 in Chapter 8, has a corresponding sub-directory 8.3.2 containing the codes (or the links to the codes), the data files and the analysis tools sufficient to obtain a solution to the problem. By supplying the tools, we hope to streamline the learning process and to give the reader a taste of the new methods and their applications without the excessive labor of writing the codes. We also hope that, while running the simulations and analyzing the results, the reader will grow curious enough to learn how a given code or a computational utility does what it does. Ideally, after working through the chapters, we would like the reader to understand the inner workings of a given set of methods sufficiently well to be able to develop her or his own simulations and to adapt the codes to running them. For this, the codes supplied on the book's web site can serve as useful templates.

As stated, our desire is to expose students who are new to computational materials sciences to a relatively wide variety of methods and concepts. At the same time, we have had to be selective in our coverage and limit the book to a reasonable size. But even limiting ourselves to dislocation simulations, the collection of relevant methods is still way too large for a single book. Thus, we further limit our focus to a smaller number of topics that stand a better chance of being explained clearly and in some detail. While our choices are inevitably biased, we hope our selection of the methods is somewhat representative of the field. When several distinct methods exist for addressing the same aspect of dislocation simulations, our choice is usually for a conceptually simpler approach.

There are likely to be various ways to go about reading this book, depending on one's level of knowledge and habits. Our suggestion is to first read the entire text, skipping the exercise problems and subsections marked with * containing more advanced material. The purpose of such a first reading is to obtain a "bird's eye view" of the computational methods presented. Some appreciation of the differences and connections between the various methods discussed in this book is the desired outcome of such a quick reading: having dislocations as the common thread for the entire book should help to appreciate these connections. Even for those more knowledgeable about dislocation simulations and interested in a specific simulation approach presented here, we still recommend reading through the whole book first and only then studying particular topics of interest in all their nitty-gritty details.

The book consists of 46 sections organized into 11 chapters. At the beginning of each chapter we outline the central ideas of the methods to be discussed. Then, each section focuses on one topic, usually a particular simulation method. We try to lay out the relevant numerical algorithms as plainly as possible, often in terms of pseudo codes. The algorithms are illustrated by examples and case studies, followed by a carefully selected set of problems. For those who are serious about learning the methods presented in this book and aspire to apply them in subsequent research, we strongly recommend working through the exercise problems given at the end of each section. The problems are designed to practice the methods just introduced on other (perhaps more complex) applications or to learn about alternative simulation methods. As already mentioned, additional materials, such as source codes and data files, are made available on the companion web site of this book to help the reader work through the case studies and exercise problems. To complement the discussion, we provide references to the literature compiled into a single bibliography at the end of the book. This bibliography is small and far from representative of the field. It also contains a number of references to our own papers that inspired some of the case studies and problem sets. At least three other books in this series, *Interatomic Forces in Condensed Matter* (2003), *Strengthening Mechanisms in Crystal Plasticity* (2007) and *Mesoscale Simulations of Dislocations and Plasticity*, contain useful material complementary to our book: these too are referenced in the text.

Except for Chapter 1, which gives a brief introduction to dislocation physics, the rest of the ten chapters of this book are organized in an ascending order of length and time scales. Chapters 2 through 7 comprise the first part of the book, covering the methods of atomistic simulations of dislocations. The second part contains Chapters 8 through 11, dealing with continuum simulation approaches.

Chapter 2 (Fundamentals of atomistic simulations) opens Part I (Atomistic models) and discusses the basic ideas of atomistic methods presented in a general context of materials simulations irrespective of dislocations. The more detailed discussion of simulation methods and their application to dislocation simulations begins in Chapter 3 (Case study of static simulation) and continues through Chapters 4 (Case study of dynamic simulation), 5 (More about periodic boundary conditions), 6 (Free energy calculations) and 7 (Finding transition pathways). All methods discussed in Part I are applicable in a wide range of situations of interest in computational materials sciences, but are discussed here in the specific context of dislocation simulations.

Part II (Continuum models) contains Chapters 8 through 11. Chapter 8 discusses the so-called Peierls–Nabarro model that is a hybrid atomistic-continuum approach to describing dislocation core properties. Developed specifically for dislocations, this model presents a transparent connection between atomistic and continuum descriptions of dislocations covered in Parts I and II, respectively. The broader appeal of the original Peierls–Nabarro approach is that its ideas inspired some of the more recent developments attempting to bridge the atomistic and continuum descriptions of materials. Chapter 9 (Kinetic Monte Carlo method) presents an approach used to model dislocation behavior over long length and time scales that are not accessible to direct atomistic simulations. Although its application to dislocations has its specific aspects, the kinetic Monte Carlo method is quite general and applicable in a much wider context of materials simulations. The purpose of the line dislocation dynamics method discussed in Chapter 10 (Line dislocation dynamics) is to model the collective behavior of many dislocations. This is a relatively new approach that has garnered enough interest in the materials modeling community to merit an entire book in this series, *Mesoscale Simulations of Dislocations and Plasticity* by Kubin and Devincre. While some of the material given in Chapter 10 is bound to be similar to that in the book by Kubin and Devincre, our approach to line dislocation dynamics is unlike theirs and is more closely connected to the other methods described in our book. The last chapter, Chapter 11 (Phase-field method) presents yet another alternative approach to modeling multiple dislocations that is especially well suited to modeling co-evolution of dislocations and alloy phases.

Acknowledgements

We would like to give special thanks to the rest of the ParaDiS code development team at the Lawrence Livermore National Laboratory, which includes Tom Arsenlis, Maria Bartelt, Gregg Hommes, Masato Hiratani, Tomas Oppestrup, Tim Pierce, Moon Rhee and Meijie Tang. Chapter 10 is mainly a report on the work done by the ParaDiS team over the past four years. We would like to thank Maurice de Koning for help with the free-energy calculations discussed in Section 6.3. We remain greatly indebted to Ali Argon and Sidney Yip for seeding into us many ideas over the years and to Elaine Chandler, Christian Mailhiot and Tomas Diaz de la Rubia for their continued support and encouragement. We also want to thank the series editors Robert Rudd, Adrian Sutton, and Sonke Adlung for their hard work, nearly infinite patience and good will. Finally, a great many thanks to our colleagues who have read the manuscript in various stages of its preparation and given us numerous and very useful suggestions; these include R. E. Rudd, A. P. Sutton, C. R. Weinberger, K. Kang, M. Victoria, W. D. Nix, C. Mailhiot, S. Bulatov, C. R. Krenn, M. Tang, G. A. Galli and F. Gygi.

CONTENTS

1

INTRODUCTION TO CRYSTAL DISLOCATIONS

Dislocations first appeared as an abstract mathematical concept. In the late 19th century, Italian mathematician Vito Volterra examined mathematical properties of singularities produced by cutting and shifting matter in a continuous solid body [1]. As happened to some other mathematical concepts, dislocations could have remained a curious product of mathematical imagination known only to a handful of devoted mathematicians. In 1934, however, three scientists, Taylor, Polanyi and Orowan, independently proposed that dislocations may be responsible for a crystal's ability to deform plastically [2, 3, 4]. While successfully explaining most of the puzzling phenomenology of crystal plasticity, crystal dislocations still remained mostly a beautiful hypothesis until the late 1950s when first sightings of them were reported in transmission electron microscopy (TEM) experiments [5]. Since then, the ubiquity and importance of dislocations for crystal plasticity and numerous other aspects of material behavior have been regarded as firmly established as, say, the role of DNA in promulgating life.

Dislocations define a great many properties of crystalline materials. In addition to a crystal's ability to yield and flow under stress, dislocations also control other mechanical behaviors such as creep and fatigue, ductility and brittleness, indentation hardness and friction. Furthermore, dislocations affect how a crystal grows from solution, how a nuclear reactor wall material is damaged by radiation, and whether or not a semiconductor chip in an electronic device will function properly. It can take an entire book just to describe the various roles dislocations play in materials behavior. However, the focus of this book is on the various computational models that have been developed to study dislocations.

This chapter is an introduction to the basics of dislocations, setting the stage for subsequent discussions of computational models and associated numerical issues. Like any other crystal defect, dislocations are best defined with respect to the host crystal structure. We begin our discussion by presenting in Section 1.1 the basic elements and common terminology used to describe perfect crystal structures. Section 1.2 introduces the dislocation as a defect in the crystal lattice and discusses some of its essential properties. Section 1.3 discusses forces on dislocations and atomistic mechanisms for dislocation motion. The same section also introduces the key tune that is repeatedly played throughout this book, which is the interplay between the atomistic and continuum aspects of dislocations. For a more comprehensive discussion of dislocation physics, we refer to the classical texts of Hull and Bacon, *Introduction to Dislocations* [6], and Hirth and Lothe, *Theory of Dislocations* [7].

1.1 Perfect Crystal Structures

1.1.1 Lattices and Bases

A crystal is a collection of atoms arranged periodically in space. A crystal can be constructed from identical building blocks by replicating them periodically. This building block is called the *basis* (motif) and the way its replicas are repeated in space is specified by the *lattice* (scaffold). Thus, it is fair to say that

$$\text{crystal structure} = \text{lattice} + \text{basis.} \qquad (1.1)$$

A basis may consist of one or more atoms. For example, Fig. 1.1 illustrates a hypothetical two-dimensional crystal structure made of two types of atoms. Its lattice is defined by two repeat vectors **a** and **b** and its basis consists of three atoms. In most crystal structures to be discussed in this book, the basis contains only one or two atoms. However, bases of protein crystals are macromolecules consisting of thousands of atoms. Obviously, an infinite variety of possible bases exist that can be used to build periodic crystal structures.

A lattice is an infinite array of mathematical points that form a periodic spatial arrangement. In 1948, French physicist Auguste Bravais proved that there are only 14 different types of lattices in three dimensions with distinct symmetry properties. These lattices have been called the *Bravais* lattices ever since.[1] For a more detailed account of the Bravais lattices, we refer to the standard textbooks on solid-state theory (e.g. [8, 9, 10]). In this book we will encounter only three Bravais lattice, which all exhibit *cubic* symmetry.

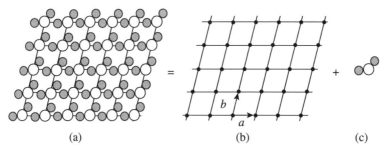

(a)   (b)   (c)

FIG. 1.1. (a) A two-dimensional crystal consisting of two types of atoms (white and gray). (b) The Bravais lattice is specified by two repeat vectors **a** and **b**. (c) The basis contains three atoms.

---

[1] In two dimensions, there are only five *Bravais* lattices.

A Bravais lattice is fully specified by its smallest repeat vectors, often called *primitive* lattice vectors, as shown in Fig. 1.1(b). In three dimensions, the position of every point in a Bravais lattice can be written as a linear combination of the *primitive* lattice vectors $\mathbf{e}_1$, $\mathbf{e}_2$, and $\mathbf{e}_3$,

$$\mathbf{R} = n_1\mathbf{e}_1 + n_2\mathbf{e}_2 + n_3\mathbf{e}_3, \qquad (1.2)$$

where n_1, n_2, n_3 are arbitrary integers. The smallest parallelepiped with a lattice point at each of its eight corners is called the *primitive* cell. Every edge of the primitive cell is a primitive lattice vector. Often, to better reflect the symmetries, certain types of Bravais lattices are specified by non-primitive lattice vectors $\mathbf{a}$, $\mathbf{b}$ and $\mathbf{c}$. The parallelepiped formed by these vectors is called the *unit* cell. Figure 1.2(a) shows the unit cell of a simple-cubic (SC) lattice. In this case, the unit cell is also the primitive cell, i.e. $\mathbf{a} = \mathbf{e}_1$, $\mathbf{b} = \mathbf{e}_2$, $\mathbf{c} = \mathbf{e}_3$. In the SC lattice, the repeat vectors $\mathbf{a}$, $\mathbf{b}$, $\mathbf{c}$ are perpendicular to each other and have the same length. The positions of all lattice points in the SC lattice can be written as

$$\mathbf{R} = i\mathbf{a} + j\mathbf{b} + k\mathbf{c}, \qquad (1.3)$$

where i, j, k are integers.

In body-centered-cubic (BCC) and face-centered-cubic (FCC) lattices, the unit cell is larger than the primitive cell, as shown in Fig. 1.2(b) and (c).[2] In these two cases, the unit cell has lattice points either on its faces (FCC) or in its interior (BCC), in addition to the lattice points at its corners. The lattice points of a BCC lattice are

$$\mathbf{R} = i\mathbf{a} + j\mathbf{b} + k\mathbf{c}$$
$$\mathbf{R} = \left(i + \tfrac{1}{2}\right)\mathbf{a} + \left(j + \tfrac{1}{2}\right)\mathbf{b} + \left(k + \tfrac{1}{2}\right)\mathbf{c}, \qquad (1.4)$$

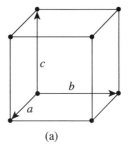
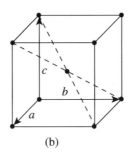
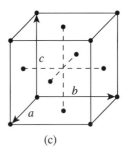

(a)   (b)   (c)

FIG. 1.2. (a) The unit cell of a simple cubic Bravais lattice. (b) The unit cell of a body-centered-cubic Bravais lattice. (c) The unit cell of a face-centered-cubic Bravais lattice.

[2] The cube-shaped unit cell reveals the cubic symmetries more clearly than the primitive cells.

and the lattice points of an FCC lattice are

$$\mathbf{R} = i\mathbf{a} + j\mathbf{b} + k\mathbf{c}$$

$$\mathbf{R} = \left(i + \tfrac{1}{2}\right)\mathbf{a} + \left(j + \tfrac{1}{2}\right)\mathbf{b} + k\mathbf{c}$$

$$\mathbf{R} = \left(i + \tfrac{1}{2}\right)\mathbf{a} + j\mathbf{b} + \left(k + \tfrac{1}{2}\right)\mathbf{c}$$

$$\mathbf{R} = i\mathbf{a} + \left(j + \tfrac{1}{2}\right)\mathbf{b} + \left(k + \tfrac{1}{2}\right)\mathbf{c}. \tag{1.5}$$

To obtain a SC, BCC or FCC crystal structure, it suffices to associate a single atom of the same chemical species with every lattice point in the SC, BCC, and FCC lattices, respectively. In all these cases, the basis consists of only one atom. It is easy to see that, for a given crystal structure, the lattice and the basis are not uniquely defined. For example, the BCC crystal structure can be equally regarded as a combination of the SC Bravais lattice with a basis consisting of two atoms (see Problem 1.1.4). To avoid confusion, the adopted convention is to associate each crystal structure with the Bravais lattice of highest possible symmetry and with the basis containing the smallest number of atoms.

Starting from the same SC, BCC or FCC lattices, more complicated crystal structures can be built using bases containing several atoms, which may be chemically distinct. For example, the diamond-cubic (DC) crystal structure, i.e. the structure of diamond, silicon and germanium crystals, is a combination of the FCC Bravais lattice with a two-atom basis. Separated by $\mathbf{a}/4 + \mathbf{b}/4 + \mathbf{c}/4$, the two atoms are shown in Fig. 1.3(a) in different colors. In the DC structure, two atoms of the basis are chemically identical (e.g. two silicon atoms). On the other hand, when the atoms in this basis are chemically distinct, the zinc-blende crystal structure results. For example, if one atom in the basis is gallium and the other is arsenic, the resulting GaAs crystal has a zinc-blende structure.

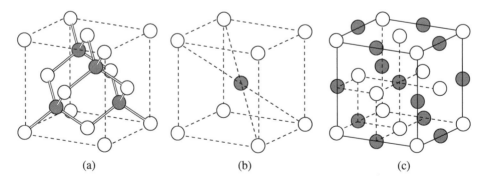

(a)   (b)   (c)

FIG. 1.3. (a) The basis of a zinc-blend structure contains two atoms that are chemically distinct. When two atoms in the basis are chemically identical, the resulting structure is diamond-cubic. (b) The CsCl structure. (c) The NaCl structure.

Similarly, the CsCl structure shown in Fig. 1.3(b) is a cubic structure with two different types of atoms in its basis. It looks similar to the BCC structure, except that two atoms in its basis are different—one is caesium and the other is chlorine. The NaCl structure of kitchen salt is also cubic, as shown in Fig. 1.3(c). At a first glance, it looks very similar to the SC structure with two types of atoms on alternating lattice sites. However, a more careful analysis reveals that the CsCl structure does not have a BCC Bravais lattice and the NaCl structure does not have an SC Bravais lattice (see Problem 1.1.5).

1.1.2 Miller indices

It is common and convenient to express the positions of atoms in a crystal in the units of repeat vectors of its lattice. Miller indices are introduced for this purpose and are frequently used as a shorthand notation for line directions and plane orientations in the crystals. The Miller indices will be very useful in our subsequent discussions, to specify a dislocation's line direction and its glide plane.

A vector l that connects one point to another in a Bravais lattice can be written as a linear combination of repeat vectors, i.e.

$$l = i\mathbf{a} + j\mathbf{b} + k\mathbf{c}, \tag{1.6}$$

where i, j and k are integers. The Miller indices notation for this vector is $[ijk]$.[3] To specify a line direction that is parallel to l, the convention is to select integer indices corresponding to the vector with the smallest length, among all vectors parallel to l. For example, the line direction along vector $\mathbf{a}$ is simply [100]. By convention, a negative component is specified by placing a bar over the corresponding index. For example, $[1\bar{2}3]$ defines direction $l = \mathbf{a} - 2\mathbf{b} + 3\mathbf{c}$. Sometimes, it is necessary to identify a family of line directions that are related to each other by crystal symmetry. Miller indices of a crystallographic family of directions are written in the angular brackets. For example, $\langle 112 \rangle$ corresponds to the set of line directions that includes $[112]$, $[11\bar{2}]$, $[1\bar{1}2]$, $[121]$, etc.

To specify a crystallographic plane, the Miller indices of the direction normal to the plane are used, but written between round brackets, i.e. (ijk). To identify a crystallographic family of planes related by symmetry, the Miller indices are written between the curly brackets, i.e. $\{ijk\}$.

Summary

- A crystal structure is a lattice plus a basis. The basis is the repeat pattern of atoms or molecules associated with every point of a Bravais lattice.

---

[3] Sometimes, fractional indices are used, such as $\left[\frac{1}{2}00\right]$, which represents vector $\mathbf{a}/2$.

- Line directions and plane orientations in a crystal are conveniently specified by their Miller indices.

Problems

1.1.1. Define Bravais lattices and bases for aluminum, iron, and silicon crystals and use MD++ code to build and display their crystal structures (MD++ can be downloaded from the book's web site [11]).

1.1.2. Find the distances from a given atom to its first, second and third nearest neighbor atoms in SC, FCC and BCC crystal structures.

1.1.3. Find the spacing between adjacent (111) planes in SC, FCC and BCC crystal structures.

1.1.4. Defining the BCC (FCC) crystal structure by the SC Bravais lattice with a basis containing two (four) atoms, write the coordinates of each atom in the basis.

1.1.5. What are the Bravais lattices of the CsCl and NaCl structures shown in Fig. 1.3(b) and (c)? Write the coordinates of each basis atom for these two crystal structures.

1.2   The Concept of Crystal Dislocations

In this section we explain how crystal dislocations are defined and describe some of their basic properties. Unless you are familiar with the concept already, be aware that, even though the dislocation is a simple object, it usually takes some time to grasp the idea and get used to it.

1.2.1   How to Make a Dislocation

A dislocation is a defect of crystal lattice topology and can be defined by specifying which atoms are dislocated or *mis-connected* with respect to the perfect, defect-free structure of the host crystal. One way to grasp the topological meaning of a dislocation is to introduce one by hand. As an example, consider a perfect crystal with the simple-cubic structure,[4] Fig. 1.4(a). Let us now imagine a horizontal plane (*A*) dividing the crystal into two halves and displace the upper half of the crystal with respect to its lower half by a lattice vector (**b**), as shown in Fig. 1.4(b). Because **b** is a lattice repeat vector, the arrangement of atoms inside the crystal remains the same as before. The only change from Fig. 1.4(a) to Fig. 1.4 (b) are two steps at the surfaces.

---

[4] The simple-cubic crystal is used here for illustration purposes whereas in reality very few crystals have this structure. A notable exception is the alpha phase of polonium, a radioactive material discovered by Marie Curie and named after her homeland Poland [12].

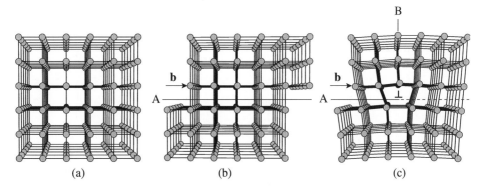

FIG. 1.4. (a) A perfect simple-cubic crystal. (b) Displacement of two half-crystals along cut plane A by lattice vector **b** results in two surface steps but does not alter the atomic structure inside the crystal. (c) The same "cut-and-slip" procedure limited to a part of cut plane A introduces an edge dislocation ⊥.

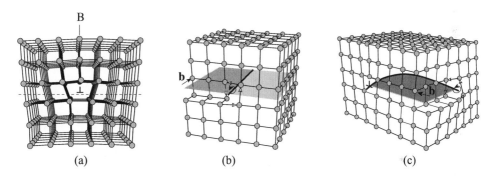

FIG. 1.5. (a) An edge dislocation created by inserting a half-plane of atoms B. (b) A screw dislocation created by a "cut-and-slip" procedure in which the slip vector is parallel to the dislocation line. The slipped area of the cut plane is shown in dark gray and the un-slipped area is shown in light gray. The dislocation line is marked by the solid line. (c) A curved dislocation line with an edge orientation at one end (on the left) and a screw orientation at the other end (on the right).

Such a "cut-and-slip" procedure produces a permanent (plastic) deformation of the crystal, but does not yet create a dislocation. A dislocation appears when two half-crystals are displaced not over the whole cut plane A, but only over a part of it marked by the solid line in Fig. 1.4(c). The remaining "un-slipped" part of the cut plane is marked by the dashed line. The boundary between the slipped and un-slipped parts of the cut plane is a dislocation line. This line runs perpendicular to the plane of the paper and is marked by symbol ⊥ on Fig. 1.4(c).

A very similar structure is created by inserting an extra half plane of atoms into a perfect SC crystal from above. Shown in Fig. 1.5(a), the new

atomic structure is essentially identical to that in Fig. 1.4(c). The only difference between the two structures is the extra surface step in Fig. 1.4(c). It
is not hard to imagine that the same dislocation can be obtained by removing a half-plane of atoms from below. Obviously, there are multiple (in fact
an infinite number of) ways to create the same dislocation. Just by looking
at the interior of a crystal, it is impossible to tell how the dislocation was
created. This simple observation underscores something important. The structure
and properties of dislocations do not depend on how they are created in the
crystal.

The dislocation shown in Fig. 1.4(c) and Fig. 1.5(a) is an *edge* dislocation,
which is located at the edge of an extra half plane of atoms. There are other types
of dislocations as well. An example is shown in Fig. 1.5(b), which is created by
a similar "cut-and-slip" operation except that the slip direction is now parallel to
the dislocation line. This dislocation is called *screw* because the atoms around
the line are arranged in a spiral, as shown by the white arrows in Fig. 1.5(b).
Dislocations with orientations between screw and edge, such as the curved line in
Fig. 1.5(c), are called *mixed* dislocations. The angle between the dislocation line
and the slip vector is called the *character angle*. The character angle is $0°$ for a
screw dislocation, $90°$ for an edge dislocation, and between $0°$ and $90°$ for a mixed
dislocation.

In real crystals, dislocations form in many different ways. For example, they can
appear by shearing along crystal planes, or by condensation of interstitials (extra
atoms in the lattice) or vacancies (empty atomic sites). These three mechanisms
can be regarded as physical realizations of the three thought experiments described
above ("cut-and-slip", insertion and removal of a half plane of atoms). Often,
dislocations appear in the process of crystal growth itself. For instance, crystal
growth from a supersaturated solution is greatly facilitated by screw dislocations
that grow simultaneously with the growing crystal. In crystals formed by cooling
from a melt, the misfit strain associated with the thermal gradients is usually large
enough to produce dislocations. Practically, with the notable exception of silicon,
it has been impossible to grow large crystals free of dislocations. Moreover, it is
very difficult to rid a crystal of the grown-in dislocations that often multiply over
the material's lifetime.

Because dislocations distort the host crystal lattice, they can be observed using
various diffraction methods, such as transmission electron microscopy (TEM). The
TEM image in Fig. 1.6(a) shows a collection of dislocation lines in BCC molybdenum. The dislocations appear curved and jagged. In more heavily dislocated
crystals, dislocations can form rather complicated sub-structures, such as braids
(Fig. 1.6(b)) or cells (Fig. 1.6(c)). Even though shapes and collective arrangements
of dislocation lines can be very complex, there is something fundamentally simple
about crystal dislocations. Each line carries with it a certain amount of crystal
distortion, quantified by its Burgers vector.

FIG. 1.6. (a) A TEM picture of dislocation structure in pure single crystal BCC molybdenum deformed at temperature 278 K (courtesy of L. L. Hsiung). (b) Dislocations formed bundles (braids) in single crystal copper deformed at 77 K (reproduced from [13] with permission from the authors and the publisher). (c) Dislocation structure formed in single crystal BCC molybdenum deformed at temperature 500 K (courtesy of L. L. Hsiung). The dark regions contain a high density of entangled dislocations lines that can no longer be distinguished individually.

1.2.2 The Burgers Vector

In addition to visual inspection, there are more precise ways to identify the presence of a dislocation. Here we will rely on a universally adopted Burgers circuit test. A Burgers circuit consists of a sequence of jumps from atoms to their neighbors. The Burgers circuit should form a complete loop when it is drawn in a perfect crystal. When the same Burgers circuit is drawn in a defective crystal, it may not end at the starting atom. When this happens, the vector pointing from the starting atom to the ending atom gives a measure of topological mis-connection of the defective crystal structure. At a first glance, such a test may appear to be rather arbitrary since there is so much freedom in choosing the Burgers circuit. Remarkably, the resulting mis-connection vector turns out to be the same (modulo sign) for any circuit that encloses the same dislocation.

Let us try the Burgers circuit test on a two-dimensional plane of atoms perpendicular to an edge dislocation, as shown in Fig. 1.7. The analysis that follows is easy to extend to more complex three-dimensional atomic structures. To set a sign convention, let us agree on the flow direction for the test circuit. The common convention is to first select a direction or *sense*, ξ, for the dislocation line. Then, the flow direction for the test circuit is defined with respect to the chosen line sense fol-

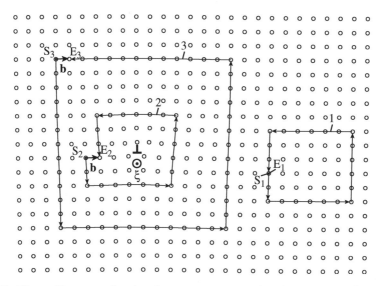

FIG. 1.7. Three Burgers circuits drawn on an atomic plane perpendicular to an edge dislocation in a SC crystal. The start and end points of the circuits are S_i and E_i, respectively. Circuit 1 does not enclose dislocation $\perp$ whereas circuits 2 and 3 do. The sense vector ξ is defined to point out of the paper so that all three circuits flow in the counterclockwise direction.

lowing the right-hand-rule.[5] The sense direction $\boldsymbol{\xi}$ is arbitrary and here we choose it to point out of the paper. According to our convention, the Burgers circuit flows counterclockwise.

As stated above, the Burgers circuit should form a complete loop if drawn in a perfect crystal. An example is circuit 1 shown in Fig. 1.7. The circuit starts at atom S_1 and consists of $N = 22$ jumps: $\downarrow, \downarrow, \rightarrow, \rightarrow, \rightarrow, \rightarrow, \rightarrow, \rightarrow, \uparrow, \uparrow, \uparrow, \uparrow, \uparrow,$ $\leftarrow, \leftarrow, \leftarrow, \leftarrow, \leftarrow, \leftarrow, \downarrow, \downarrow, \downarrow$, where each arrow corresponds to a translation along one of the primitive vectors of the SC lattice, $\mathbf{v}_\rightarrow = [1\,0]$, $\mathbf{v}_\leftarrow = [\bar{1}\,0]$, $\mathbf{v}_\uparrow = [0\,1]$, $\mathbf{v}_\downarrow = [0\,\bar{1}]$. The end point E_1 coincides with the starting point S_1, because circuit 1 does not contain a dislocation. However, when the very same sequence starts from atom S_2, the resulting circuit encloses a dislocation and fails to close. The vector connecting starting point S_2 to end point E_2 of test circuit 2 is the Burgers vector $\mathbf{b}$ associated with the enclosed dislocation. An important point we would like to to emphasize once again is that the direction (or sign) of the Burgers vector is meaningful only when the sense of the dislocation line is either explicitly defined or implied by context. As an illustrative example, let us reverse the direction of line sense vector $\boldsymbol{\xi}$ and make it point into the plane of the paper. Our right-hand-rule convention then dictates that, to obtain the Burgers vector, circuit 2 should now run clockwise, from atom E_2 to atom S_2, which obviously reverses the direction of the resulting Burgers vector.

The following algorithm constructs Burgers circuits in a crystal. The required input is an index n_S of the starting atom and a sequence of translations for the test circuit $\mathbf{v}_i$, $i = 1, \ldots, N$. The algorithm returns index n_E of the end atom of the circuit and a set of N vectors $\Delta\mathbf{u}_i$, which are the differences between the actual (distorted) and expected (perfect) relative positions of atom pairs connected by translations $\mathbf{v}_i$.

Algorithm 1.1

1. Define the starting position $\mathbf{r}_0$ of the circuit to be the position of atom n_S. Set $i := 1$.

2. Find atom n_i whose position $\mathbf{r}_i$ is nearest to point $\mathbf{r}_{i-1} + \mathbf{v}_i$.

3. Compute the difference between the actual and perfect relative positions of the atom pair connected by the current translation, $\Delta\mathbf{u}_i := \mathbf{r}_i - (\mathbf{r}_{i-1} + \mathbf{v}_i)$.

4. Increment the step counter $i := i + 1$. If $i \leq N$, go to step 2.

5. Define the index of the end atom, $n_E := n_N$.

---

[5] Pointing the right thumb along the chosen line direction $\boldsymbol{\xi}$, the remaining four fingers curve in the direction of flow of the Burgers circuit.

The Burgers vector is obtained either as the difference $\mathbf{r}_N - \mathbf{r}_0$ or, equivalently, as the sum of vectors $\Delta\mathbf{u}_i$:

$$\mathbf{b} = \mathbf{r}_N - \mathbf{r}_0 = \sum_{i=1}^{N} \Delta\mathbf{u}_i. \tag{1.7}$$

The sum in this equation is a discrete analogue of the equation used in continuum mechanics to define the Burgers vector of a Volterra dislocation, i.e.

$$\mathbf{b} = \oint_C \frac{\partial\mathbf{u}}{\partial l}\, dl, \tag{1.8}$$

where C is any contour enclosing the dislocation line and $\partial\mathbf{u}/\partial l$ is the elastic displacement gradient along the contour [7]. Another, more practical reason to retain $\Delta\mathbf{u}_i$ is because they can be used for computing the Burgers vector in the slightly more complicated case of *partial* dislocations (see Problem 1.2.2).

As stated earlier, all Burgers circuits flowing in the same direction and enclosing the same dislocations give rise to the same Burgers vector. As an example, consider circuit 3 in Fig. 1.7, starting at S_3 and ending at E_3. This circuit yields the same Burgers vector $\mathbf{b}$ as circuit 2. More generally, the resulting Burgers vector is unaffected by any deformation and/or translation of the test circuit as long as such deformation and/or translation does not make the circuit "cut" through a dislocation line. In particular, translation of the Burgers circuit along the enclosed line leaves the Burgers vector unchanged, meaning that the Burgers vector is conserved along any given dislocation. Therefore, the Burgers vector is an intrinsic property of the dislocation line that can be regarded as the dislocation's *topological charge*. Another consequence of the Burgers vector conservation is that a dislocation cannot end within an otherwise perfect crystal. If this were possible, the Burgers vector would have to change abruptly from a non-zero value to zero as the test circuit is moved along the line beyond its termination point.

As illustrated in Fig. 1.8, a test circuit drawn around two dislocation lines reveals a Burgers vector equal to the vector sum of the Burgers vectors of two enclosed dislocations, $\mathbf{b}_2 + \mathbf{b}_3$. Imagine now that two dislocation lines $\boldsymbol{\xi}_2$ and $\boldsymbol{\xi}_3$ merge into a single line at junction node P. The Burgers vector of the resulting dislocation $\boldsymbol{\xi}_1$ can be obtained from Burgers circuit q drawn on cross-section C. Because circuit q can be obtained from circuit p by deformation and translation without cutting through the dislocation lines, the Burgers vector revealed by both circuits must be the same, that is

$$\mathbf{b}_1 = \mathbf{b}_2 + \mathbf{b}_3. \tag{1.9}$$

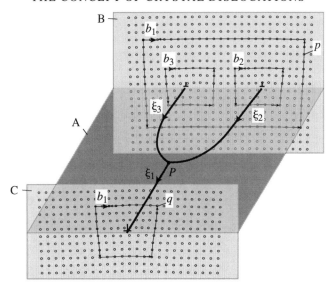

FIG. 1.8. The larger circuit p on cross-section B encloses two smaller circuits, each of which encircles a single dislocation. Dislocations $\boldsymbol{\xi}_2$ and $\boldsymbol{\xi}_3$ merge into dislocation $\boldsymbol{\xi}_1$ at junction node P. Conservation of the Burgers vector requires that $\mathbf{b}_1 = \mathbf{b}_2 + \mathbf{b}_3$, as illustrated by the Burgers circuits drawn on the two cross-sections B and C.

This condition is analogous to the conservation of current in an electric circuit, except that now the conserved "current" is a vector rather than a scalar. This conservation law can be generalized to situations where an arbitrary number of dislocations merge at a node. In such cases, it is often more convenient to define all line directions to *flow out* of the node. Following this convention, the directions of $\boldsymbol{\xi}_2$, $\boldsymbol{\xi}_3$, $\mathbf{b}_2$, $\mathbf{b}_3$ in Fig. 1.8 must be reversed and the conservation of Burgers vector can be conveniently rewritten as,

$$\sum_{i=1}^{n} \mathbf{b}_i = 0, \qquad (1.10)$$

where n is the total number of dislocation lines connected to node P ($n = 3$ in Fig. 1.8).

Summary

- A dislocation is a line defect in an otherwise perfect crystal that carries a certain amount of distortion quantified by the Burgers vector. A variety of mathematical operations can be used to introduce a dislocation into a perfect crystal. Likewise, a number of physical processes produce dislocations in real materials.

- A dislocation's Burgers vector is defined with respect to the sense (direction of flow) of the line. When the line sense is inverted, the Burgers vector changes sign.

- The Burgers vector is conserved along the dislocation lines. When several dislocations merge at a node, the sum of their Burgers vectors must be zero, provided all the line directions are selected to flow out of the node.

Problems

1.2.1. Modify the Matlab code available at the book web site to perform the Burgers circuit analysis on a two-dimensional triangular lattice. Create a two-dimensional crystal structure with the triangular lattice and insert a single edge dislocation inside the crystal. Test the Matlab code by performing the Burgers vector analysis to see if the Burgers vector revealed by a test circuit matches that of the inserted dislocation.

1.2.2. The Burgers circuit test implemented in Algorithm 1.1 may not work correctly when the crystal contains a *partial* dislocation, i.e. a dislocation with a Burgers vector that is only a fraction of the primitive lattice vector. One such situation is illustrated in Fig. 1.9, where a partial edge dislocation was created by a "cut-and-slip" operation but the slip vector was

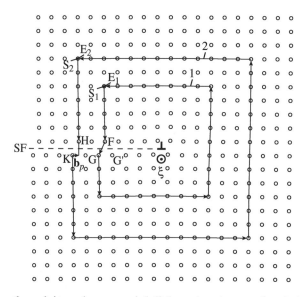

FIG. 1.9. Even though it encloses partial dislocation ⊥, test circuit 1 does not reveal any closure defect. The partial Burgers vector $\mathbf{b}_p$ can be detected if large jumps $\Delta\mathbf{u}_i$ are excluded from the sum in eq. (1.11), as illustrated by circuit 2.

only half of lattice vector [1 0]. The atomic arrangement to the left of the dislocation is not the same as that in the perfect crystal but forms an extended defect called a *stacking fault*.

A correct Burgers circuit analysis should reveal the partial Burgers vector $\mathbf{b}_p = \left[\frac{1}{2}0\right]$. To see if this is indeed the case, subject the configuration shown in Fig. 1.9 to a Burgers circuit test (the file containing the atom positions is available at the book's web site). Select a starting atom and a test circuit enclosing the dislocation and use Algorithm 1.1 to compute the Burgers vector. If you do everything right, the result can be a little surprising. The Burgers vector detected by the test will be either zero or a full lattice vector [1 0]. The reason for this failure can be seen in the same figure, where test circuit 1 finds a large lattice distortion when trying to connect atom F to a neighbor atom in the [0 $\bar{1}$] direction. At this step, atom G is selected as the neighbor because it happens to be a little closer to atom F than atom G': the resulting circuit reveals no closure defect. A slight distortion of the crystal would result in G' being selected in this step. Then the Burgers circuit would predict a Burgers vector that is equal to the full lattice vector.

It turns out that Algorithm 1.1 can be fixed to make it capable of detecting partial dislocations. The fix is to simply exclude from the sum in eq. (1.7) all distortion terms $\Delta\mathbf{u}_i$ that are suspiciously large. In the modified algorithm, the test circuit is drawn exactly the same way as before, but the Burgers vector is now computed as

$$\mathbf{b} = \sum_{i=1}^{N} \Delta\mathbf{u}_i H(\epsilon - |\Delta\mathbf{u}_i|) \tag{1.11}$$

where $H(\cdot)$ is the step function, i.e. $H(x) = 1$ when $x \geq 0$ and $H(x) = 0$ when $x < 0$. ϵ is a numerical parameter chosen by trial-and-error to limit the magnitude of distortions between two neighboring atoms that is acceptable in the sum. Test the modified algorithm and find a range of values for parameter ϵ suitable for detecting the Burgers vector of the partial dislocation shown in Fig. 1.9.

1.3 Motion of a Crystal Dislocation

While dislocations define many properties of crystalline materials, the numerical models in this book are mostly oriented towards mechanical properties, and how they are controlled by the motion of dislocations. In this section, we illustrate how the driving force for dislocation motion can be obtained from continuum elasticity theory, whereas the dislocation's response to this force is governed by discrete atomistic mechanisms.

1.3.1 Driving Forces for Dislocation Motion

Dislocation motion provides a mechanism for a crystal to deform plastically, or *yield*, when the crystal is subjected to an applied stress. Consider the mechanism illustrated in Fig. 1.10(a)–(e). Let the dimension of the crystal be $L_x \times L_y \times L_z$. Apply a traction force T_x in the x direction to its top surface and fix the position of the bottom surface, so that the crystal is subjected to a stress,

$$\sigma_{xy} = \frac{T_x}{L_x L_z}. \tag{1.12}$$

Imagine that an edge dislocation nucleates from the left side of the crystal (b), then moves to the right (c)–(d), and finally exits the crystal from the right surface (e). The dislocation line sense $\boldsymbol{\xi}$ is along the positive z direction and its Burgers vector is $\mathbf{b} = [b_x\, 0\, 0]$. During this process, the dislocation travels along the x direction for a distance L_x, and the net result is that the top half of the crystal is displaced by $\mathbf{b}$ with respect to the lower half (Fig. 1.10(e)). Because the top surface moves by b_x, the total work done by the surface traction is

$$W = b_x \cdot T_x. \tag{1.13}$$

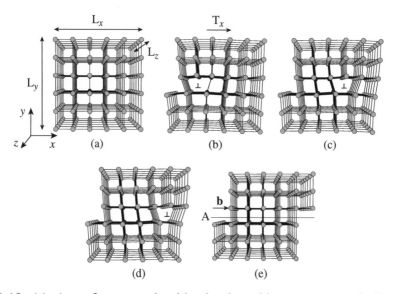

FIG. 1.10. (a) A perfect crystal with simple cubic structure and dimension $L_x \times L_y \times L_z$. (b) The top surface is subjected to a traction force T_x while the bottom surface is fixed. An edge dislocation nucleates from the left surface. In (c) and (d) the dislocation moves to the right. In (e) the dislocation finally exits the crystal from the right surface. The net result is that the upper half of the crystal is displaced by $\mathbf{b}$ with respect to the lower half.

At the same time, we may interpret W as the work done by a driving force f_x per unit length on the dislocation line. Since the dislocation length is L_z and it moves by L_x, we have,

$$W = L_x \cdot f_x L_z. \qquad (1.14)$$

Hence

$$f_x = \frac{W}{L_x L_z} = b_x \sigma_{xy}. \qquad (1.15)$$

Thus the force on a dislocation is the product between the Burgers vector and an appropriate stress component. This identity can be generalized to the force per unit length $\mathbf{f}$ at an arbitrary point P on a (possibly curved) dislocation line. The result is the Peach–Koehler (PK) formula [7],

$$\mathbf{f} = (\boldsymbol{\sigma} \cdot \mathbf{b}) \times \boldsymbol{\xi}, \qquad (1.16)$$

where $\boldsymbol{\sigma}$ is the local stress field and $\boldsymbol{\xi}$ is the local line tangent direction at point P. The cross product with $\boldsymbol{\xi}$ ensures that the PK force is always perpendicular to the line itself.

The significance of the PK formula is that the force acting on a dislocation is fully defined by the local stress $\boldsymbol{\sigma}$ on the dislocation, regardless of the origin of this stress. In addition to surface traction forces, local stresses can be induced by other dislocations or any other strain-producing defects. The total stress is then a linear superposition of all such contributions (Chapter 10). Continuum elasticity theory of solids supplies the equations for computing stress induced by various sources, and hence the driving forces on dislocation lines.

The process of dislocation motion is not only plausible, as illustrated in Fig. 1.10, but it also provides the correct microscopic picture for plastic deformation. Pre-dating the initial proposal for dislocations in 1934 [2, 3, 4], plastic deformation of single crystals was known to proceed by sliding (or slipping) along well-defined crystallographic planes resulting in shifting of the atoms on one side of the plane with respect to the atoms on the other side. However, laboratory experiments presented ample evidence that something was missing in this picture. The theoretical stress required to force a perfect crystal to slide was found to be on the order of 0.1μ while the measured yield stress ranged between $10^{-4}\mu$ to $10^{-2}\mu$ (μ is the elastic shear modulus of the crystal). The presence of dislocations explained this fundamental discrepancy by offering an easier path for crystallographic slip. Sliding by dislocation motion only requires significant atomic rearrangements near the dislocation core, as opposed to over an entire plane. The level of stress required to make a dislocation move is usually orders of magnitudes lower than the critical stress to break all bonds on a crystallographic plane, and is much closer to experimentally measured yield stress of crystals.

Plastic strain produced by dislocation motion* Another important property of dislocation motion is that the *area* swept out by a moving dislocation is proportional to the plastic *strain* ε^p it introduces to the crystal. This concept is closely related to the already established fact that the *force* on a dislocation is proportional to the local *stress* σ. Again let us use Fig. 1.10 as an example. The total area swept out by the dislocation is $\Delta A = L_x L_z$, whereas the total plastic strain is

$$\varepsilon^p_{xy} = \frac{b_x}{2L_y} = \frac{b_x \Delta A}{2\Omega}, \tag{1.17}$$

where $\Omega \equiv L_x L_y L_z$ is the volume of the crystal. This result can be generalized as well, by considering a dislocation with Burgers vector $\mathbf{b} = [b_1\ b_2\ b_3]$, which sweeps out an area of ΔA on a plane with normal vector $\mathbf{n} = [n_1\ n_2\ n_3]\ (n_1^2 + n_2^2 + n_3^2 = 1)$. The plastic strain it introduces is

$$\varepsilon^p_{ij} = \frac{(b_i n_j + b_j n_i)\Delta A}{2\Omega}. \tag{1.18}$$

Suppose that the total length of all dislocations is L and their average velocity is v; then the total area swept out by dislocations during a period Δt is $\Delta A = vL\Delta t$. This quickly leads to the famous *Orowan's* formula,

$$\dot{\varepsilon}^p = \rho b v, \tag{1.19}$$

which relates the plastic strain rate $\dot{\varepsilon}^p$ to the dislocation density ρ (line length per unit volume), Burgers vector magnitude b, and average dislocation velocity v.

1.3.2   Conservative versus Non-conservative Motion

While the PK force on a dislocation can be obtained from continuum elasticity theory, the speed at which a dislocation moves in response to the PK force may be sensitive to the details of the atomic structures and interactions at the dislocation core. This level of detail is beyond the reach of continuum elasticity theory. Before we discuss this in detail, it is important to first observe the fundamental difference between the motion of edge and screw dislocations.

Consider an edge dislocation moving to the right by one lattice spacing, as shown in Fig. 1.11(a) and (b). All it requires is for the atoms near the dislocation core to exchange their neighbors. However, when this dislocation moves downward by one lattice spacing, as in Fig. 1.11(c), a row of atoms has to be inserted into the crystal. The difference between the two cases is that, in (b), the dislocation is moving along the *glide plane*, plane A, whereas in (c) the dislocation is moving perpendicular to it. The glide plane is defined as the plane that simultaneously contains the dislocation line and the Burgers vector. Not surprisingly, motion along the glide plane is called *glide*, whereas motion perpendicular to it is called *climb*.

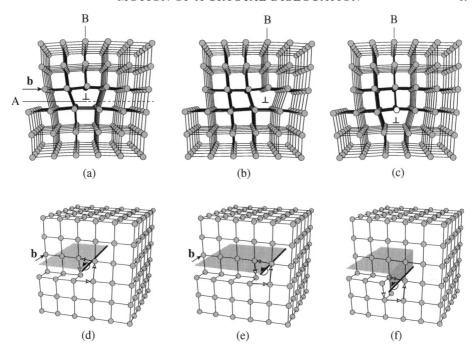

FIG. 1.11. (a) An edge dislocation in a simple cubic crystal. When it moves on its glide plane by one lattice spacing (b), no atom needs to be inserted or removed. When it moves perpendicular to its glide plane (c), a row of atoms (white spheres) needs to be inserted. (d) A screw dislocation in a simple cubic crystal. Regardless of which direction the dislocation moves, (e) and (f), no atom needs to be inserted or removed. The grey area indicates regions where atoms on both sides have slipped with respect to one another.

Dislocation glide is often called *conservative motion*, meaning that the total number of atoms is conserved, whereas dislocation climb is *non-conservative*.

For a screw dislocation, the Burgers vector is parallel to the line direction; hence the glide plane is not uniquely defined. The consequence of this is that screw dislocation motion does not require atoms to be inserted or removed regardless of the plane on which it moves, as illustrated in Fig. 1.11(e) and (f). In other words, a screw dislocation always glides and never climbs.

For a mixed dislocation, the edge component of the Burgers vector determines how many atoms need to be inserted or removed during climb motion. Consider a general dislocation with Burgers vector $\mathbf{b}$, which moves on a plane with normal vector $\mathbf{n}$ and sweeps out an area ΔA (ΔA is positive if the dislocation moves in the direction of $\mathbf{n} \times \boldsymbol{\xi}$). Let $\Delta V = (\mathbf{b} \cdot \mathbf{n}) \Delta A$. If $\Delta V > 0$, then material of volume ΔV has to be inserted. If $\Delta V < 0$, then material of volume $|\Delta V|$ has to be removed.

The number of atoms to be inserted or removed can then be obtained by dividing ΔV by the volume of the primitive cell and multiplying by the number of atoms in the basis.

In real crystals, nobody inserts or removes the atoms by hand. Instead, climb motion involves atomic diffusion and takes place when dislocations absorb or emit atoms or vacancies. Generally, dislocations move with some combination of glide and climb, where one of these two mechanisms can dominate, depending on stress, temperature and other conditions. At low temperatures, climb is usually difficult and glide is often dominant. However, at high temperatures or under conditions of vacancy super-saturation, climb can become dominant instead.

Consequently, the mobility of non-screw dislocations is usually highly anisotropic with respect to their glide planes. At low temperatures, when climb is difficult, the motion of non-screw dislocations is essentially confined to their glide planes. On the other hand, screw dislocations are truly special among all dislocations, because of their ability to glide on multiple planes. Even at low temperatures, screw dislocations are not confined to any particular glide plane, but can move in three dimensions.[6]

1.3.3   Atomistic Mechanisms of Dislocation Motion

A dislocation's response to PK force $\mathbf{f}$ can be often described by a mobility function $\mathbf{M}(\cdot)$ that relates dislocation velocity $\mathbf{v}$ to force $\mathbf{f}$, i.e. $\mathbf{v} = \mathbf{M}(\mathbf{f})$. Both $\mathbf{f}$ and $\mathbf{v}$ are two-dimensional vectors in the plane perpendicular to the local dislocation line orientation. The mobility function $\mathbf{M}(\cdot)$ is an important material-specific input for dislocation simulations based on continuum theory (Chapter 10). $\mathbf{M}(\cdot)$ is usually highly anisotropic for glide and climb orientations of non-screw dislocations. Because climb is very difficult at low temperature, we can often make a simplifying approximation that the dislocation velocity is entirely confined to the glide plane (i.e. the climb mobility is zero). In this case $\mathbf{M}(\cdot)$ will be determined if we obtain the glide mobility.

Dislocation glide mobility may be influenced by both *extrinsic* factors, such as impurities acting as obstacles, and *intrinsic* factors, such as the interatomic interactions at the dislocation core [14]. Intrinsic factors dominate when the crystal is sufficiently pure. Most of this book deals with dislocations in an idealized pure single crystal, whose mobility is only affected by intrinsic factors. An exception is Chapter 11, which contains an example of dislocation interaction with alloy

---

[6] In face-centered-cubic (FCC) crystals (see Section 1.1), even the screw dislocations appear to be confined to move in a plane. This apparent confinement is specific to these and other similar materials where the screws spontaneously dissociate into two non-screw (partial) dislocations with a common glide plane. This behavior does not contradict our statement here. Screw dislocations in other materials such as body-centered-cubic (BCC) crystals, where no such planar dissociation takes place, are observed to move in a variety of available glide planes.

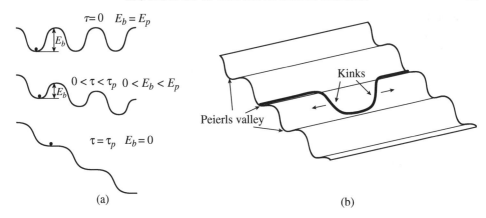

FIG. 1.12. (a) A schematic of the periodic Peierls potential for dislocation motion through a discrete lattice, viewed along the dislocation line (the positions of the lines are marked by the small filled circles). E_b is the energy barrier that the dislocation sees when it moves from left to right. The energy barrier decreases with the increasing stress τ and disappears completely when the stress reaches the Peierls threshold $\tau = \tau_P$. (b) A three-dimensional sketch of the Peierls potential. When $\tau < \tau_P$, the dislocation moves by nucleating a kink pair and propagating the kinks sideways.

obstacles. In the following, we briefly introduce two fundamental parameters that characterize intrinsic lattice resistance to dislocation motion: the *Peierls barrier* and the *Peierls stress*. These parameters can be computed using the atomistic models to be described in the following chapters.

Consider a straight dislocation moving in its glide plane. The effect of the crystal lattice on this motion can be represented by an energy function of the dislocation position, which has the periodicity of the lattice, as illustrated in Fig. 1.12(a). Such periodic energy variations cannot be explained from the standpoint of linear elasticity theory (Section 8.1). Instead, they reflect changes in the dislocation core energy (Section 5.2) as the dislocation translates from one lattice position to the next. Periodicity of this energy is a consequence of the translation symmetry of the crystal.

The minima of this function mark the preferred dislocation positions. These are sometimes called *Peierls valleys*. The energy barrier (per unit length) that a dislocation must surmount to move from one Peierls valley to an adjacent one under zero stress, is called the *Peierls barrier* E_p. In the presence of a non-zero local stress, the PK force acting on a dislocation modifies the periodic energy function. This modification is represented by a constant slope superimposed on the periodic potential, as shown in Fig. 1.12(a); the slope is equal to the PK force. As a result, the actual energy barrier E_b experienced by the dislocation becomes lower than the Peierls barrier E_p. A further increase in the local stress can make the energy barrier E_b vanish completely. The magnitude of such critical stress is called the *Peierls stress* τ_P.

While the Peierls stress is an idealized concept—the minimum stress required to make a straight dislocation move at zero temperature—it is still very useful. First, it provides a quantitative measure of lattice resistance to dislocation motion that can be directly computed in an atomistic simulation (Section 5.3). Second, it is related to the yield stress measured at (or extrapolated to) vanishingly low temperatures. Third, the very magnitude of the computed Peierls stress often provides valuable insights into the mechanisms of dislocation motion.

For example, the Peierls stress usually delineates two different regimes of dislocation motion. When the local stress is lower than the Peierls stress, a dislocation cannot move at zero temperature,[7] but can move at a finite temperature with the help of thermal fluctuations. In the latter case, rather than moving the whole straight dislocation at once, motion begins by throwing a short dislocation segment into the next Peierls valley, as shown in Fig. 1.12(b). Such an event creates a pair of *kinks*. Then, with the assistance of stress and/or thermal fluctuations, the kinks run away from each other in two opposite directions. In the end, the entire dislocation line finds itself in the next Peierls valley, and the process may repeat itself. In this regime, the dislocation mobility increases at higher temperatures where kink pair nucleation is more frequent. A kinetic Monte Carlo (kMC) model for dislocation motion based on the kink mechanism is discussed in Chapter 9.

On the other hand, when local stress is higher than the Peierls stress, thermal activation is no longer necessary for dislocation motion. In this so-called "viscous drag" regime, dislocation velocity becomes a linear function of stress and is usually limited by the viscosity due to dislocation interaction with lattice vibrations, i.e. sound waves. Dislocation mobility now decreases with increasing temperature, when the amplitude of lattice vibrations becomes larger. In this regime, molecular dynamics (MD) simulations in Chapter 4 can be used to compute dislocation mobility.

There are discernible trends in the lattice resistance and the mechanisms of dislocation motion among different materials. In semiconductors, dislocation motion requires breaking and re-forming of covalent bonds, resulting in a very high Peierls stress, of the order of few GPa, much higher than the typical stress seen by dislocations moving in a crystal. Hence, the kMC model of kinks in Chapter 5 is well suited here. On the other hand, the Peierls stress for ordinary dislocations in the face-centered-cubic (FCC) metals is low, so that they usually move in the viscous drag regime. Hence, the MD model in Chapter 4 is well suited to this case. The resistance to dislocation motion in the body-centered-cubic (BCC) metals rests somewhere in between these two extremes. In particular, screw dislocations in

---

[7] Here we ignore the effect of quantum fluctuations.

BCC metals usually move by kink mechanisms, whereas non-screw dislocations usually move in the viscous drag regime, so that different models are needed to study dislocations with different orientations.

Summary

- The linear-elasticity theory is fairly accurate in predicting the forces acting on dislocations. How a dislocation moves in response to these forces is defined by the atomistic core mechanisms.

- The plane containing the Burgers vector and the local line direction vector of a dislocation is its glide plane. Dislocation motion in the glide plane is conservative. However, motion of non-screw dislocations out of their glide planes requires inserting or removing material and is difficult at low temperature when diffusion is slow. In principle, a screw dislocation can glide in any plane containing its line direction.

- At a stress below the Peierls threshold, dislocations move by thermally assisted kink pair mechanisms. At a stress above the threshold, viscous drag mechanisms usually limit the dislocation mobility.

PART I

ATOMISTIC MODELS

2

FUNDAMENTALS OF ATOMISTIC SIMULATIONS

Fundamentally, materials derive their properties from the interaction between their constituent atoms. These basic interactions make the atoms assemble in a particular crystalline structure. The same interactions also define how the atoms prefer to arrange themselves in the dislocation core. Therefore, to understand the behavior of dislocations, it is necessary and sufficient to study the collective behavior of atoms in crystals populated by dislocations. This chapter introduces the basic methodology of atomistic simulations that will be applied to the studies of dislocations in the following chapters. Section 1 discusses the nature of interatomic interactions and introduces empirical models that describe these interactions with various degrees of accuracy. Section 2 introduces the significance of the *Boltzmann distribution* that describes statistical properties of a collection of interacting atoms in thermal equilibrium. This section sets the stage for a subsequent discussion of basic computational methods to be used throughout this book. Section 3 covers the methods for energy minimization. Sections 4 and 5 give a concise introduction to Monte Carlo and molecular dynamics methods.

2.1 Interatomic Interactions

When put close together, atoms interact by exerting forces on each other. Depending on the atomic species, some interatomic interactions are relatively easy to describe, while others can be very complicated. This variability stems from the quantum mechanical motion and interaction of electrons [15, 16]. Henceforth, rigorous treatment of interatomic interactions should be based on a solution of Schrödinger's equation for interacting electrons, which is usually referred to as the first principles or *ab initio* theory. Numerical calculations based on first principles are computationally very expensive and can only deal with a relatively small number of atoms. In the context of dislocation modelling, relevant behaviors often involve many thousands of atoms and can only be approached using much less sophisticated but more computationally efficient models. Even though we do not use it in this book, it is useful to bear in mind that the first principles theory provides a useful starting point for constructing approximate but efficient models that are needed to study large-scale problems involving many atoms. The usual way to construct a model of interatomic interactions is to postulate a relatively simple, analytical functional form for the potential energy of a set of atoms,

$$V(\{\mathbf{r}_i\}) \equiv V(\mathbf{r}_1, \mathbf{r}_2, \ldots, \mathbf{r}_N), \tag{2.1}$$

where $\mathbf{r}_i$ is the position vector of atom i and N is the total number of atoms. The force on an atom is the negative derivative of the potential function with respect to its position, i.e.

$$\mathbf{f}_j = -\frac{\partial V(\{\mathbf{r}_i\})}{\partial \mathbf{r}_j}. \tag{2.2}$$

The hope is that approximate forms can be developed that capture the most essential physical aspects of atom–atom interaction. Such functions are commonly called *interatomic potentials*. The parameters in the interatomic potential function are usually fitted to experimental or *ab initio* simulation data.

2.1.1   Interatomic Potential Models

The most obvious physical feature of interatomic interactions is that atoms do not like to get too close to each other. This is why all solid materials assume a certain volume and resist compression. While this effect is quantum mechanical in nature (due to repulsion between the electron clouds around each atom), it is simple to account for by making the potential energy increase when the distance between any two atoms becomes small. Probably the simplest model that accounts for the short-range repulsion is the "hard-sphere" model, where the energy becomes infinite whenever the distance between two atoms gets smaller than σ_0 (the diameter of the spheres), i.e.

$$V(\{\mathbf{r}_i\}) = \sum_{i=1}^{N-1} \sum_{j=i+1}^{N} \phi(|\mathbf{r}_i - \mathbf{r}_j|), \tag{2.3}$$

where

$$\phi(r) = \begin{cases} +\infty, & r < \sigma_0 \\ 0, & r \geq \sigma_0 \end{cases}, \tag{2.4}$$

as shown in Fig. 2.1. Even though the hard-sphere potential is not a very realistic model for atoms, computer simulations based on this model have contributed much to the understanding of the atomistic structure of liquids.

   The other important aspect of the interatomic interaction is that atoms attract each other at longer distances. This is why atoms aggregate into various bulk forms, such as a solid crystal of silicon or a liquid droplet of water. A well-known model

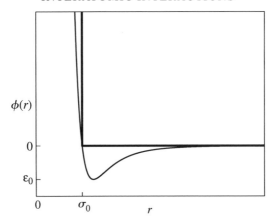

FIG. 2.1. Interaction energy $\phi(r)$ as a function of the distance between two atoms. The thick line is the hard-sphere model and the thin line is the Lennard-Jones model.

that describes both long-range attraction and short-range repulsion between atoms is the Lennard-Jones (LJ) potential,

$$\phi(r) = 4\epsilon_0 \left[\left(\frac{r}{\sigma_0}\right)^{-12} - \left(\frac{r}{\sigma_0}\right)^{-6} \right], \tag{2.5}$$

also plotted in Fig. 2.1. Here, ϵ_0 is the depth of the energy well and $2^{1/6}\sigma_0$ is the distance at which the interaction energy between two atoms reaches the minimum.

The two potential energy functions considered so far are constructed under the assumption that the potential energy can be written as a sum of interactions between pairs of atoms, eq. (2.3). Model potential functions of this type are called *pair potentials*. Relatively few materials, among them the noble gases (He, Ne, Ar, etc.) and ionic crystals (e.g. NaCl), can be described by pair potentials with reasonable accuracy. For most other materials pair potentials do a poor job, especially in the solid state. For example, it is well known that all pair potential models are bound to produce equal values for C_{12} and C_{44}, which are two different elastic constants for cubic crystals.[1] This is certainly not true for most cubic semiconductors and metals.

There are several well established ways to go beyond pair potentials. One approach is to represent the many-body potential energy as a sum of two-body,

---

[1] Elastic properties of crystals with cubic symmetry are fully characterized by three elastic constants: C_{11}, C_{12} and C_{44}. The first two describe the response to tension while C_{44} describes the response to shear.

three-body, four-body, and all the way up to N-body terms, i.e.

$$V(\{\mathbf{r}_i\}) = \sum_{i<j} \phi(|\mathbf{r}_i - \mathbf{r}_j|) + \sum_{i<j<k} V_3(\mathbf{r}_i, \mathbf{r}_j, \mathbf{r}_k)$$

$$+ \sum_{i<j<k<l} V_4(\mathbf{r}_i, \mathbf{r}_j, \mathbf{r}_k, \mathbf{r}_l) + \cdots . \tag{2.6}$$

The hope is that such a series converges fast enough to justify truncation of the expansion at some low-order term while retaining an accurate description of inter-atomic interactions.[2] As an example, the Stillinger–Weber (SW) potential [18] for semiconductor silicon contains two-body and three-body terms. The three-body terms of the SW potential are proportional to

$$\left(\cos\theta_{jik} + \tfrac{1}{3}\right)^2, \tag{2.7}$$

where θ_{jik} is the angle between bond ij and bond ik, as shown in Fig. 2.2. Such terms penalize the deviation of the bond angle away from the tetrahedral angle (109.74°) and help to stabilize the diamond-cubic structure. Physically, this three-body term reflects the strong bond directionality of most tetrahedrally bonded covalent semiconductors.

While electrons in dielectric and semiconductor solids form well-localized covalent bonds, in metals they are more diffuse and shared by many atoms. This different behavior of bonding electrons suggests that many-body effects in the inter-atomic interactions in metals should be accounted for in a different manner. The embedded-atom model (EAM) is a simple and widely used interatomic potential

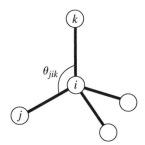

FIG. 2.2. In the diamond-cubic structure, every atom is bonded with four other atoms. The angle between any two bonds involving the same atom, such as θ_{jik}, equals 109.47°, i.e. $\cos\theta_{jik} = -1/3$.

[2] A low-order truncation is not always feasible for an arbitrary many-body function $V(\{\mathbf{r}_i\})$ [17].

for metals. Its functional form is

$$V(\{\mathbf{r}_i\}) = \sum_{i<j} \phi(|\mathbf{r}_i - \mathbf{r}_j|) + \sum_i F(\rho_i), \tag{2.8}$$

$$\rho_i = \sum_{j\neq i} f(r_{ij}), \tag{2.9}$$

where $r_{ij} = |\mathbf{r}_i - \mathbf{r}_j|$. The first term in eq. (2.8) is the usual pairwise interaction accounting for the effect of core electrons. ρ_i is the local density of bonding electrons supplied by the atoms neighboring with atom i. Function $f(r)$ specifies the contribution of an atom to the electron density field. Finally, $F(\rho_i)$ is an embedding function defining the energy required to embed atom i into an environment with electron density ρ_i. For example, the form $F(\rho) = -A\sqrt{\rho}$ is used in the Finnis–Sinclair (FS) potential [19]. Other similar potentials, such as effective medium theory (EMT) models, use somewhat different forms for the electron density contribution function $f(r)$ and the embedding function $F(\rho)$ [17].

Because the embedding function is non-linear, the EAM-like potentials include many-body effects that cannot be expressed by a superposition of pair-wise interactions. As a result, EAM potentials can be made more realistic than pair potentials. For example, EAM potentials give rise to non-zero values of the Cauchy pressure $(C_{12} - C_{44})$ and can be fitted to accurately reproduce the elastic constants of metals.

Numerous interatomic potentials have been developed for atomistic simulations of various types of materials. For some materials there is a choice of potentials of varying degrees of accuracy and computational complexity. For example, there are more than 30 interatomic potentials for silicon alone and new potentials are likely to be developed in the future. In such a situation, it is not always obvious how to choose a potential that best fits the research objectives. In order to make reasonable choices, it is useful to appreciate what goes into the development of an interatomic potential.

Development of an interatomic potential is not a straightforward process. The first step is usually the selection of a functional form that can capture the underlying physics of interatomic interaction in the material of interest. Although inspired by physical considerations, this choice is often intuitive and hence subjective. The functional form should contain a sufficient number of adjustable parameters so that the potential can be fitted to reproduce a selected set of material properties. The fitting database of the material properties can include experimental data, or *ab initio* calculation results, or both. The selection of material properties to fit to is also a matter of subjective judgement and often reflects which material behaviors the developer is most interested in. Because fitting to empirical data is often involved in their development, the interatomic potential functions are sometimes referred to as *empirical potentials*, to reflect their less-than-rigorous physical foundation.

The key issue in developing a good interatomic potential is its *transferability*, the ability to accurately describe or predict a material's property that is *not* included in its fitting database. Transferability of an empirical potential is by no means guaranteed. As a rule of thumb, the further away from the fitted behavior the potential is used, the poorer the description it provides. It is not surprising that some users regard interatomic potentials as nothing more than interpolation functions between the data points they were fitted to. In general, it is wise to be cautious in the interpretation of simulation results obtained with the use of interatomic potentials, especially if the potential is used far from its demonstrated applicability range.

In this book we are mostly interested in the dynamics of crystal dislocations and dislocation-related mechanical properties of crystals. In this context, the physical parameters that matter most are the relative energies of the most stable crystal structures (Section 1.1), elastic constants, point defect energies and stacking fault energies. Since these properties are related to atomic structures associated with dislocations, an interatomic potential fitted to these parameters stands a better chance of describing dislocation behaviors accurately.

2.1.2   Locality of Interatomic Interactions

Interatomic interactions are usually short range (exceptions include Coulomb interactions in ionic crystals such as NaCl). Intuitively, displacement of one atom from its initial position should only cause appreciable forces in a more or less localized neighborhood of this atom, even though the agents of this interaction, the electrons, could be completely delocalized in space. Take the Lennard-Jones (LJ) potential (Fig. 2.1) as an example. The magnitude of pair interaction energy already decreases down to $\sim 10^{-3}\epsilon_0$ when the distance between two atoms becomes larger than $4\sigma_0$. To improve numerical efficiency, it is common to truncate interatomic potentials, i.e. by setting the interaction energy to zero whenever the distance between two atoms exceeds a cut-off radius r_c. However, a simple truncation like this would lead to a (small) discontinuity of energy at $r = r_c$. This discontinuity can lead to undesirable artifacts in simulations. It is a simple matter to "smooth out" the LJ pair potential function so that both the energy and its derivative become zero at the cut-off distance (see Problem 2.1.1).

For short-range potentials, calculation of the force on a given atom requires the positions of its neighbors only within the cut-off radius. Taking the truncated LJ potential as an example, the force on atom i is

$$\mathbf{f}_i = \sum_j -\frac{\partial \phi(|\mathbf{r}_i - \mathbf{r}_j|)}{\partial \mathbf{r}_i}, \qquad (2.10)$$

where the sum is over all atoms within radius r_c around atom i. Depending on the potential model, the average number of interacting neighbors (per atom) within the cut-off sphere ranges from a few to a few hundred atoms. In solids, this is a relatively constant number, which is related to the atomic density of the material. Hence, the time required to compute the potential energy and atomic forces for short-range potential scales as $\mathcal{O}(N)$. This linear scaling is an important advantage of using the interatomic potentials.[3]

Another useful property of interatomic potential models is that it is usually straightforward to partition the potential energy into local energy contributions for each atom. Although largely artificial and somewhat arbitrary (only the potential energy of the entire system has physical meaning), this partitioning is useful in data analysis and visualization of localized lattice defects (see Section 3.3). For potentials constructed as cluster expansions, e.g. eq. (2.6), the convention is to add half of every pair-wise interaction term (such as in the LJ potential) and one third of every three-body term (such as in the SW potential) to the local energy of each participating atom. For the EAM-like potentials, the local energy of atom i can be defined as

$$E_i = \sum_j \frac{1}{2} \phi(|\mathbf{r}_i - \mathbf{r}_j|) + F(\rho_i). \qquad (2.11)$$

It is easy to check that the sum of local energies of all atoms is equal to the total potential energy, eq. (2.8), as it should be.

2.1.3 Computational Cost of Interatomic Interaction Models

As has already been mentioned, the first principles theory describes the interatomic interactions more accurately than the interatomic potentials but at a much higher computational cost. Among first principles methods, those based on the density functional theory (DFT) have become very popular in the computational materials science community [21]. The tight binding (TB) models present another popular alternative whose accuracy and efficiency lie somewhere between DFT and empirical potentials. In the following, we compare the computational cost among three models of silicon that are representative of DFT, TB and interatomic potential models. We hope that even a rough comparison will be helpful in better appreciating the computational limits of various models. For this estimate, we use reasonably optimized codes, specifically VASP [22] for DFT calculations, TBMD [23] for TB, and MD++ [11] with the SW potential for silicon.

The results of a series of benchmark calculations are presented in Fig. 2.3, showing the amount of time required in these three models to compute the potential energy and atomic forces for silicon crystals with various numbers of atoms. To

---

[3] In practice, we need to implement a *neighbor list* in order to achieve the $\mathcal{O}(N)$ scaling. A neighbor list provides the indices of all atoms that fall within the cut-off distance of any atom. To retain the overall $\mathcal{O}(N)$ scaling, the neighbor list itself should be constructed in $\mathcal{O}(N)$ time [20].

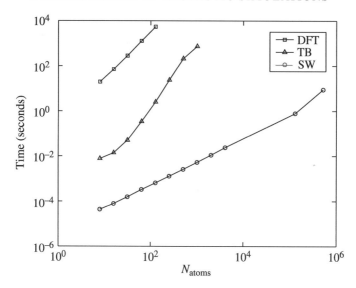

FIG. 2.3. Computer time (in seconds) expended on evaluation of the total energy and atomic forces for a silicon crystal with different numbers of atoms using DFT, Tight Binding (TB) and Stillinger–Weber (SW) interatomic potential models.

avoid uncertainties associated with parallel machines, all calculations were performed on a single-processor Linux Alpha workstation, with 666 MHz clock speed and 1.33 GFlop peak performance.

As shown in Fig. 2.3, for systems larger than a few hundred atoms DFT calculations become prohibitively expensive on our benchmark workstation. The feasibility limit for TB calculations is higher, over a thousand atoms.[4] On the other hand, the computing time for the empirical potential model is not only much shorter but also scales linearly with the total number of atoms. Thus simulations of millions of atoms are feasible on a desktop workstation. Feasibility limits can be pushed significantly upward by using many processors in parallel. The gain due to parallel computing is again most impressive for the interatomic potential models because of their linear scaling. Molecular dynamics simulations of billions of atoms have been performed on parallel machines with thousands of processors [24].

---

[4] The actual time for one step in real simulations is likely to be different from this benchmark data. For example, in DFT and TB methods, the electronic wave functions need to be relaxed to their ground states to compute potential energy and atomic forces. The relaxation of wave functions usually take more time in the first step (recorded here) than in subsequent steps because they do not change much from one step to the next. DFT usually scales as $\mathcal{O}(N^3)$ at large N on the more powerful parallel computers, whereas on single CPU workstations an $\mathcal{O}(N^2)$ scaling is usually observed. To extend their range of applicability, $\mathcal{O}(N)$ methods are being developed for DFT and TB calculations.

To choose an appropriate model for atomistic simulations several factors need to be considered. The final selection is most often a compromise between computational expediency and theoretical rigor. If the problem of interest involves a large number of atoms far exceeding the feasibility limit of DFT and TB models, classical potential models may be the only choice. For some other materials, such as novel alloys for which no empirical potential or semi-empirical TB model exists, DFT becomes the method of choice. However, if the situation of interest demands both high accuracy and a large number of atoms simultaneously, one may need to develop a new potential, which is usually a very tedious task. The lack of computationally expedient approaches to modelling alloys and other "realistically dirty" materials is probably the greatest limitation for atomistic modelling in general and dislocation simulations in particular.

To improve the accuracy and to expedite atomistic simulations, different models, such as DFT and empirical potential, can be used in combination. For example, a "cheap" empirical potential model can be used to screen a large family of crystal defect configurations and select candidates that are most likely to control the material behavior of interest. These observations often call for a "heavy artillery fire", suggesting targets for further calculations using more accurate and expensive methods such as TB and DFT. It is also possible to use inexpensive interatomic potential models to estimate and correct errors in TB and DFT calculations because of their limited number of atoms (Chapter 5). Yet another very interesting possibility is to apply different models to different regions in the same simulation. For example, in atomistic simulations of cracks, a few atoms near the crack tip can be treated by more accurate DFT or TB methods, to obtain an accurate description of the bond breaking process. At the same time, atoms farther away from the crack tip experience much smaller deformations and can be well described by an empirical potential model. How to handle interfaces between regions described by different models is a subject of active research (see [25] as an example). The models available for atomistic simulations are becoming more plentiful and their accuracy continues to improve. Along with the ever increasing computing power, this assures that atomistic simulations will be playing an increasingly important role in future materials research.

Summary

- The intricacies of interaction between atoms in molecules and solids are rooted in the quantum mechanics of electrons.

- First principles approaches, such as DFT, provide accurate descriptions of interatomic interactions but are computationally demanding.

- Empirical potential models are less accurate but allow simulations of a much larger number of atoms. These models are fitted to experimental data and/or first principles calculations.

- Models with different levels of complexity and accuracy can be combined to provide a more accurate description at a more reasonable computational cost.

Problem

2.1.1. It is easy to make the Lennard-Jones potential strictly short range by setting the energy to zero beyond the cut-off distance, $r > r_c$. However, the resulting modified LJ potential will be discontinuous at $r = r_c$. To avoid artifacts in atomistic simulations, it is best to make sure that both the energy and its first derivative vanish smoothly at $r = r_c$. Specifically, for the LJ potential this can be done by adding a linear term to $\phi(r)$ when $r < r_c$, i.e.,

$$\phi(r) = \begin{cases} 4\epsilon_0[(r/\sigma_0)^{-12} - (r/\sigma_0)^{-6}] + A + Br, & r < r_c \\ 0, & r \geq r_c \end{cases}. \quad (2.12)$$

Find values for A and B such that the truncated LJ potential and its first derivative with respect to r become zero at $r = r_c$.

2.2 Equilibrium Distribution

Consider a system containing N atoms of the same mass m interacting via an interatomic potential $V(\{r_i\})$. The total energy of such a system is

$$H(\{\mathbf{r}_i, \mathbf{p}_i\}) = \sum_{i=1}^{N} \frac{|\mathbf{p}_i|^2}{2m} + V(\{\mathbf{r}_i\}). \quad (2.13)$$

Function H is called the Hamiltonian. The first term on the right hand side is the sum of the kinetic energies of all atoms in which $\mathbf{p}_i = m\mathbf{v}_i$ is the momentum vector of atom i and $\mathbf{v}_i = d\mathbf{r}_i/dt$ is its velocity. In classical mechanics, the instantaneous state of the system is fully specified by the positions and momenta of all atoms $\{\mathbf{r}_i, \mathbf{p}_i\}$, and is usually called a *microstate*. In three dimensions, this corresponds to $6N$ variables. Therefore, the microstate of the system can be regarded as a point in the $6N$-dimensional space spanned by $\{\mathbf{r}_i, \mathbf{p}_i\}$, called the *phase space*.

The theoretical basis of most atomistic simulation methods is the fundamental law of classical statistical mechanics—*Boltzmann's law* [26, 27]. This states that, when the system is in thermal equilibrium at temperature T, the probability density of finding the system near a specific point $\{\mathbf{r}_i, \mathbf{p}_i\}$ in the phase space is $f(\{\mathbf{r}_i, \mathbf{p}_i\})$,

$$f(\{\mathbf{r}_i, \mathbf{p}_i\}) = \frac{1}{Z} \exp\left[-\frac{H(\{\mathbf{r}_i, \mathbf{p}_i\})}{k_B T}\right], \quad (2.14)$$

where

$$Z = \int \prod_{i=1}^{N} \mathrm{d}\mathbf{r}_i \, \mathrm{d}\mathbf{p}_i \exp\left[-\frac{H(\{\mathbf{r}_i, \mathbf{p}_i\})}{k_B T}\right].$$

Here, $k_B = 8.6173 \times 10^{-5}$ eV/K is Boltzmann's constant. Equation (2.14) is the famous *Boltzmann's distribution* for classical particles at thermal equilibrium and Z is called the partition function. Z ensures the proper normalization of the probability density f. Boltzmann's distribution is suitable when the temperature is not too low, so that quantum effects can be ignored[5] (see Problem 2.2.3).

An important concept in statistical mechanics is the *statistical ensemble* of systems. Consider a very large number of replicas of the system, each with N atoms and described by the same Hamiltonian. At any instance of time, each system's state is represented by a point $\{\mathbf{r}_i, \mathbf{p}_i\}$ in the $6N$-dimensional phase space. An *ensemble* is the collection of all these replicas or, equivalently, of points in the $6N$-dimensional phase space. Each replica (a member of the ensemble) may evolve with time, which is reflected by the motion of a corresponding point in the phase space. The replicas do not interact, i.e. the points in the phase space move independently of each other. When the replicas are distributed according to Boltzmann's distribution, i.e. eq. (2.14), the ensemble is called *canonical*.

Boltzmann's law makes it possible to express macroscopic quantities of the system at thermal equilibrium as a statistical average of microscopic functions over the canonical ensemble. Given a microscopic function $A(\{\mathbf{r}_i, \mathbf{p}_i\})$, which depends on the positions and momenta of all atoms, its (canonical) ensemble average is

$$\langle A \rangle \equiv \frac{1}{Z} \int \prod_{i=1}^{N} \mathrm{d}\mathbf{r}_i \, \mathrm{d}\mathbf{p}_i \, A(\{\mathbf{r}_i, \mathbf{p}_i\}) \exp\left[-\beta H(\{\mathbf{r}_i, \mathbf{p}_i\})\right], \qquad (2.15)$$

where $\beta \equiv 1/(k_B T)$. For example, the average potential energy U of the system is

$$U = \langle V \rangle \equiv \frac{1}{Z} \int \prod_{i=1}^{N} \mathrm{d}\mathbf{r}_i \, \mathrm{d}\mathbf{p}_i \, V(\{\mathbf{r}_i\}) \exp\left[-\beta H(\{\mathbf{r}_i, \mathbf{p}_i\})\right]$$

$$= \frac{\int \prod_{i=1}^{N} \mathrm{d}\mathbf{r}_i \, V(\{\mathbf{r}_i\}) \exp\left[-\beta V(\{\mathbf{r}_i\})\right]}{\int \prod_{i=1}^{N} \mathrm{d}\mathbf{r}_i \exp\left[-\beta V(\{\mathbf{r}_i\})\right]}. \qquad (2.16)$$

Because the Hamiltonian has a simple dependence on atomic momenta, the momentum part of the integral for the Boltzmann's partition function Z can be

---

[5] Quantum-mechanical particles satisfy different types of distributions, such as Bose–Einstein and Fermi–Dirac distributions.

separated out, i.e.

$$f(\{\mathbf{r}_i, \mathbf{p}_i\}) = \frac{1}{Z} \prod_{i=1}^{N} \exp\left[-\frac{|\mathbf{p}_i|^2}{2mk_BT}\right] \exp\left[-\frac{V(\{\mathbf{r}_i\})}{k_BT}\right]. \qquad (2.17)$$

Physically, this means that at thermal equilibrium the momentum of each atom obeys an independent and identical distribution. The average kinetic energy of an atom j is

$$\langle E_{\text{kin},j} \rangle = \left\langle \frac{|\mathbf{p}_j|^2}{2m} \right\rangle \equiv \frac{1}{Z} \int \prod_{i=1}^{N} \mathrm{d}\mathbf{r}_i \, \mathrm{d}\mathbf{p}_i \, \frac{|\mathbf{p}_j|^2}{2m} \exp\left[-\beta H(\{\mathbf{r}_i, \mathbf{p}_i\})\right]$$

$$= \frac{\int \mathrm{d}\mathbf{p}_j \frac{|\mathbf{p}_j|^2}{2m} \exp\left[-\frac{|\mathbf{p}_j|^2}{2mk_BT}\right]}{\int \mathrm{d}\mathbf{p}_j \exp\left[-\frac{|\mathbf{p}_j|^2}{2mk_BT}\right]} = \frac{3}{2} k_B T. \qquad (2.18)$$

The average kinetic energy of the entire system is,

$$\langle E_{\text{kin}} \rangle = \sum_{j=1}^{N} \langle E_{\text{kin},j} \rangle = \frac{3}{2} N k_B T. \qquad (2.19)$$

Notice that these results are independent of the atomic mass m. They are the direct consequence of the quadratic dependence of the Hamiltonian on atomic momentum. Similarly, if the system is harmonic, meaning that the potential $V(\{\mathbf{r}_i\})$ is a quadratic function of the atomic positions, the average potential energy of the system is $U = \frac{3}{2} N k_B T$. Therefore, the average total energy of a harmonic system is $E_{\text{tot}} = E_{\text{kin}} + U = 3 N k_B T$, to which the kinetic and potential energies contribute equally. For a crystal well below its melting temperature, the potential energy function can be approximated by a set of harmonic (quadratic) terms through a Taylor expansion around the energy minimum (see Section 6.2). In such a case, we would expect the total energy of the system to divide approximately equally between kinetic and potential energy parts.

Summary

- Boltzmann's distribution describes statistical properties of a classical system of atoms at thermal equilibrium.

- Some of the statistical averages over the canonical ensemble are particularly simple. The average kinetic energy of a system of N atoms is $\frac{3}{2} N k_B T$. When the potential energy is a quadratic function of atomic coordinates, the average potential energy is also $\frac{3}{2} N k_B T$.

Problems

2.2.1. The Hamiltonian of a one-dimensional harmonic oscillator is

$$H(x, p_x) = \frac{p_x^2}{2m} + \frac{1}{2}kx^2. \tag{2.20}$$

Assuming that an ensemble of such oscillators satisfies Boltzmann's distribution, find the average kinetic energy and the average total energy over this ensemble at temperature T.

2.2.2. The Hamiltonian of a three-dimensional Harmonic oscillator is

$$H(\mathbf{r}, \mathbf{p}) = \frac{|\mathbf{p}|^2}{2m} + \frac{1}{2}k|\mathbf{r}|^2. \tag{2.21}$$

Find the distribution function $f(E)$ for the energy of an ensemble of such oscillators in thermal equilibrium at temperature T.

2.2.3. In quantum mechanics, every particle possesses a de Broglie wavelength $\lambda = h/p$, where $h = 6.626068 \times 10^{-34}$ m^2 kg s^{-1} is Planck's constant, $p = mv$ is the momentum, and v is the velocity. The motion of interacting particles can be treated classically when the typical distance between them exceeds λ. In thermal equilibrium, the average kinetic energy of a classical particle is $\langle \frac{1}{2}m|\mathbf{v}|^2 \rangle = \frac{3}{2}k_B T$. Taking a copper crystal as an example, estimate the temperature below which the motion of its atoms will have to be treated quantum mechanically. Copper has the FCC crystal structure with a lattice constant $a = 3.61$ Å and atomic mass of $m = 63.546$ a.u. (1 a.u. $= 1.6604 \times 10^{-27}$ kg).

2.3 Energy Minimization

According to Boltzmann's distribution, the probability of finding the system at a microstate $\{\mathbf{r}_i, \mathbf{p}_i\}$ decreases exponentially with increasing energy $H(\{\mathbf{r}_i, \mathbf{p}_i\})$. In the low-temperature limit, this exponential dependence means that the system is most likely to be found in the neighborhood of the global minimum of $H(\{\mathbf{r}_i, \mathbf{p}_i\})$. From eq. (2.13), the minimum of $H(\{\mathbf{r}_i, \mathbf{p}_i\})$ corresponds to $\mathbf{p}_i = 0$ for all i and $\{\mathbf{r}_i\}$ at the minimum of the potential function $V(\{\mathbf{r}_i\})$. Therefore, the minimum of the potential energy function $V(\{\mathbf{r}_i\})$ provides a good description of the atomic structure of the system at low temperatures. Searching for minima of a function of many variables is an extensively studied subject in computational sciences. Many algorithms exist [28], and the development of still more efficient methods remains an active area of research.

To simplify notation for the following discussions, let us define a $3N$-dimensional vector $\mathbf{R} = (x_1, y_1, z_1, x_2, y_2, z_2, \ldots, x_N, y_N, z_N)^\mathrm{T}$, where x_i, y_i, z_i are the cartesian coordinates of atoms in a system of N atoms. The space spanned by $\mathbf{R}$ is called the *configurational space*. Now we can rewrite the potential energy $V(\{\mathbf{r}_i\})$ as $V(\mathbf{R})$. The $3N$-dimensional vector of force is then $\mathbf{F} = -\partial V(\mathbf{R})/\partial \mathbf{R}$ or, in the component form, $\mathbf{F} = (f_{x_1}, f_{y_1}, f_{z_1}, \ldots, f_{x_N}, f_{y_N}, f_{z_N})^\mathrm{T}$, where $f_{x_i} = -\partial V/\partial x_i$ is the force on atom i along the x direction. Let us define the magnitude of force $\mathbf{F}$ as $|\mathbf{F}| = (f_{x_1}^2 + f_{y_1}^2 + \cdots + f_{z_N}^2)^{1/2}$. These definitions will simplify the following discussions.

2.3.1 The Steepest-descent Method

Steepest descent is a simple, although not very efficient, iterative algorithm for finding a (local) minimum of $V(\mathbf{R})$ starting from an arbitrary initial configuration $\mathbf{R}$. At every iteration, the force vector $\mathbf{F}$ is computed and $\mathbf{R}$ is displaced by a small step along $\mathbf{F}$. The iterations continue until $|\mathbf{F}|$ becomes smaller than a prescribed tolerance ϵ. A simple steepest-descent algorithm is given below, which requires the step size Δ to be specified as an input.

Algorithm 2.1

1. $\mathbf{F} := -\partial V(\mathbf{R})/\partial \mathbf{R}$

2. If $|\mathbf{F}| < \epsilon$, exit.

3. $\mathbf{R} := \mathbf{R} + \mathbf{F}\Delta$. Go to 1.

The idea of the steepest-descent algorithm is that, as long as $|\mathbf{F}|$ is non-zero, $V(\mathbf{R})$ can be further reduced by displacing $\mathbf{R}$ in the direction of $\mathbf{F}$. This algorithm is equivalent to a numerical integration of an over-damped equation of motion,

$$\mathbf{f}_i - \gamma \mathbf{v}_i = 0, \tag{2.22}$$

where $\mathbf{v}_i$ is the velocity of atom i and γ is the friction coefficient. This equation describes the dynamics of a set of interacting particles in a highly viscous medium.[6] After a sufficiently long time, the particles will eventually arrive at a structure corresponding to a local energy minimum for which all forces vanish.

Although very simple, the *steepest-descent* algorithm is rarely used in atomistic simulations because it is numerically inefficient, often requiring many steps to converge to a minimum. The second problem with this method is that, even when a minimum energy state is found, it is not guaranteed to be the global minimum. In

---

[6] To make the analogy exact, Δ in Algorithm 2.1 corresponds to $\Delta t/\gamma$, where Δt is the time step for the numerical integration of eq. (2.22).

fact, potential energy functions constructed from the interatomic potentials typically have multiple local minima. A steepest-descent search can converge to any one of them, depending on the initial atomic configuration. In the following, we discuss slightly more sophisticated approaches that address these two problems.

2.3.2 Conjugate Gradient Relaxation

Closely related to the steepest descent method is the *conjugate gradient relaxation* (CGR) algorithm. CGR relies on exactly the same information as steepest descent, i.e. atomic forces, but uses it in a more intelligent way. CGR goes through a series of search directions. The (local) minimum energy point along each search direction is reached before CGR proceeds to the next search direction. The search sequence is constructed in such a way that subsequent search directions "avoid" (i.e. are conjugate to) all previously searched directions. This is the key to the greater efficiency of the CGR algorithm compared to the steepest-descent method.

The CGR algorithm works best in an idealized situation when the potential energy is a quadratic function, i.e. $V(\mathbf{R}) = \frac{1}{2}\mathbf{R}^T \cdot \mathbf{G} \cdot \mathbf{R}$, where $\mathbf{G}$ is a $3N \times 3N$ symmetric matrix. The *conjugate* condition means that any two search directions $\mathbf{d}^{(n)}$ and $\mathbf{d}^{(m)}$ ($n \neq m$) must satisfy

$$\mathbf{d}^{(n)T} \cdot \mathbf{G} \cdot \mathbf{d}^{(m)} = 0. \tag{2.23}$$

One can prove that that the minimum (in this case it is $\mathbf{R} = 0$) is reached in no greater than $3N$ searches provided that the consecutive search directions are all conjugate to each other.[7] Making sure that the current search direction is conjugate to all previous search directions seems to be a daunting task, especially if the relaxation requires hundreds of search directions. The key to the success of CGR is to make sure that the current search direction $\mathbf{d}^{(n)}$ is conjugate to the previous one $\mathbf{d}^{(n-1)}$. Then, it is possible to show that $\mathbf{d}^{(n)}$ is also conjugate to all the previous search directions $\mathbf{d}^{(m)}$ [28]. The algorithm can be described as follows.

Algorithm 2.2

1. Initialize iteration counter $n := 1$

2. Compute forces $\mathbf{F} := -\partial V(\mathbf{R})/\partial \mathbf{R}$. If $|\mathbf{F}| < \epsilon$, exit.

3. If $n = 1$, $\mathbf{d}^{(n)} := \mathbf{F}^{(n)}$, i.e. the search direction is the same as the force direction; otherwise, $\gamma^{(n)} := [\mathbf{F}^{(n)} \cdot \mathbf{F}^{(n)}]/[\mathbf{F}^{(n-1)} \cdot \mathbf{F}^{(n-1)}]$, $\mathbf{d}^{(n)} := \mathbf{F}^{(n)} + \gamma^{(n)}\mathbf{d}^{(n-1)}$;

4. Find an energy minimum along direction $\mathbf{d}^{(n)}$, i.e. find x_0 that is a local minimum of the one-dimensional function $V(x) \equiv V(\mathbf{R} + x\mathbf{d}^{(n)})$.

---

[7] In practice, the search brings the system close to the minimum much earlier than that.

5. $\mathbf{R} := \mathbf{R} + x_0 \mathbf{d}$.

6. $n := n + 1$. Go to 2.

Given its computational efficiency CGR is often used in atomistic simulations, including several examples in this book. The CGR algorithm has been implemented in MD++ [11], which works well in simulations of up to about half a million atoms.

2.3.3   Global Minimization

Neither the CGR nor the steepest-descent algorithm guarantees finding the global minimum of an energy function. To better understand the nature of this problem, Fig. 2.4 illustrates the difference between local minima and the global minimum. Function $V(x)$ has three local minima, E_0, E_1 and E_2, at $x = x_0$, x_1 and x_2 respectively. Among the three, E_0 is the lowest or global minimum. Depending on the choice of the initial point, both minimization algorithms discussed so far can converge to one of the local minima E_1 or E_2. If the global minimum is the target, a brute-force approach is to run minimization many times starting from randomly selected initial configurations. Then, the lowest local minimum found from a series of relaxations provides an upper bound on the global minimum. Obviously, such an approach is very inefficient.[8]

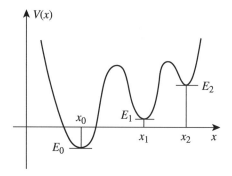

FIG. 2.4. A function $V(x)$ with multiple local minima: E_0, E_1 and E_2. The lowest of the minima, E_0, is the global minimum.

---

[8] On the other hand, we are not always interested in the global energy minimum of the potential energy function in the entire configurational space. The global energy minimum state of a large collection of atoms is usually the perfect crystal. Thus, the system is by definition not in its global energy minimum if we are studying dislocations. Therefore, what we are usually interested in is a "global" minimum within a certain domain of the configurational space, such as the minimum energy structure of a dislocation (Chapter 3).

Some of the more efficient methods for finding the global minimum of functions of many variables are motivated by processes occurring in nature. For example, genetic algorithms invoke mutation and natural selection to evolve a collection of systems in search for the fittest (having the lowest energy) [29, 30, 31]. Another useful method is *simulated annealing* (SA), which mimics the thermal annealing of a material, as its name suggests [32, 33].

SA requires methods to simulate atomistic systems at non-zero temperatures. Both Monte Carlo (MC) and molecular dynamics (MD) methods described in following sections can be used for this purpose. At low temperatures, the system can become trapped in the neighborhood of a local energy minimum that depends on the choice of the initial configuration. To lessen this unwanted dependence, SA simulation starts at a high temperature, making it easier for the system to escape from the traps of local minima and explore a larger volume in the configurational space. The SA search then slowly reduces the temperature to zero. At the end of this procedure, the system converges to a local minimum that is likely to be less dependent on the initial configuration, and is more likely to be the global minimum. To further reduce the statistical uncertainty, it is advisable to run SA several times, starting from different initial configurations. In this way a tight upper bound on the global minimum can be obtained, provided that the temperature varies with time sufficiently slowly. How long the SA simulation runs is usually dictated by the available computing resources. At the same time, there should exist an optimal annealing schedule, i.e. temperature T as a function of time t, for a given total annealing time. The optimal form of $T(t)$ is system dependent and generally not known in advance. In practice, annealing functions of the form $T(t) = T_0 \exp(-t/\tau)$ are commonly used, where τ is a prescribed time scale of temperature reduction.

While the MC and MD methods can be combined with SA to search for the global minimum of the potential function, they are often used by themselves to study the finite temperature properties of atomistic systems. The relaxed atomistic configuration from local minimization algorithms described in this section (such as CGR) is usually used as initial conditions for MC and MD simulations.

Summary

- An atomistic configuration with the lowest possible potential energy is the most likely state in the limit of zero temperature.

- The steepest-descent and the conjugate gradient relaxation algorithms are used for finding local energy minima. Finding the global energy minimum is a much more difficult task. The method of simulated annealing can provide a tight upper bound on the global minimum.

2.4 Monte Carlo

According to Boltzmann's distribution, eq. (2.14), a system at finite temperature does not spend all of its time in the minimum energy state. Instead, the system may be found in any state in the phase space, with a finite probability exponentially dependent on the state's energy. Various thermodynamic quantities of the system can be written as ensemble averages of the appropriate microscopic functions over all possible states, as in eq. (2.15). This is a powerful statement that connects experimentally measurable, macroscopic properties of materials to the fundamental interactions between atoms. However, direct integration of eq. (2.15) to obtain ensemble averages is not practical except for highly idealized systems. Numerical integration (i.e. quadrature) in $6N$-dimensional space becomes prohibitively expensive even for relatively small N (e.g. $N > 10$).

This is where simulation methods, such as Monte Carlo (MC) and molecular dynamics (MD) come to the rescue. The idea is to simulate the motion of atoms at a finite temperature in such a way that the fraction of time the system spends in each state satisfies Boltzmann's distribution. When this is the case, the ensemble average in eq. (2.15) can be replaced by a time average over the simulation period t_{sim}, i.e.

$$\langle A \rangle_t = \frac{1}{t_{sim}} \int_0^{t_{sim}} A(\{\mathbf{r}_i(t), \mathbf{p}_i(t)\}) \, dt. \qquad (2.24)$$

2.4.1 Average over the Configurational Space

If the microscopic function depends only on the atomic configuration $\mathbf{R}$, i.e. $A = A(\mathbf{R})$, the ensemble average can be written in terms of an integral over the configurational space,

$$\langle A \rangle = \int d\mathbf{R} \, A(\mathbf{R})\rho(\mathbf{R}), \qquad (2.25)$$

where $\rho(\mathbf{R})$ is Boltzmann's distribution function in the configurational sub-space of the phase space and $Z_\mathbf{R}$ is the configurational part of the partition integral,

$$\rho(\mathbf{R}) = \frac{1}{Z_\mathbf{R}} \exp[-\beta V(\mathbf{R})] \qquad (2.26)$$

$$Z_\mathbf{R} = \int d\mathbf{R} \exp[-\beta V(\mathbf{R})], \quad \beta \equiv 1/(k_B T). \qquad (2.27)$$

The idea of the Monte Carlo method is to generate a stochastic process in the configurational space such that the states visited during the simulation satisfy the equilibrium distribution $\rho(\mathbf{R})$.[9] If this is the case, then the ensemble average $\langle A \rangle$ can be replaced by an average of $A(\mathbf{R})$ over the trajectory of the Monte Carlo simulation. The stochastic process is artificial and its sole purpose is to reproduce a desired distribution function $\rho(\mathbf{R})$. Hence "time" in such MC simulations does not have much physical meaning. However, a kinetic version of the Monte Carlo method exists in which each MC step corresponds to a certain physical time interval. The kinetic Monte Carlo (kMC) method will be discussed in Chapter 9.

2.4.2 Designing a Stochastic Monte Carlo Process

What kind of stochastic process can generate an ensemble of configurations that would satisfy Boltzmann's distribution? Most Monte Carlo methods simulate a Markov process, meaning that the probability of moving to a specific state in the next step is only a function of the current state and is independent of the states visited in the preceding steps. A Markov process is specified by its *transition probability matrix*, $\pi(\mathbf{R}, \mathbf{R}')$, which is the probability density of visiting state $\mathbf{R}'$ in the next step if the system is currently at $\mathbf{R}$. π satisfies the normalization condition,

$$\int d\mathbf{R}'\pi(\mathbf{R}, \mathbf{R}') = 1. \tag{2.28}$$

If the probability distribution of the system at step n is $\rho^{(n)}(\mathbf{R})$, then the probability distribution at step $n+1$ is

$$\rho^{(n+1)}(\mathbf{R}) = \int d\mathbf{R}'\rho^{(n)}(\mathbf{R}')\pi(\mathbf{R}', \mathbf{R}). \tag{2.29}$$

Hence, to sample the equilibrium distribution $\rho(\mathbf{R})$, the Markov process should satisfy

$$\rho(\mathbf{R}) = \int d\mathbf{R}'\rho(\mathbf{R}')\pi(\mathbf{R}', \mathbf{R}). \tag{2.30}$$

Therefore, the task of designing a Monte Carlo algorithm is equivalent to designing a transition probability matrix $\pi(\mathbf{R}, \mathbf{R}')$ such that the solution of eq. (2.30) for $\rho(\mathbf{R})$ is exactly Boltzmann's distribution eq. (2.26).

Many transition probability matrices can reproduce the same equilibrium distribution. To narrow down our choices, let us consider a more strict condition than eq. (2.30), namely

$$\rho(\mathbf{R})\pi(\mathbf{R}, \mathbf{R}') = \rho(\mathbf{R}')\pi(\mathbf{R}', \mathbf{R}). \tag{2.31}$$

Obviously, π and ρ that satisfy condition (2.31) will also satisfy eq. (2.30) but the reverse is not true (Problem 2.4.1). Equation (2.31) is called the *detailed balance*

---

[9] Therefore, a Monte Carlo simulation *samples* the equilibrium distribution.

condition, it states that the net rate of transitions from state $\mathbf{R}$ to state $\mathbf{R}'$ is exactly the same as the net rate of the reverse transitions from $\mathbf{R}'$ to $\mathbf{R}$, when the stationary distribution ρ is reached. Given eq. (2.26), the detailed balance condition leads to the following restriction on the transition probabilities

$$\frac{\pi(\mathbf{R}, \mathbf{R}')}{\pi(\mathbf{R}', \mathbf{R})} = \frac{\rho(\mathbf{R}')}{\rho(\mathbf{R})} = \exp\left[-\beta(V(\mathbf{R}') - V(\mathbf{R}))\right]. \tag{2.32}$$

2.4.3   Metropolis Algorithm

A widely used version of Monte Carlo is the Metropolis algorithm corresponding to the following choice of π,[10]

$$\pi(\mathbf{R}, \mathbf{R}') = \alpha(\mathbf{R}, \mathbf{R}') P_{\text{acc}}(\mathbf{R}, \mathbf{R}'), \quad \text{for } \mathbf{R}' \neq \mathbf{R} \tag{2.33}$$

$$P_{\text{acc}}(\mathbf{R}, \mathbf{R}') \equiv \min\{1, \exp\left[-\beta(V(\mathbf{R}') - V(\mathbf{R}))\right]\}, \tag{2.34}$$

where $\alpha(\mathbf{R}, \mathbf{R}')$ is a symmetric but otherwise arbitrary probability density matrix, i.e.

$$\alpha(\mathbf{R}, \mathbf{R}') = \alpha(\mathbf{R}', \mathbf{R}). \tag{2.35}$$

The two factors on the right-hand side of eq. (2.33) correspond to two sub-steps of a single Monte Carlo step. First, a trial move from the current configuration $\mathbf{R}$ to another configuration $\mathbf{R}'$ is selected with probability $\alpha(\mathbf{R}, \mathbf{R}')$. Usually, the trial moves displace one or several atoms in a random direction, $\mathbf{R}' = \mathbf{R} + \delta\mathbf{R}$. If the potential energy of the trial state $\mathbf{R}'$ is lower than the energy of the current state $\mathbf{R}$, the move is accepted. However, if the potential energy of the new state is higher by $dV = V(\mathbf{R}') - V(\mathbf{R})$, the move is accepted with probability $P_{\text{acc}} = \exp(-\beta dV)$. Otherwise, the move is rejected and the system remains in the old state $\mathbf{R}$ for yet another step. The following is a Metropolis algorithm that computes the average of function $A(\mathbf{R})$, for a system of N atoms with initial positions r_i, $i = 1, 2, \ldots, N$.

Algorithm 2.3

1. Pick an atom i at random, from 1 to N.

2. Draw three random numbers dx, dy, dz uniformly distributed in $[-\Delta, \Delta]$ to form a trial displacement vector $\mathbf{dr} = (dx, dy, dz)$.

3. Displace atom i by $\mathbf{dr}$, $\mathbf{r}_i := \mathbf{r}_i + \mathbf{dr}$.

4. Compute the energy change dV caused by the trial displacement.

5. If $dV \leq 0$, accept the move, compute the new value of function $A(\mathbf{R})$ and include it in the average. Go to 1.

---

[10] Normalization condition of π is satisfied by letting $\pi(\mathbf{R}, \mathbf{R}) = 1 - \int_{\mathbf{R}' \neq \mathbf{R}} d\mathbf{R}' \pi(\mathbf{R}, \mathbf{R}')$.

6. Otherwise draw a random number ξ uniformly distributed in $[0, 1]$.

7. If $\xi \leq \exp(-dV/k_BT)$, accept the move, include the new value of $A(\mathbf{R})$ in the average. Go to 1.

8. Otherwise reject the move, return to the old state by $\mathbf{r}_i := \mathbf{r}_i - d\mathbf{r}$ and include the old value of $A(\mathbf{R})$ one more time in the average. Go to 1.

How to select the amplitude Δ for the trial displacements is somewhat arbitrary.[11] On one hand, it is desirable to use a large Δ to let the simulation explore a large volume in the configurational space as quickly as possible. On the other hand, if Δ is too large, the energy of the trial states will be almost always much higher than the energy of the current state, leading to a low acceptance probability of the trial moves. As a result, the system may stay in the same state for many MC steps. It is a common practice to adjust Δ empirically so that the fraction of accepted moves stays close to 50 per cent during the simulation.

Algorithm 2.3 is a common implementation of the Monte Carlo method in atomistic simulations. However, various modifications exist that are equally applicable. In constructing an MC procedure, the most important requirement is that the trial moves proceed unbiased, i.e. the probability of a trial move from state $\mathbf{R}$ to state $\mathbf{R}'$ equals the probability of a trial move from $\mathbf{R}'$ to $\mathbf{R}$. The symmetry condition, eq. (2.35), eliminates the unwanted bias. It is also necessary to make sure that the simulation trajectory can span, in principle, the entire configurational space. In addition to Algorithm 2.3, many other procedures satisfy these two requirements. For example, the rule for generating the trial moves can involve displacement of several or even all of the atoms at once. Alternatively, it is also possible to modify only one coordinate of one atom in each step.

Summary

- In the Monte Carlo simulations, a sequence of random moves is constructed to generate micro-states whose distribution obeys Boltzmann's statistics in thermal equilibrium.

- The Metropolis algorithm generates the desired equilibrium distribution by making trial moves that are accepted with a probability that depends on the potential energy of the trial state.

Problems

2.4.1. With the help of the normalization condition, eq. (2.28), show that the detailed balance condition eq. (2.31) is a sufficient condition for eq. (2.30).

---

[11] This reflects the arbitrariness of $\alpha(\mathbf{R}, \mathbf{R}')$ in eq. (2.33).

2.4.2. Show that, when $\pi(\mathbf{R}, \mathbf{R}')$ satisfies eq. (2.33), it satisfies eq. (2.32) as well.

2.4.3. The Barker algorithm is an alternative to the Metropolis algorithm. In the former, the acceptance probability in eq. (2.34) is

$$P_{\text{acc}}(\mathbf{R}, \mathbf{R}') = \{1 + \exp[\beta(V(\mathbf{R}') - V(\mathbf{R}))]\}^{-1}. \qquad (2.36)$$

Show that the resulting transition matrix $\pi(\mathbf{R}, \mathbf{R}')$ satisfies eq. (2.32) so that, just as in the Metropolis algorithm, its equilibrium distribution is Boltzmann's distribution. Because the acceptance probability P_{acc} in the Barker algorithm is always smaller than that in the Metropolis algorithm, it is not as efficient.

2.5 Molecular Dynamics

Unlike the Monte Carlo method, which generates artificial trajectories spanning the configurational space and complying with Boltzmann's distribution, molecular dynamics (MD) attempts to simulate the "true" dynamics of atoms while also preserving Boltzmann's statistics.[12] At moderate to high temperatures, quantum effects can be ignored (see Problem 2.2.3). The classical equations of motion for a system with Hamiltonian $H(\{\mathbf{r}_i, \mathbf{p}_i\})$ are [34]

$$\frac{d\mathbf{r}_i}{dt} = \frac{\partial H}{\partial \mathbf{p}_i}, \qquad (2.37)$$

$$\frac{d\mathbf{p}_i}{dt} = -\frac{\partial H}{\partial \mathbf{r}_i}. \qquad (2.38)$$

Given the form of $H(\{\mathbf{r}_i, \mathbf{p}_i\})$ in eq. (2.13), the equation of motion can be written as

$$m\frac{d^2\mathbf{r}_i}{dt^2} = -\frac{\partial V(\{\mathbf{r}_i\})}{\partial \mathbf{r}_i}, \qquad (2.39)$$

which is simply Newton's third law $\mathbf{f}_i = m\mathbf{a}_i$, where $\mathbf{f}_i = -\partial V(\{\mathbf{r}_i\})/\partial \mathbf{r}_i$ is the force on atom i and $\mathbf{a}_i = d^2\mathbf{r}_i/dt^2$ is its acceleration. Molecular dynamics, at its heart, is simply the numerical integration of eq. (2.39).

---

[12] The Boltzmann's statistics are preserved when a thermostat is used in the simulation to sample the canonical ensemble (see Section 4.4). Without a thermostat, MD trajectories sample the microcanonical ensemble, in which the total energy remains constant.

2.5.1 The Verlet Algorithm

Several algorithms for numerical integration of ordinary differential equations are widely used in MD simulations [20, 35, 36]. Among them, the *Verlet algorithm* is one of the simplest and most stable. It is based on the following symmetric finite-difference approximation to the acceleration,

$$\mathbf{a}_i(t) = [\mathbf{r}_i(t + \Delta t) - 2\mathbf{r}_i(t) + \mathbf{r}_i(t - \Delta t)]/(\Delta t)^2, \tag{2.40}$$

which leads to

$$\mathbf{r}_i(t + \Delta t) = 2\mathbf{r}_i(t) - \mathbf{r}_i(t - \Delta t) + \mathbf{f}_i(t)\frac{(\Delta t)^2}{m}. \tag{2.41}$$

Because $\mathbf{f}_i(t)$ is a function of $\{\mathbf{r}_i(t)\}$, eq. (2.41) can be used to compute atomic positions at the next time step $(t + \Delta t)$, based on the atomic positions at two consecutive time steps (t and $t - \Delta t$). Using this equation repeatedly, atomic positions are computed step by step. Notice that atomic velocity does not appear explicitly in the Verlet algorithm but can be evaluated, if necessary, at time t through

$$\mathbf{v}_i(t) = [\mathbf{r}_i(t + \Delta t) - \mathbf{r}_i(t - \Delta t)]/(2\Delta t) \tag{2.42}$$

after Eq. (2.41) is solved for atomic positions at time $t + \Delta t$.

2.5.2 The Velocity Verlet Algorithm

The trajectory of a Hamiltonian system satisfying equations of motion (2.37) and (2.38) is fully determined by the initial conditions, $\{\mathbf{r}_i(0), \mathbf{v}_i(0)\}$. It would be natural for the integration algorithm to use atomic velocities as initial conditions for computing the trajectory. Obviously, this is not what the Verlet algorithm does—it uses $\{\mathbf{r}_i(0), \mathbf{r}_i(-\Delta t)\}$ as initial conditions instead. At the same time, the numerical estimation of the atomic velocities obtained in the Verlet algorithm is not very accurate (see Problem 2.5.2). Several variants of the original Verlet algorithm have been proposed to address these problems. One of the variants is the velocity Verlet algorithm. It can be summarized by the following equations.

$$\mathbf{r}_i(t + \Delta t) = \mathbf{r}_i(t) + \mathbf{v}_i(t)\Delta t + \mathbf{a}_i(t)\frac{\Delta t^2}{2} \tag{2.43}$$

$$\mathbf{a}_i(t + \Delta t) = -\frac{1}{m}\frac{\partial V(\{\mathbf{r}_i(t + \Delta t)\})}{\partial \mathbf{r}_i(t + \Delta t)} \tag{2.44}$$

$$\mathbf{v}_i(t + \Delta t) = \mathbf{v}_i(t) + [\mathbf{a}_i(t) + \mathbf{a}_i(t + \Delta t)]\frac{\Delta t}{2}. \tag{2.45}$$

The initial conditions for the velocity Verlet algorithm are $\{\mathbf{r}_i(0), \mathbf{v}_i(0)\}$ and the atomic velocities obtained in this algorithm are more accurate than in the original Verlet algorithm (see Problem 2.5.3). Interestingly, despite their apparent differences, the velocity Verlet and the original Verlet algorithms generate identical atomic trajectories (see Problem 2.5.4).

2.5.3 Energy Conservation

The equations of motion of a Hamiltonian system should conserve the system's energy. This becomes immediately obvious by taking the time derivative of the Hamiltonian and using eqs. (2.37) and (2.38),

$$\frac{dH}{dt} = \sum_{i=1}^{N} \frac{\partial H}{\partial \mathbf{r}_i} \frac{d\mathbf{r}_i}{dt} + \frac{\partial H}{\partial \mathbf{p}_i} \frac{d\mathbf{p}_i}{dt} = \sum_{i=1}^{N} \frac{\partial H}{\partial \mathbf{r}_i} \frac{\partial H}{\partial \mathbf{p}_i} - \frac{\partial H}{\partial \mathbf{p}_i} \frac{\partial H}{\partial \mathbf{r}_i} = 0. \qquad (2.46)$$

Among the vast number of integrators developed for solving the ordinary differential equations (ODE), the Verlet and the velocity Verlet algorithms belong to a class of the *symplectic* integrators [36] that preserve the energy particularly well. In contrast, if the system is evolved by non-symplectic integrators, the energy may drift away significantly from the initial value over a long time. We will rely on Verlet algorithms in our subsequent MD simulations throughout this book.

Selection of a proper time step Δt is an important issue in MD simulations and the conservation of system's energy is often used to check if the simulation is running correctly. Naturally, it is desirable to use a large Δt to maximize simulation efficiency, but too large Δt may result in an inaccurate trajectory (see Problem 2.5.1) and may even lead to numerical instabilities. In particular, the total energy of the system,

$$E_{\text{tot}}(t) = \sum_{i=1}^{N} \frac{1}{2} m |\mathbf{v}_i(t)|^2 + V(\{\mathbf{r}_i(t)\}), \qquad (2.47)$$

may "drift away" or even diverge when the selected Δt is too large.[13] For MD simulations of solids, Δt around 0.01 of the inverse *Debye* frequency ν_D is usually a safe choice. ν_D is defined as the maximum frequency of atomic vibrations in a given solid and is typically of the order of 10^{13} Hz [9]. Therefore Δt of the order of 10^{-15} s (one femtosecond) is common.

Numerical implementation of MD is not very complicated and writing a basic code for simple simulations from scratch (such as Problem 2.5.5) can greatly help in understanding the fundamentals of the method. In the subsequent chapters, we

---

[13] Especially if non-symplectic integrators are used.

use MD and other simulation methods to study dislocation properties. These simulations are carried out using MD++ [11], which has molecular dynamics, Monte Carlo and conjugate gradient relaxation capabilities, and has mostly been used to simulate crystal defects. While relatively efficient for simulations with up to half a million atoms, MD++ is still a serial code, i.e. it only runs on one processor. For larger scale simulations, several excellent parallel codes are available in the public domain. MD++ and links to parallel MD codes can be accessed from the book web site [11].

Summary

- Molecular dynamics simulations rely on the numerical integration of Newton's equation of motion for the interacting atoms.

- The total energy of a closed system of atoms (i.e. interacting only among themselves) should be conserved when the MD simulation is running correctly.

Problems

2.5.1. In this exercise, we examine the order of accuracy of the Verlet algorithm. Let $\mathbf{r}(t)$ be the trajectory of an atom. Taking a Taylor expansion of $\mathbf{r}(t + \Delta t)$ around time t, we obtain

$$\mathbf{r}(t + \Delta t) = \mathbf{r}(t) + \frac{d\mathbf{r}(t)}{dt} \Delta t + \frac{1}{2} \frac{d^2\mathbf{r}(t)}{dt^2} \Delta t^2$$

$$+ \frac{1}{3!} \frac{d^3\mathbf{r}(t)}{dt^3} \Delta t^3 + \mathcal{O}(\Delta t^4), \qquad (2.48)$$

where $\mathcal{O}(\Delta t^4)$ means that the rest of the expansion terms are of order Δt^4 or higher. In other words, the order of accuracy of the expansion truncated at the third term is 4. Similarly, take a Taylor expansion of $\mathbf{r}(t - \Delta t)$ around time t and find the order of accuracy for eq. (2.41). This order quantifies the numerical error produced over a single time step of the Verlet algorithm and is called its *local* order of accuracy. If it is intended to integrate the equations of motion up to a specified time in the future, it is important to know how the local errors accumulate over time. The *global* order of accuracy tells us how this accumulated error depends on the value of the time step Δt. Because a smaller Δt requires more steps to integrate to the specified time, the global order of accuracy is equal to the local order of accuracy minus one. What is the global order of accuracy of the Verlet algorithm?

2.5.2. Find the local order of accuracy of velocity for the Verlet algorithm, as given by eq. (2.42), assuming that $\mathbf{r}(t)$ and $\mathbf{r}(t - \Delta t)$ (but not $\mathbf{r}(t + \Delta t)$) are known exactly.

2.5.3. What is the local order of accuracy for the velocity in the velocity Verlet algorithm, as given by eq. (2.45), assuming that $\mathbf{r}(t)$ and $\mathbf{v}(t)$ are known exactly?

2.5.4. In this exercise, show that the original Verlet algorithm and the velocity Verlet algorithm generate identical trajectories $\mathbf{r}(t)$ when they both use the same time step Δt. Consider a simulation using the velocity Verlet algorithm with initial conditions $\mathbf{r}(0)$ and $\mathbf{v}(0)$. Let $\mathbf{r}(n\Delta t)$, $n = 1, 2, \ldots$ be the trajectory generated by this simulation. Show that, if $\mathbf{r}(0)$ and $\mathbf{r}(\Delta t)$ are used as the initial conditions for the original Verlet algorithm, it will produce the same sequence $\mathbf{r}(n\Delta t)$ for $n = 2, 3, \ldots$.

2.5.5. Write a Matlab program that simulates the orbit of planet Earth around the Sun using the velocity Verlet algorithm. Assuming that the Sun is fixed at the origin, the Earth is moving in a potential function $V(\mathbf{r}) = -GmM/|\mathbf{r}|$, where $m = 5.9736 \times 10^{24}$ kg is the mass of the Earth, $M = 1.9891 \times 10^{30}$ kg is the mass of the Sun and $G = 6.67259 \times 10^{-11}$ N m^2 kg^{-2} is the gravitational constant. We will use an arbitrary coordinate system such that the position of the Earth on 27 September 2004, 12:00 am is $x = 0.97878$, $y = -0.17895$, $z = 0.12004$ in Astronomical Units (AU, 1 AU $= 1.49598 \times 10^{11}$ m). The velocity of the Earth is $v_x = 2.8838 \times 10^{-3}$, $v_y = 1.6897 \times 10^{-2}$, $v_z = -7.1847 \times 10^{-4}$ in units of AU/day [37]. Predict the Earth's position 36 500 days after this moment.

3

CASE STUDY OF STATIC SIMULATION

Having discussed the basic concepts of atomistic simulations, we now turn to a case study that demonstrates the use of the static simulation techniques and, along the way, reveals some of the realistic aspects of the dislocation core structure and highlights the coupling between continuum and atomistic descriptions of dislocations. Section 3.1 explains how to use simple solutions of the continuum elasticity theory for setting up initial positions of atoms. The important issue of boundary conditions is then discussed in Section 3.2. Section 3.3 presents several practical methods for visualization of dislocations and other crystal defects in an atomistic configuration. This first case study sets the stage for a subsequent exploration of more complex aspects of dislocation behavior, which demands more advanced methods of atomistic simulations to be discussed in Chapters 4 through 7.

3.1 Setting up an Initial Configuration

In Section 1.2 we already considered the atomistic structure of dislocations in simple cubic crystals. In other crystals, the atomistic structure of dislocations is considerably more complicated but can be revealed through an atomistic simulation. This is the topic of this chapter.

An atomistic structure is specified by the positions $\mathbf{x}_i$ of all atoms. In a perfect crystal, $\mathbf{x}_i$'s are completely determined by the crystal's Bravais lattice, its atomic basis and its lattice constant (Section 1.1). Now assume a dislocation or some other defect is introduced, distorting the crystal structure and moving atoms to new positions $\mathbf{x}_i'$. A good way to describe the new structure is by specifying the displacement vector $\mathbf{u}_i \equiv \mathbf{x}_i' - \mathbf{x}_i$ for every atom i.

The relationship between $\mathbf{u}_i$ and $\mathbf{x}_i$ can be obtained analytically if we approximate the crystal as a continuum linear elastic solid. This is certainly an approximation, but it works well as long as crystal distortions remain small. Let us define $\mathbf{x}$ as the position of a material point in the continuum before the dislocation is introduced and $\mathbf{x}'$ as the position of the same material point after the dislocation is introduced. Here, $\mathbf{x}$ is a continuous variable and the displacement vector $\mathbf{u}(\mathbf{x}) \equiv \mathbf{x}' - \mathbf{x}$ expressed as a function of $\mathbf{x}$ is the displacement field. The continuum representation is effective because, in a state of mechanical equilibrium, $\mathbf{u}(\mathbf{x})$ satisfies a set of simple partial differential equations [7, 38] that allow analytic solutions for various interesting dislocation configurations.[1] For our purpose here, we can use the continuum solution $\mathbf{u}(\mathbf{x})$ to approximate the displacement of atoms in a crystal, i.e., $\mathbf{u}_i \approx \mathbf{u}(\mathbf{x}_i)$. As we will see shortly, this is a very good approximation for atoms far away from the dislocation center. On the other hand, the same approximation becomes inaccurate near the dislocation center where atomic positions are strongly affected by lattice discreteness and non-linear interactions among the atoms.

As an example, consider an infinitely long, straight screw dislocation in an infinite continuum medium, as shown in Fig. 3.1(a). Similar to the thought experiments considered in Section 1.2, this dislocation can be created by the following sequence. First, cut the solid along the horizontal plane $y = 0$ from $x = -\infty$ to $x = 0$. Second, displace the material points just above the cut $(y > 0)$ by b along the z axis with respect to the material points just below the cut $(y < 0)$. Third, reconnect the points across the cut and let the solid reach mechanical equilibrium. For the dislocation so created, let us choose its line sense in the positive z direction, $\xi = \mathbf{e}_z$, where $\mathbf{e}_z$ is the unit vector along the z axis. Following the Burgers circuit analysis described in Section 1.2, the Burgers vector of this dislocation is $\mathbf{b} = b\,\mathbf{e}_z$.

When the medium is elastically isotropic, the displacement field of this dislocation has a particularly simple analytic solution. The only non-zero component of the displacement vector $\mathbf{u} = (u_x, u_y, u_z)$ is u_z and it depends only on the x, y coordinates of the field point $\mathbf{x} = (x, y, z)$

$$u_z(x, y) = b\frac{\theta}{2\pi}, \tag{3.1}$$

where $\theta \in (-\pi, \pi]$ is the angle between the vector connecting the origin to $(x, y, 0)$ and the horizontal x axis, Fig. 3.1(a). This solution is plotted in Fig. 3.1(b). Consistent with the "cut-and-shift" operation just described, θ jumps from $-\pi$ and π across the cut plane (from $y < 0$ to $y > 0$) so that the displacement jumps from $-b/2$ to $b/2$.

This continuum solution has a problem right on the dislocation line where $x = y = 0$ and θ and u_z are not defined. The same solution predicts that elastic distortions related to the spatial derivatives of $\mathbf{u}(\mathbf{x})$ become infinite as $\mathbf{x}$ approaches the dislocation line. Consequently, solution (3.1) predicts that the elastic energy of the same dislocation should be infinite. These artifacts are manifestations of the well-known failure of the linear elasticity theory to describe lattice distortion near the dislocation core, the region close to the dislocation center. We will be dealing with this problem repeatedly in this book (Chapters 5 and 10).

There are several reasons for the continuum linear elasticity theory to break down in the dislocation core. First, in a discrete crystal structure, relative displacements

---

[1] Continuum solutions for $\mathbf{u}(\mathbf{x})$ depend only on the elastic constants of the crystal and do not explicitly depend on the crystal structure.

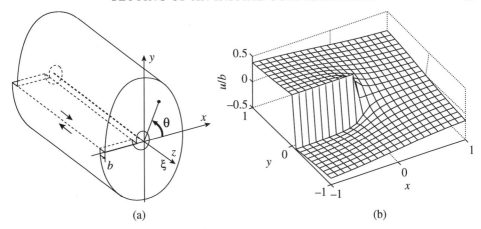

(a)   (b)

FIG. 3.1. (a) A straight screw dislocation is introduced in a solid cylinder along its
axis (thick dashed line). (b) z-component of the displacement field produced by
the dislocation in the units of b.

between neighboring atoms in the dislocation core can be very large and may
vary rapidly from one atom to its neighbor, making the continuum approximation
problematic. Second, the interatomic interactions in the core are highly non-linear,
invalidating the linear elastic assumption. When the linear elastic solutions cannot
be used, the equilibrium atomic positions should be determined by minimizing the
interatomic potential energy of the entire system using the methods described in
Section 2.3.

The total number of atoms in a crystal is often very large, making it impractical
to minimize the potential energy as a function of the coordinates of all atoms in
the crystal. Fortunately, linear elasticity theory is very accurate far away from the
dislocation core. In other words, the farther away from the dislocation center, the
more accurate the continuum solution becomes in describing the atom positions in
the crystal. This observation is constructive since we can fix the positions of atoms
away from the dislocation line on the positions predicted by the linear elasticity
solution. At the same time, all other atoms are allowed to adjust their positions
freely during the energy minimization. When the interatomic potential is short
range (Section 2.1), only a small number of fixed atoms are needed that are within
the interaction range of the "free" atoms. In such a way, the atomistic simulation
only deals with atoms close to the dislocation core subjected to boundary conditions
provided by the linear elasticity solution.

As an example, consider a screw dislocation in a BCC crystal with Burgers
vector $\mathbf{b} = \frac{1}{2}[111]$. As shown in Fig. 3.2, x and y axes are along $[11\bar{2}]$ and $[1\bar{1}0]$
directions, respectively. To obtain the core structure of this dislocation in an atom-
istic simulation, we start by creating a perfect BCC crystal structure contained

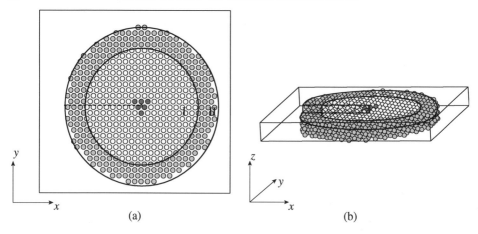

FIG. 3.2. An atomistic simulation of a screw dislocation in BCC tantalum. The atoms are arranged in a cylindrical slab and initially positioned according to an appropriate linear elasticity solution. The atoms in region I (open circles) are are free to adjust their positions while the atoms in region II (grey) remain fixed during the energy minimization. Periodic boundary conditions are applied in z direction. The shape and the physical spread of the relaxed dislocation core are illustrated by the atoms shown in dark grey. (a) The view along the z axis. The cut plane used to insert the dislocation is shown as a dashed line. (b) A 3-dimensional perspective view.

inside a cylindrical slab. The dislocation is inserted by displacing all atoms in the slab according to the linear elastic solution, eq. (3.1). To be able to use this solution, it is necessary to decide where in the slab to position the origin (0, 0, 0) and the cut. Here, it is important to avoid the ambiguity created when the origin and/or the cut coincide with the position of one or several atoms. Here we define the origin to be in the center of the triangle formed by three central rows of atoms. Then, the cut plane connects to the origin between two horizontal layers of atoms (dashed line on Fig. 3.2(a)). The resulting atomistic structure provides an initial atomic configuration for a subsequent conjugate gradient relaxation (Section 2.3).

The simulation volume is divided into two regions, as shown in Fig. 3.2(a). Atoms in the inner region I are free to change their positions during the relaxation while atoms in the outer region II are fixed in their initial positions. Since the thickness of outer region II is greater than the cut-off distance of the interatomic potential, any further increase in this thickness should have no effect on the behavior of atoms in region I. This is a good *fixed* boundary condition for our simulation because, by fixing atoms in region II according to the continuum solution of the dislocation displacement field, we have effectively extended the dimensions of the system to infinity in the x and y directions.

As stated, it is our intention here to simulate an infinitely long dislocation, but what we have created so far is a very short screw dislocation threading an ultrathin metal foil with two surfaces of the foil exposed to a vacuum. It is possible to make our slab dimension effectively infinite in the z direction by using a periodic boundary condition (PBC) to be discussed in more detail in the next section. With this choice of boundary conditions, fixed in x and y and periodic in z, our setup is appropriate for simulations of dislocations in an infinite crystal.

In this particular simulation, the interatomic interaction is described by the Finnis–Sinclair potential for BCC tantalum. The final atomic configuration after a conjugate gradient relaxation is plotted in Fig. 3.2. As expected, the atoms close to the dislocation center adjust their positions significantly more than the atoms farther away from the center. In Fig. 3.2, atoms with local energy E_i higher than that in a perfect crystal by 0.05 eV are shown in dark grey. The positions of high-energy atoms depict the physical size and shape of the dislocation core. Other methods for identifying crystal defects in an atomistic configuration will be described in Section 3.3.

Summary

- Linear elasticity theory provides simple solutions for the displacement fields of dislocations. These solutions can be used for setting up initial conditions for atomistic simulations.

Problem

3.1.1. Use the methods discussed in Section 3.1 to examine the core structure of an edge dislocation in BCC tantalum. First, use MD++ code to create a rectangular slab of a perfect BCC crystal of tantalum with dimensions $x = 20[111]/2$, $y = 16[1\bar{1}0]$, $z = [11\bar{2}]$. Displace the atoms according to the linear elasticity solution for an edge dislocation with Burgers vector $\mathbf{b} = [111]/2$. The displacement field of an infinite edge dislocation in an isotropic elastic medium is [7]

$$u_x(x, y) = \frac{b}{2\pi}\left[\theta(x, y) + \frac{xy}{2(1 - v)(x^2 + y^2)}\right],$$

$$u_y(x, y) = -\frac{b}{2\pi}\left[\frac{1 - 2v}{4(1 - v)}\ln(x^2 + y^2) + \frac{x^2 - y^2}{4(1 - v)(x^2 + y^2)}\right],$$

$$u_z(x, y) = 0,$$

where the dislocation line direction is $\boldsymbol{\xi} = \mathbf{e}_z$ and its Burgers vector is $\mathbf{b} = b\,\mathbf{e}_x$. Here, $\theta(x, y)$ is the angle between $\mathbf{e}_x$ and the vector connecting the origin to point $(x, y, 0)$.

The cut plane is defined by selecting the range for angle $\theta(x, y)$, as in $\theta_0 \leq \theta \leq \theta_0 + \pi$. First, create a dislocation with a $(1\bar{1}0)$ cut connecting to the origin. Second, create the same dislocation using a (111) cut. In the latter case, atoms will need to be removed. Relax both configurations using the CGR algorithm and the FS potential for tantalum. Observe that the resulting atomic structure of the core does not depend on the choice of the cut plane.

3.2   Boundary Conditions

Suitable boundary conditions are key to a successful atomistic simulation. The simple simulation in the preceding section is a good example in which fixed and periodic boundary conditions are combined to model a dislocation in an infinite crystal. To appreciate the importance of boundary conditions for atomistic simulations, consider Avogadro's number, $N_A = 6.022 \times 10^{23}$, which is the number of molecules in one mole of any substance.[2] Roughly speaking, the number of atoms contained in a cubic centimeter of solid or liquid is of the order of N_A. In comparison, a typical simulation on a desktop workstation can only handle 10^3 to 10^6 atoms. Even in the billion-atom simulations on massively parallel computers [24], the total number of atoms is still very small compared with Avogadro's number. Therefore, unless one is specifically interested in isolated nano-sized atomic clusters, the actual simulation volume is only a very small portion of the material of interest. The behavior of atoms in the simulation volume is affected by a large number ($\sim 10^{23}$) of surrounding atoms, which cannot be explicitly included in the simulation. The influence of these surrounding atoms can only be accounted for implicitly and approximately, through special treatment of boundary conditions discussed below and external couplings to be discussed in Section 4.4.

Unless a different boundary condition is specifically defined, "no-boundary" boundary condition or, equivalently, "free-surface" boundary condition means that there is no constraint on the motion of any atom. In such cases, the atoms usually form a compact cluster in which some of the atoms are exposed to vacuum. Such a boundary condition is not a good description of bulk materials since it completely ignores the effects of atoms outside the simulation volume and introduces unnecessary surfaces. One way to reduce the unwanted surface effects is to fix atoms on the periphery of the simulation volume in equilibrium positions that they would occupy in an infinite solid (Section 3.1). Such a fixed boundary condition is relatively simple to implement but has its own artifacts, especially in dynamical simulations since in a real material all atoms should be able to move. As an improvement, flexible boundary conditions allow the atoms in the boundary layer to adjust their positions

---

[2] N_A is formally defined as the number of atoms in 12 grams of carbon-12.

in response to the motion of inner atoms [39, 40]. The search for better boundary conditions continues to the present day. The goal is to reduce surface artifacts as much as possible, while maximizing the numerical efficiency by not including too many atoms explicitly in the simulation. Periodic boundary condition (PBC) is a special case among many types of boundary conditions available. The idea of PBC has a long history, but it remains a very competitive method today. The particular advantage of PBC is that it completely eliminates surface effects and maintains translational invariance of the simulation volume, which is especially important in simulations of crystals. PBC can be applied along one (Section 3.1), two or three directions of the simulation cell. In the following, we describe how to implement PBC along all three directions.

3.2.1   Periodic Boundary Conditions

The idea of periodic boundary conditions (PBC) is to embed the simulation volume or simulation *cell* into an infinite, periodic array of replicas or *images*. This is illustrated in Fig. 3.3 for a two-dimensional simulation. The atoms in the replicas are assumed to behave exactly in the same way as the atoms in the original or *primary* simulation cell [20]. Because the primary and image cells are identical, it is irrelevant which one of them is regarded as primary and which ones are the images. This type of periodic arrangement of atoms is fully specified by a set of repeat vectors, c_i, $i = 1, 2$ in two dimensions and $i = 1, 2, 3$ in three dimensions. The repeat vectors relate positions of atoms in the periodic replicas. Whenever there

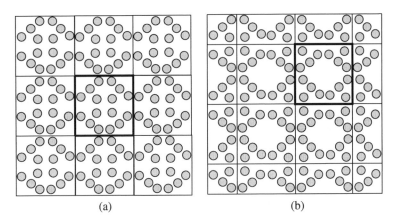

(a)   (b)

FIG. 3.3. (a) A simulation supercell with 16 atoms replicated periodically in two-dimensional space. (b) Shifting supercell boundaries produces a different set of atoms in each supercell but does not alter the overall periodic arrangement.

is an atom at position $\mathbf{r}$, there are also atoms at positions $\mathbf{r} + n_1\mathbf{c}_1 + n_2\mathbf{c}_2 + n_3\mathbf{c}_3$, where n_1, n_2 and n_3 are arbitrary integers.[3]

A remarkable property of PBC is that no point in space is treated any more specially than others. To better understand this feature, let us first consider a simulation under free-surface boundary condition. In this case, the atoms closer to the surface experience a different environment than the atoms further away from the surface. In other words, the presence of the surface breaks the translational invariance. At a first glance, even when PBC is applied one might think that the border of the simulation supercell (solid line in Fig. 3.3(a)) creates artificial interfaces that break the translational invariance. But this is not the case. The boundary of the primary cell can be shifted arbitrarily, as shown in Fig. 3.3(b). Such a shift has no effect on the dynamics of any atom, even though each periodic replica now apparently contains a different arrangement of atoms. In other words, translational invariance of space is fully preserved.[4] Hence, there are no boundaries and no artificial surface effects to speak of when PBC is used. This fundamental property has made PBC a very popular boundary condition for simulations of condensed-matter systems. In particular, PBC is a standard boundary condition for first-principles calculations (see Section 2.1) that rely on the plane-wave basis functions requiring translational invariance of space.

In an atomistic simulation with only short-range interactions, implementation of PBC is straighforward. All it takes is a careful enforcement of the *minimum image convention* that states that the relative displacement vector between atoms i and j is taken to be the shortest of all vectors that connect atom i to all periodic replicas of atom j. Since the energy and forces depend only on the relative positions of the atoms, this convention avoids any ambiguity, provided that the potential cut-off distance is sufficiently small so that no more than one replica of atom j falls within the cut-off radius of atom i (see Problem 3.2.1).

Enforcement of the minimum image convention is simplest in *scaled coordinates*. For an atom at $\mathbf{r} = (x, y, z)$, its scaled coordinates $s = (s_x, s_y, s_z)$ are defined through

$$s = \mathbf{h}^{-1} \cdot \mathbf{r}, \tag{3.2}$$

$$\mathbf{r} = \mathbf{h} \cdot s, \tag{3.3}$$

---

[3] An atomistic model under three-dimensional PBC is equivalent to a crystal structure whose primitive vectors are the same as the PBC repeat vectors and whose basis is the total collection of atoms inside the primary cell (see Section 1.1 for the definition of crystal structures). The unit cell of such a super-crystal is sometimes called a *supercell*.

[4] If PBC is applied only in one direction, then the translational invariance in that direction is preserved.

where $\mathbf{h}$ is a 3×3 matrix whose columns are the repeat vectors of the simulation cell, $\mathbf{h} \equiv (\mathbf{c}_1 | \mathbf{c}_2 | \mathbf{c}_3)$. Or, in the component form,

$$\begin{pmatrix} x \\ y \\ z \end{pmatrix} = \begin{pmatrix} c_{1x} & c_{2x} & c_{3x} \\ c_{1y} & c_{2y} & c_{3y} \\ c_{1z} & c_{2z} & c_{3z} \end{pmatrix} \cdot \begin{pmatrix} s_x \\ s_y \\ s_z \end{pmatrix}. \tag{3.4}$$

The manifestation of PBC in scaled coordinates is such that, whenever there is an atom at position $\mathbf{s} = (s_x, s_y, s_z)$, there are also atoms at positions $\mathbf{s} = (s_x + n_1, s_y + n_2, s_z + n_3)$, where n_1, n_2 and n_3 are arbitrary integers. That is, in the space of scaled coordinates periodic images of every atom in the supercell form a simple cubic lattice with the unit lattice constant. Under the minimum image convention, the distance between two atoms with scaled coordinates $\mathbf{s}_1 = (s_{x1}, s_{y1}, s_{z1})$ and $\mathbf{s}_2 = (s_{x2}, s_{y2}, s_{z2})$ is $\Delta\mathbf{s} = (\Delta s_x, \Delta s_y, \Delta s_z)$, where

$$\Delta s_x = f(s_{x1} - s_{x2}), \tag{3.5}$$
$$\Delta s_y = f(s_{y1} - s_{y2}),$$
$$\Delta s_z = f(s_{z1} - s_{z2}).$$
$$f(t) \equiv t - [t]. \tag{3.6}$$

Here, function $[t]$ returns the nearest integer to t and is called rint() in the C programming language, dnint() in Fortran and round() in Matlab. The useful property of $f(t)$ is that its values lie within $[-0.5, 0.5)$, meaning that the nearest image of atom j is certain to stay within a unit cube with atom i at its center. Therefore, the separation vector in the real space can be calculated simply as[5]

$$\Delta\mathbf{r} = \mathbf{h} \cdot \Delta\mathbf{s}. \tag{3.7}$$

Even though in PBC the simulation cell has no boundaries, it has finite physical dimensions and occupies volume

$$\Omega = \det(\mathbf{h}). \tag{3.8}$$

The use of PBC does not completely eliminate the artifacts caused by the inevitably small number of atoms. Not surprisingly, the only way to reduce the unwanted small-size effects is to increase the size of the simulation cell. In fact, PBC can bring in its own artifacts. One relevant example is when the simulation cell contains a crystal defect and is subjected to PBC. Such a simulation does not exactly

---

[5] Because the transformation between the real and the scaled coordinates does not necessarily preserve the relative length of segments in different orientations, it is possible that two images that are nearest in the real space are not nearest images in the scaled space. This will not create a problem if the cut-off distance of the interatomic potential is sufficiently small, so that, whenever this situation occurs, atoms i and j are separated by a distance larger than the cut-off radius.

reproduce the behavior of a single defect in an infinite crystal. Instead, it will be more representative of an infinite periodic array of defects. Unless the simulation domain is sufficiently large, the interaction between primary and image defects may "pollute" the simulation results. This is an important issue given the wide applicability and robustness of PBC in atomistic simulations of materials. How to eliminate or mitigate the artifacts introduced by PBC will be discussed in Chapter 5.

As a result of the prevalence of PBC, many simulation codes, such as *ab initio* codes [22] and MD++ [11], have PBC built into them as a default boundary condition. Other types of boundary conditions can still be constructed "inside" PBC. For example, in the simulation shown in Fig. 3.2, we used fixed boundary conditions in x and y and a periodic boundary condition in z. This corresponds to an isolated, infinitely long cylinder containing a screw dislocation in the center. These same boundary conditions can be imposed within the periodic supercell provided it is long enough to house the whole assembly. In fact, this is exactly how we did it in MD++. First, the z vector of the supercell was defined to coincide with the axis of the cylindrical slab. Then, to be able to enforce the fixed boundary conditions along two other directions, x and y dimensions of the supercell were selected large enough to enclose the whole cylinder. The resulting simulation corresponds to a two-dimensional array of infinitely long cylinders. As long as the neighboring cylinders do not overlap, the behavior of their constituent atoms will not be affected even if, nominally, the supercell is periodic along all three directions.

Summary

- The total number of atoms in a simulation is typically many orders of magnitude smaller than required in most situations of interest for materials modelling. Well chosen boundary conditions mimic the effects of atoms outside the simulation volume and help to eliminate or mitigate the unwanted artifacts associated with the unavoidably small size of the simulation domain.

- Periodic boundary conditions remove artificial surface effects and preserve translational invariance in atomistic simulations.

Problem

3.2.1. The minimum image convention states that an atom interacts with at most one image of any other atom. To be able to enforce this convention, the size and shape of the periodic box under PBC should be selected with some care. The allowable size and shape can be deduced from the condition that no two images of the same atom can fit inside a sphere with a radius equal to the cut-off radius of the potential. Given three periodicity vectors c_1, c_2, c_3 and a cut-off radius r_{cut} and assuming pairwise interactions, derive inequalities that determine when the box becomes too small or too skewed for the minimum image convention to work safely. (Hint: Consider a sphere of radius r_{cut} inscribed in the supercell.)

3.3   Data Analysis and Visualization

In Fig. 3.2 the atoms were shaded according to their local energies computed after the atomistic structure was relaxed to a local energy minimum. This helped us identify the location and spread of the dislocation core. There are other ways to identify crystal defects in an atomistic configuration, such as the Burgers circuit test described in Section 1.2. In this section, we discuss another useful approach somewhat similar to local energy filtering but with several important advantages. It will be used in the subsequent chapters to track the location of dislocations.

In addition to analyzing the data at the end of a simulation, it is often necessary to perform data filtering during the simulation. This is because atomistic simulations often generate huge amounts of data. For example, in an MD simulation, the instantaneous atomic coordinates and velocities are available at every time step. In the early days, it was possible to save the entire MD trajectory for future analysis. Today, large-scale atomistic simulations present serious challenges for data handling, storage and analysis. For example, an MD simulation of one billion atoms [24] produces over 10^{10} bytes of data (atomic coordinates and velocities) at every time step. In one million time steps, the total amount of data will exceed 10^{16} bytes or 10 petabytes. Such enormous amounts of data are overwhelming and can quickly fill up even the largest arrays of hard disks. Furthermore, this data is quite useless in its "raw" form. Processing or reduction of the raw data on-the-fly, during the simulation, is necessary to make it more useful.

While there may be many atoms in a simulation cell, some of them are more important than others in terms of their information content. In atomistic simulations of crystalline solids, it is often desirable to find where crystal defects are and how the atoms are arranged around these defects. The remaining atoms further away from the defects are usually much less interesting. Because defects typically occupy only a small fraction of the total volume of the crystal, focusing on the local regions in the defect *cores* significantly reduces the amount of data that needs to be stored. For this purpose, it is desirable to have robust algorithms that can efficiently identify crystal defects in a given atomistic configuration.

Because a perfect crystal structure usually corresponds to a state of minimum energy for a set of interacting atoms, an obvious way to identify crystal defects is to find locations where the local energy exceeds the energy per atom of the perfect crystal. How to partition the potential energy among individual atoms (Section 2.1) is not unique, but reasonable partitioning should work for visualization purposes. Once the local energy of every atom is computed, it is useful to plot its histogram, such as in Fig. 3.4(a). Distinct energy intervals in the histogram may correspond

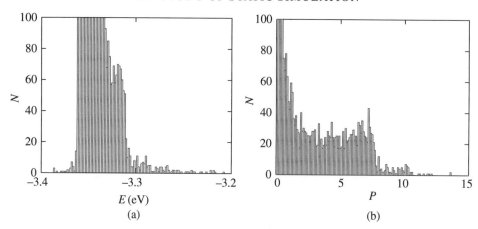

FIG. 3.4. Histograms of (a) the local energy and (b) the centro-symmetry deviation parameter for the atomistic configuration shown in Fig. 3.5. Both histograms are truncated above 100.

to different features of the atomistic configuration. In particular, the single highest peak in the histogram accounts for the majority of atoms in near-perfect crystal environments. There is no point in displaying these atoms. After some experimentation, it is usually possible to find an optimal window in the histogram that corresponds to interesting, "defect" atoms. Atoms with local energy inside this interval can be plotted to highlight the crystal defects [41]. This approach was used in Section 3.1, where atoms with local energy E_i exceeding that of perfect crystal atoms by 0.05 eV were highlighted to show the dislocation core.

While this method of energy filtering is simple and often used, it has limitations. The most serious one is that its signal-to-noise ratio tends to be low especially when applied to molecular dynamics simulations. At finite temperatures, thermal fluctuations smear out the energy differences between the atoms so that the signatures of "defect" atoms seen very clearly at zero temperature may be completely washed out by the thermal noise. For example, in FCC Cu represented by an EAM potential [42], it becomes very difficult to identify dislocations by local energy filtering at temperatures over 200 K.

Inherent structures and data filtering* It is possible to suppress much of the unwanted thermal noise using a simple procedure of partial steepest descent. In this approach, the snapshots of an MD simulation are used to initiate steepest descent paths towards underlying local minima. Eventually, after many steepest-descent iterations, each descent path will arrive at a local energy minimum. However, just a few, typically 5–10, descent steps are usually sufficient for the thermal part of energy to subside significantly, making it again possible to apply energy filtering for identification of crystal defects. If desired, the steepest descent paths can continue to the very bottom of the local energy basin. By observing which minimum the system descends to starting from every point of its dynamic trajectory, it is possible to map the system's dynamics onto a discrete sequence of energy basins, also known as *inherent structures* [43] visited by the system. This *steepest-descent mapping* can be a useful reduced representation of the system's dynamics as a sequence of resident intervals separated by instant transitions from one energy basin to the next. In cases when time intervals between transitions are considerably longer than characteristic times of atomic vibrations, such an interpretation makes considerable physical sense (for more on this see Chapter 7). When combined with energy filtering, partial steepest descent can drastically reduce the amount of data to store, without much increase in the computational cost. As a word of caution, in certain situations the system's configuration may change significantly during a steepest-descent relaxation. This makes the interpretation of the inherent structure questionable. Therefore, it is usually a good idea to limit each steepest-descent relaxation to a small number of (say 10) iterations. To further reduce the computational effort, this analysis can be applied at every 10th or every 20th step of the simulation. For a given simulation, the frequency of quenches and the number of steepest-descent steps can be adjusted empirically.

Another limitation of the energy filtering approach is that it is not necessarily true that atoms in the defect cores must have a distinctly higher local energy than the "bulk" atoms in the perfect crystal. For example, the excess local energy of atoms in a stacking fault (Fig. 3.5) in an FCC metal can be very small or even zero for some interatomic potential models, making it virtually impossible to differentiate the stacking fault atoms from the bulk atoms based on the local energies alone. In many cases, the so-called *centro-symmetry deviation* (CSD) parameter is a better indicator of crystal defects than the local energy. In the following, we describe when the CSD analysis can be used and how to implement it.

In many crystal structures, every atom is a center of inversion symmetry, meaning that the positions of its neighboring atoms are centro-symmetric relative to this atom. Simple cubic, FCC, BCC, and other crystal structures possess this symmetry (see Section 1.1), while hexagonal-close-packed (HCP), diamond cubic, and some other structures do not. Specifically, an atom possesses local centro-symmetry when for each of its neighbors at the relative position $\mathbf{r}$, there is another neighbor at the relative position $-\mathbf{r}$. These two neighboring atoms form a centro-symmetric pair. Whenever the local atomic arrangement deviates from the perfect crystal structure, the same two atoms may no longer be centro-symmetric. The CSD parameter of a given atom is a quantitative measure of the deviation from the ideal centro-symmetry in the atom's neighborhood. It can be defined as [44]

$$P = \sum_{i=1}^{N_p} |\mathbf{r}_i + \mathbf{r}_{i+N_p}|^2, \qquad (3.9)$$

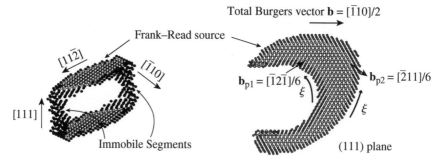

FIG. 3.5. (a) A dislocation loop created by removing a rectangular plate of atoms on a $(\bar{1}10)$ plane of FCC aluminum, followed by conjugate gradient relaxation (from [41] with publisher's permission). Two dislocation segments lie on (111) planes and are easy to move while the other two segments lie on $(11\bar{2})$ planes and are difficult to move. The immobile segments provide pinning points to the mobile segments, making them Frank–Read sources. This configuration is also simulated in Chapter 10 using the dislocation dynamics model (see Fig. 10.4). (b) Under applied stress, the Frank–Read segments bow out on (111) planes (only the top segment is shown). The dislocation is dissociated into two partial dislocations (black atoms) bounding an area of stacking fault (white atoms). These defects are identified according to the CSD parameter of every atom. The histogram of the CSD parameter is shown in Fig. 3.4(b).

where $\mathbf{r}_i$ and $\mathbf{r}_{i+N_p}$ are positions of the ith pair of neighbors relative to the atom under consideration. In this sum, the number of nearest-neighbor pairs N_p depends on the host crystal structure: N_p is 6 for FCC and 4 for BCC. For perfect FCC (or BCC) structures, vectors $\mathbf{r}$ and $\mathbf{r}_{i+N_p}$ in each neighbor pair cancel each other so that P vanishes for every atom. On the other hand, because crystal defects destroy local symmetry, P is usually non-zero near a defect. The following algorithm computes the CSD parameter for a given atom 0. In this algorithm, $\mathbf{r}_j$ ($j = 1, 2, \ldots, n$) are the relative positions of neighboring atoms with respect to atom 0 and n is a number equal to or larger than $2N_p$.[6]

Algorithm 3.1

1. Select appropriate N_p for the crystal structure considered.

2. Initialize marker array $f(j) := 0$, $j = 1, 2, \ldots, n$. Initialize $P := 0$.

3. Find k such that $|\mathbf{r}_k|^2$ is the smallest among $|\mathbf{r}_j|^2$ and its marker is still zero, $f(k) = 0$.

---

[6] The array $\mathbf{r}_j$ is usually available during the calculation of interatomic potential energy and atomic forces. For every atom subjected to CSD analysis, the dimension of array $\mathbf{r}_j$ is equal to the number of its neighbors falling within the cut-off radius.

4. Assign $f(k) := 1$ to mark the atom as no longer available for subsequent evaluations.

5. Find l such that $|\mathbf{r}_k + \mathbf{r}_l|^2$ is the smallest among $|\mathbf{r}_k + \mathbf{r}_j|^2$ and $f(l) = 0$.

6. Mark $f(l) := 1$.

7. $P := P + |\mathbf{r}_k + \mathbf{r}_l|^2$.

8. If $\sum_{j=1}^{n} f(j) = 2N_p$ stop; otherwise go to step 3.

For each neighbor of atom 0, this algorithm finds another neighbor that minimizes the sum of two position vectors of the pair. As an example, here we apply this algorithm to an atomic configuration containing a dislocation loop. The dislocation loop is created by removing a rectangular plate of atoms in FCC aluminum, followed by conjugate gradient relaxation [41]. The histogram of the CSD parameter distribution for all atoms is shown in Fig. 3.4(b). Similar to the local energy distribution, shown in Fig. 3.4(a), the CSD distribution shows a large peak near zero corresponding to "bulk" atoms far away from the defect. However, the spread of the CSD distribution is wider than that of the local energies, suggesting that CSD analysis may have a better resolution and signal-to-noise ratio. From the distribution in Fig. 3.4(b), two relevant ranges of the CSD parameter are established by trial and error that are very useful in illustrating the fact that perfect dislocations on {111} planes of FCC crystals dissociate into partial dislocations. The atoms with CSD parameter falling in the range of $1 \text{ Å}^2 < P < 3 \text{ Å}^2$ correspond to the cores of the partial dislocations. They are shown as filled circles in Fig. 3.5. The atoms with $P > 3 \text{ Å}^2$ correspond to the stacking fault area between the two partial dislocations. These are shown as open circles. It would be difficult to observe all such details using the method of energy filtering described earlier.

In addition to its ability to resolve defects that have weak energy signatures, such as stacking faults, CSD analysis is considerably more robust than energy filtering with respect to thermal noise. The defects retain their distinct features in the CSD distribution up to relatively high temperatures. Another useful feature for the CSD analysis is that it is independent of the interatomic potential function, unlike the energy filtering method. However, CSD analysis can only be applied to crystals in which every atom is a center of inversion symmetry in the perfect crystal structure.

Summary

• Atomistic simulations generate large amounts of data. On-the-fly data analysis and post-processing are often necessary to reduce the amount of simulation data and to generate images to facilitate analysis.

- Energy filtering and centro-symmetry deviation (CSD) analyses are two useful methods for data processing. Of the two, CSD analysis offers more robust identification and recognition of crystal defects in those crystals where each atom is the center of inversion symmetry.

Problems

3.3.1. Build a perfect BCC crystal in a periodic box defined by vectors 10[100], 10[010], and 10[001]. Create a vacancy–interstitial pair by taking one of the atoms from its regular position in the lattice and moving it to another remote interstitial position. Relax the configuration using the FS potential for molybdenum. Construct histograms of local energies and CSD parameters for all atoms in the box. What are the signatures of the vacancy and interstitial defects in these histograms? Select a window in each of the two histograms and plot the atoms near the defects.

3.3.2. Build a perfect BCC crystal of molybdenum atoms in a periodic box with periodicity vectors 16[112], 8[$\bar{1}\bar{1}$1], and 16[1$\bar{1}$0] along x, y, z directions respectively. Build another perfect BCC crystal, but now in a periodic box with periodicity vectors 16[11$\bar{2}$], 8[111], and 16[1$\bar{1}$0] along x, y, z. The two crystals have exactly the same shape and dimensions but different crystal orientations. Stack one crystal on top of the other along the z-axis and relax the resulting configuration using CGR and the FS potential. This creates a perfect grain boundary (GB) that materials scientists refer to as a pure-twist $\Sigma 3$ boundary. Plot the histograms of local energies and CSD parameters for all atoms. Identify the signatures of the grain boundary in both histograms.

3.3.3. Similar to Problem 3.3.2, build two crystals with dimensions $c_1 =$ [50, 50, 48], $c_2 = [\bar{1}\bar{1}, \bar{1}\bar{1}, 26]$, $c_3 = 16[1\bar{1}0]$, and $c_1 = [49, 49, \bar{50}]$, $c_2 = [13.5, 13.5, 23.5]$, $c_3 = 16[1\bar{1}0]$. Relative rotation between the two crystals is close to but slightly different from that in the $\Sigma 3$ boundary considered in Problem 3.3.2. Stack two crystals on top of each other and relax the resulting configuration. The extra rotation is now accommodated by two grain boundary dislocations superimposed on top of an otherwise perfect $\Sigma 3$ grain boundary. Identify the signatures of the grain boundary dislocations in the histograms of local energy and CSD parameter. Plot the atoms in the grain boundary using two different colors for the atoms in the boundary plane and in the dislocation cores.

4

CASE STUDY OF DYNAMIC SIMULATION

The preceding chapter focused on the dislocation core structure at zero temperature obtained by energy minimization. In this chapter we will discuss a case study of dislocation motion at finite temperature by molecular dynamics (MD) simulations. MD simulations offer unique insights into the mechanistic and quantitative aspects of dislocation mobility because accurate measurements of dislocation velocity are generally difficult, and direct observations of dislocation motion in full atomistic detail are still impossible. The discussion of this case study is complete in terms of relevant details, including boundary and initial conditions, temperature and stress control, and, finally, visualization and data analysis.

4.1   Setting up an Initial Configuration

In Section 3.1 we discussed a method for introducing a dislocation into a simulation cell. It relies on the linear elasticity solutions for dislocation displacement fields. To expand our repertoire, let us try another method here. The idea is to create a planar misfit interface between two crystals, such that subsequent energy minimization would automatically lead to dislocation formation.

For example, let us start by creating a rectangular slab of BCC crystal molybdenum with dimensions $L_{x1} = 30[111]/2$, $L_{y1} = 8[\bar{1}01]$, $L_{z1} = 8[1\bar{2}1]$, and lattice constant $a = 3.1472$ Å. Label it crystal 1. Next, build crystal 2 such that $L_{x2} = 29[111]/2$, $L_{y2} = L_{y1}$ and $L_{z2} = L_{z1}$. Notice that crystal 2 has one atomic plane fewer in the x direction than crystal 1. We now elongate crystal 1 and compress crystal 2 along the x direction so that both crystals have the same length, $L_x = 29.5[111]/2$. Joining two crystals along their x–z faces produces a bi-crystal with a misfit interface on the horizontal mid-plane shown in Fig. 4.1. At this point, the misfit is uniformly distributed over the interface. Let us now employ the Finnis–Sinclair potential [19] and use the conjugate gradient option of MD++ to relax the system's energy. As always, MD++ embeds our bi-crystal in a supercell that is periodic in all three directions. However, as shown in Fig. 4.1, here we define the supercell dimension along the y axis to exceed the total height of the bi-crystal,

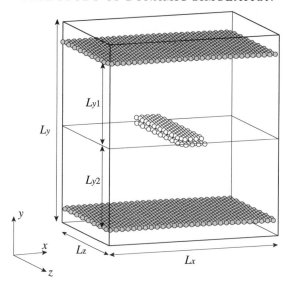

FIG. 4.1. An edge dislocation obtained by relaxing a bi-crystal obtained by joining
    together two crystals. The lower crystal contains one atomic plane fewer than
    the upper crystal. Only the atoms with an elevated value of the CSD parameter
    ($>1.5\,\text{Å}^2$) are plotted, showing an edge dislocation in the center and two atomic
    planes at the free surfaces.

$L_y > L_{y1} + L_{y2}$. As a result, a "vacuum gap" separates the top x–z surface of crystal
1 from the bottom x–z surface of the periodic replica of crystal 2. Therefore, the
bi-crystal has two free surfaces.

The atomistic configuration reached at the end of the conjugate gradient
relaxation is shown in Fig. 4.1, where the misfit between the two crystals has
condensed into a dislocation.[1] To visualize this dislocation, we computed the CSD
parameter for every atom in the bi-crystal (Section 3.3) and selected two intervals in
the CSD distribution to show the "interesting" atoms. Atoms with CSD parameter
larger than $10\,\text{Å}^2$ are plotted in grey—these are the atoms at the top and bottom
surfaces of the bi-crystal. Atoms with CSD parameter between $1.5\,\text{Å}^2$ and $10\,\text{Å}^2$
are plotted in white—these are the atoms in the core of an edge dislocation at the
center of the bi-crystal. Obviously, the line direction of this dislocation is parallel
to the z axis while its Burgers vector (obtained from the Burgers circuit analysis
described in Section 1.2) is $\mathbf{b} = [111]/2$. The dislocation core appears planar and
spreading on the x–z glide plane.

This configuration is designed for a subsequent MD simulation of dislocation
motion. The periodic boundary condition along z effectively makes the dislocation length infinite and eliminates undesirable end effects. The surfaces at the top

---

[1] This is similar to the formation of misfit dislocations in heteroepitaxial thin films.

and bottom can be used to apply surface tractions and to force the dislocation to move. Finally, PBC along the x direction allows the dislocation to traverse the simulation volume multiple times without ever leaving the simulation cell. As soon as it exits the supercell on one side, one of its periodic images enters from the other side. This "treadmill" boundary condition is useful for obtaining high-quality statistics because the dislocation can keep moving for distances much longer than the dimension of the simulation cell.

Summary

- Dislocations sometimes form as a result of atomic relaxation in misfit interfaces. This behavior can be exploited to introduce dislocations in the atomistic models.

4.2   Initializing Atomic Velocities

The atomistic configuration created in the previous section corresponds to a dislocation at zero temperature since the atom positions are at a potential energy minimum and the atom velocities are zero. To prepare this configuration for a subsequent MD simulation, we would like first to bring the system to a state of thermal equilibrium at a finite temperature, say $T = 300$ K. In the language of statistical mechanics, we would like to assign our system a microstate that is representative of a *canonical ensemble* of microstates obeying Boltzmann's distribution (see Section 2.2). This section describes a method that can be used for this purpose.

In Section 2.3 we said that, when the entire system obeys Boltzmann's distribution, the velocity of each atom obeys the following Gaussian distribution,

$$f(\mathbf{v}_i) = \frac{1}{Z_v} \exp\left(-\frac{m|\mathbf{v}_i|^2}{2k_B T}\right), \tag{4.1}$$

$$Z_v = \int \exp\left(-\frac{m|\mathbf{v}|^2}{2k_B T}\right) d\mathbf{v}. \tag{4.2}$$

It is certainly possible to use random numbers to initialize velocities to satisfy this probability distribution [45]. By itself, however, this will not bring the system to thermal equilibrium because the atomic positions will also have to be initialized according to Boltzmann's distribution. The method explained below simultaneously initializes both atomic velocities and atomic positions. The approach relies on the system's own ability to reach thermal equilibrium. It has been observed that an MD simulation starting from an arbitrary initial configuration can bring the system to a state of thermal equilibrium after some period of time. This preliminary MD simulation is sometimes called an *equilibration* run. Its purpose is to bring the system to a state ready for a subsequent MD simulation of real interest. Over the duration of the equilibration run, the system's progress towards equilibrium is monitored but no statistical or dynamic data is collected.

While we can hope that the system will reach thermal equilibrium after some time, it remains unclear what would be the temperature of the resulting equilibrium state. Let us now figure out how to initialize the velocities in such a way that the temperature of the equilibrium state is equal to the temperature we desire to use in the subsequent MD simulation. In Section 2.2 we observed that the average kinetic energy of a system in an equilibrium state is related to its temperature,

$$\langle E_{\text{kin}} \rangle = \frac{3}{2} N k_B T. \tag{4.3}$$

It is useful to define an "instantaneous temperature" T^{inst} in terms of the instantaneous kinetic energy of the system,

$$T^{\text{inst}} = \frac{2}{3Nk_B} E_{\text{kin}} = \frac{2}{3Nk_B} \sum_{i=1}^{N} \frac{1}{2} m |\mathbf{v}_i|^2. \tag{4.4}$$

In equilibrium, T^{inst} should fluctuate around the "true" temperature T. Therefore, a simple procedure to make the simulated system match a desired temperature T is to scale the atomic velocities such that $T^{\text{inst}} = T$. This velocity rescaling may have to be repeated if T^{inst} is observed to drift to another value during the equilibration run. The following algorithm describes how to initialize atomic velocities in such a way that the instantaneous temperature of the system matches a specified value T^*.

Algorithm 4.1

1. Assign a random number uniformly distributed from -0.5 to 0.5 to each component of atom velocities, $3N$ components in all.

2. Compute the average velocity $\langle \mathbf{v} \rangle = \frac{1}{N} \sum_{i=1}^{N} \mathbf{v}_i$.

3. Subtract the average velocity from each atom velocity to make sure that the velocity of the center of mass of all atoms is zero, i.e. $\mathbf{v}_i := \mathbf{v}_i - \langle \mathbf{v} \rangle$, for all $i = 1, \ldots, N$.

4. Compute the instantaneous temperature T^{inst} according to eq. (4.4).[2]

---

[2] Step 3 of this algorithm reduces the total number of degrees of freedom from $3N$ to $3(N-1)$. Hence, the average kinetic energy of the system at thermal equilibrium should be $\frac{3}{2}(N-1)k_B T$ and the instantaneous temperature should be defined as $T^{\text{inst}} = 2E_{\text{kin}}/[3(N-1)k_B]$ instead. This correction is negligible at large N and is ignored here.

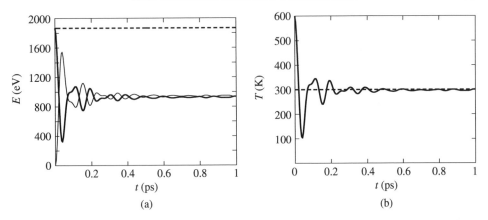

FIG. 4.2. (a) Kinetic energy E_{kin} (thick line), potential energy E_{pot} (thin line), and total energy $E_{tot} = E_{kin} + E_{pot}$ (dashed line), as a function of simulation time (in picoseconds). (b) The instantaneous temperature as a function of time (solid line) gradually settles to 300 K (dashed line) towards the end of the equilibration run.

5. $\mathbf{v}_i := \mathbf{v}_i \sqrt{T^*/T^{inst}}$, for all $i = 1, \ldots, N$.

6. Compute T^{inst} using eq. (4.4) again. By construction, T^{inst} should match the specified value T^*.

As an example, let us use Algorithm 4.1 with $T^* = 600$ K to assign velocities to the atomistic configuration obtained in the preceding section. Using the resulting velocities as the initial condition, let us run an MD simulation in which the equations of motion are integrated using the velocity Verlet algorithm (Section 2.5) with time step $\Delta t = 0.5$ fs (0.5×10^{-15} s). Fig. 4.2(a) shows the evolution of kinetic energy E_{kin}, potential energy E_{pot}, and total energy $E_{tot} = E_{kin} + E_{pot}$, during the equilibration run. The plot reveals a vigorous exchange between the kinetic and the potential energies over the first 0.4 picosecond and correspondingly large fluctuations of the instantaneous temperature shown in Fig. 4.2(b). This behavior is a clear indication that the initial state is far from thermal equilibrium. After about $t = 0.6$ ps, the amplitude of fluctuations gradually decreases and the instantaneous temperature begins to settle down closer to $T = 300$ K, indicating that the system is equilibrating. Over the course of the entire simulation, including its "wild" early period, the total energy is conserved within high accuracy, indicating that the selected time step is adequately small. In contrast, excessive fluctuations and/or "drift" in the total energy would be a warning sign that either the time step was too large or that there is a bug in the code.

It is interesting to note that the system eventually settles down to a temperature T that is very close to half of the instantaneous temperature T^* at the beginning of the equilibration run. This is not an accident but a manifestation of energy equipartition. At the beginning of the equilibration run, the atomic positions are at a local potential energy minimum. During equilibration, the average potential energy increases while the kinetic energy decreases by the same amount so that the total energy is conserved. As was discussed in Section 2.2, the average potential energy (counted from the local minimum) should be equal to the average kinetic energy in thermal equilibrium provided that the potential energy is a harmonic function of the atomic coordinates. In this case, the instantaneous temperature should drop by exactly one half from its initial value because the potential energy is initially zero. The Finnis–Sinclair potential function used in this simulation is obviously more complicated than a harmonic function, yet it can be approximated (through a Taylor expansion) by a harmonic function provided that the atomic positions are not too far away from the local minimum.[3] This condition is satisfied when the temperature is much lower than the melting temperature T_m of the crystal ($T_m = 2896$ K for molybdenum), which is certainly the case in the simulation considered here.

The equilibration run discussed in this section is a common procedure for obtaining the initial conditions for subsequent MD simulations. Anticipating that the instantaneous temperature will drop by approximately half during the equilibration run, it is common to set the initial instantaneous temperature T^* in Algorithm 2.1 to twice the desired temperature. If it is necessary to bring the equilibrated temperature even closer to the desired value, the atomic velocities may need to be rescaled a few more times, each time followed by an equilibration run.

Summary

- To set up a finite temperature MD simulation, atom velocities can be assigned at random, followed by an equilibration run.

- During the equilibration run, the instantaneous temperature drops to about half of its initial value provided that the atoms initially occupy positions in a local minimum of potential energy.

Problem

4.2.1. Use MD simulations to make a crystal melt. Generate the coordinates of an FCC crystal (see Section 1.3) made of argon atoms with the following dimensions expressed in the units of lattice constant $a = 5.256$ Å:

$$L_x = [8, 0, 0]$$
$$L_y = [0, 8, 0]$$
$$L_z = [0, 0, 8].$$

---

[3] This harmonic approximation (HA) will be used again in Section 6.2.

Use a simulation box (supercell) that is larger than the crystal itself, e.g.

$$\mathbf{c}_1 = [10, 0, 0]$$
$$\mathbf{c}_2 = [0, 10, 0]$$
$$\mathbf{c}_3 = [0, 0, 10]$$

so that all six faces of the crystal are free surfaces. To describe the interactions among argon atoms use the Lennard-Jones potential (eq. (2.5)) with parameters $\epsilon_0 = 1.03235 \times 10^{-2}$ eV and $\sigma_0 = 3.405$ Å. The mass of an argon atom is $m = 39.948$ a.u. (1 a.u. $= 1.6605 \times 10^{-27}$ kg).

Perform a series of MD simulations using a time step of $\Delta t = 0.1$ fs. Start by randomly setting the initial atomic velocities to instantaneous temperatures $T^* = 10$ K, 50 K, 100 K, 150 K, 200 K, and 250 K, respectively. Equilibrate the system for 1 ps and observe it settle down to an equilibrium temperature. Continue each simulation for another 100 ps beyond the initial equilibration. Observe the crystal melt and estimate its melting temperature.

4.3 Stress Control

Having obtained an atomistic configuration equilibrated at $T = 300$ K, here we discuss how to apply stress to make the dislocation move in an MD simulation. According to the Peach–Koehler formula, eq. (1.16), the stress component that produces force on the dislocation in the x direction is σ_{xy}. One way to induce this stress is to exert external forces along x- and $-x$-directions on the atoms in the top and bottom surface layers, respectively. Let the total force on the top surface be F_x and that on the bottom surface be $-F_x$, then

$$\sigma_{xy} = \frac{F_x}{A}, \tag{4.5}$$

where $A = L_x L_z$ is the area of the surface (see Fig. 4.3). To avoid stress concentrations, these forces should be evenly distributed on all atoms on the top and bottom surfaces. Now is a good time to remember that the number of atoms on the top and bottom surfaces are different (720 versus 696). Because of this, the external force to be exerted on the surface atoms should be different for the two surfaces. To apply stress $\sigma_{xy} = 300$ MPa, a total force $F_x = \sigma_{xy} A = 9.287$ eV Å^{-1} should be applied to each of the two surface layers. On a per atom basis, this translates into force $f_x^{\text{top}} = F_x/720 = 0.01290$ eV Å^{-1} on each atom in the top layer and force $f_x^{\text{bot}} = -F_x/696 = -0.01334$ eV Å^{-1} on each atom in the bottom layer. These forces are added to the forces on each surface atom due to its interaction with other atoms.

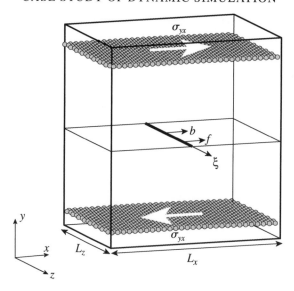

FIG. 4.3. The edge dislocation first shown in Fig. 4.1 is represented here by the thick line in the center of the simulation supercell. The atoms in two surface layers experience external forces along the x direction so that stress σ_{xy} is maintained during the simulation.

Applying stress in simulations with PBC in all three directions* The simple procedure for stress control described above is sufficient for the purpose of our case study in this chapter. It is feasible here because the crystal has two free surfaces on which to exert the surface tractions. It is possible to control stress even when no free surfaces are present. Here we describe the Parrinello–Rahman method that can be used for applying stress in such situations.

The so-called virial equation expresses the average *internal* stress in a simulation cell under three-dimensional PBC in terms of atomic velocities and interatomic forces [46, 47, 48, 49],

$$\Omega\sigma = \sum_{i} -m_i \mathbf{v}_i \otimes \mathbf{v}_j + \sum_{(ij)} \frac{\partial V(\{\mathbf{r}_{ij}\})}{\partial \mathbf{r}_{ij}} \otimes \mathbf{r}_{ij}, \qquad (4.6)$$

where V is the potential energy written as a function of relative distances between the atoms $\mathbf{r}_{ij} \equiv \mathbf{r}_j - \mathbf{r}_i$. It is possible to write the potential energy in this way because it is translationally invariant, i.e. it remains unchanged when all atoms are simultaneously translated by the same distance. Accordingly, the term

$$\mathbf{f}_{ij} \equiv \frac{\partial V(\{\mathbf{r}_{ij}\})}{\partial \mathbf{r}_{ij}} \qquad (4.7)$$

can be interpreted as the force on atom i due to atom j. This interpretation of $\mathbf{f}_{ij}$ is obvious when V is a pair potential. At the same time, the applicability of the above virial equation for stress is not limited to models described by the pair potentials.

The Parrinello–Rahman (PR) method [50] "applies" stress to the periodic super-cell by dynamically adjusting its shape so that the virial stress fluctuates around a specified value. In the PR approach, the atomistic model acquires nine additional degrees of freedom, which are the components of matrix $\mathbf{h} = (\mathbf{c}_1|\mathbf{c}_2|\mathbf{c}_3)$ made up of three repeat vectors $\mathbf{c}_1$, $\mathbf{c}_2$ and $\mathbf{c}_3$. Thus, for a system with N atoms, the total number of degrees of freedom in the PR approach is $3N + 9$. The equations of motion for these degrees of freedom are best specified in terms of the scaled coordinates $\mathbf{s}_i$ (Section 3.2) and matrix $\mathbf{h}$,

$$\ddot{\mathbf{s}}_i = -\mathbf{h}^{-1}\frac{1}{m}\frac{\partial V}{\partial \mathbf{r}_i} - \mathbf{G}^{-1}\dot{\mathbf{G}}\dot{\mathbf{s}}_i, \quad i = 1, \ldots, N, \tag{4.8}$$

$$W\ddot{\mathbf{h}} = (-\sigma - p\,\mathbf{I})(\mathbf{h}^{-1})^{\mathrm{T}}\Omega - \mathbf{h}\Sigma, \tag{4.9}$$

where

$$\mathbf{G} = \mathbf{h}^{T}\mathbf{h}, \tag{4.10}$$

$$\Sigma = \mathbf{h}_0^{-1}(-\sigma^{\mathrm{ext}} - p\,\mathbf{I})(\mathbf{h}_0^{-1})^{\mathrm{T}}\Omega_0. \tag{4.11}$$

Here, σ^{ext} is the stress one wishes to maintain during the simulation, $p = -\frac{1}{3}\mathrm{Tr}(\sigma^{\mathrm{ext}})$ is its hydrostatic pressure component, and $\mathbf{I}$ is the identity matrix. $\mathbf{h}_0$ is the matrix describing the shape of the periodic cell at the beginning of the simulation—this matrix defines the reference frame in which the orientation of the applied stress is expressed. $\Omega = \det(\mathbf{h})$ and $\Omega_0 = \det(\mathbf{h}_0)$ are the current and initial volumes of the simulation cell, respectively. Equation (4.9) provides a feedback mechanism by which $\mathbf{h}$ is adjusted whenever the virial stress σ becomes different from the desired value σ^{ext}. W is a "mass" parameter that determines how fast the box changes its shape in response to the imbalance between the desired and the actual stress. The value of W can be adjusted so that the stress fluctuations in the simulation volume are comparable to those in a volume of similar size in an infinite solid. In addition to molecular dynamics, the PR method can be used to apply stress in Monte Carlo simulations and during conjugate gradient relaxation.

Summary

- When free surfaces are available, stress can be applied by exerting additional forces on the atoms in the surface layers.

- In a simulation with no surfaces, the Parrinello–Rahman method can be used to maintain the internal virial stress close to a preselected value.

4.4   Temperature Control

Once the system has reached equilibrium, its temperature will remain close to the equilibrium value forever unless external forces are applied. But external forces are exactly what we need in order to "convince" the dislocation to move. During dislocation motion the work done by the external stress is dissipated as heat and, unless some mechanism to remove this heat is used, this would inevitably cause the temperature to rise (see Problem 4.4.1).

A obvious way to remove heat and maintain a constant temperature is to scale the atomic velocities at periodic intervals during the simulation, each time resetting the instantaneous temperature to the desired value. However, this approach is known to disrupt the dynamics of atoms and may lead to serious artifacts in the simulation results. A less disruptive approach to temperature control is the Nose–Hoover thermostat, which mimics heat exchange between a simulation volume and its surroundings.

4.4.1   Ensembles and Extended Systems

In Section 3.2 we introduced boundary conditions as a way to mimic the interaction between the atoms in a (small) simulation volume with atoms in the (much larger) surrounding volume. In addition to special treatments of the boundary atoms, other methods exist that mimic various effects of the surrounding environment on the atoms inside the simulation cell. Heat, momentum, and mass exchange with the environment can be all simulated using special coupling procedures, each of which corresponds to a specific *statistical ensemble* in the language of statistical mechanics.

For instance, MD simulations under fixed boundary conditions (Section 3.1) or under PBC with fixed repeat vectors correspond to the *micro-canonical* or *NVE* ensemble. The abbreviation *NVE* means that the total number of particles N, volume Ω and total energy E are all conserved during the simulations.[4] On the other hand, a system exchanging heat with an external thermostat will maintain a constant temperature T resulting in an isothermal or *canonical* (NVT) ensemble. If, in addition, the same system interacts with an external barostat, its total volume will adjust so as to maintain a constant pressure, corresponding to an isothermal isobaric, or *NPT* ensemble. Furthermore, mass exchange with a large external reservoir will cause the total number of atoms in the system to fluctuate so that its chemical potential μ remains close to that of the *massostat*. Such simulations correspond to a grand-canonical or μVT ensemble [51]. Different statistical ensembles can mimic various effects of the surrounding environment on the behavior of atoms inside the simulation cell.

---

[4] V in this abbreviation stands for volume. It is used here because this abbreviation is so common. Everywhere else in this book V stands for potential energy and Ω stands for volume.

The Nose–Hoover thermostat to be used in this case study is designed to reproduce the *canonical* ensemble. It is an example of the *extended-system* approach, in which one or several fictitious degrees of freedom are introduced to mimic various aspects of the system's interaction with the surroundings. The total system is assumed to consist of two interacting sub-systems, the atomistic system itself plus the additional fictitious degrees of freedom. The total system is still closed and its total energy is conserved. However, the atomic sub-system is open and its energy is not conserved. Through its interaction with the fictitious degrees of freedom, the temperature of the atomistic sub-system can be controlled.

4.4.2 Nose–Hoover Thermostat

The Nose–Hoover thermostat uses a single degree of freedom to mimic a large reservoir that exchanges heat with the atomistic sub-system. The specific implementation of this idea is due to Nose and starts by introducing a new dynamical variable s that serves as a scaling factor of time [52]. This time scaling distorts the atomic trajectories, but, in return, maintains the temperature of the atomistic sub-system close to a preselected value. The following equations of motion describe the joint evolution of atomistic and fictitious sub-systems, in the form suggested by Hoover [36, 53]:

$$\frac{d\mathbf{r}_i(t)}{dt} = \mathbf{v}_i(t), \tag{4.12}$$

$$\frac{d\mathbf{v}_i(t)}{dt} = \frac{\mathbf{f}_i(t)}{m} - \zeta(t)\mathbf{v}_i(t), \tag{4.13}$$

$$\frac{d\zeta(t)}{dt} = \frac{1}{M_s}\left[\sum_{i=1}^{N} m|\mathbf{v}_i(t)|^2 - 3Nk_BT\right], \tag{4.14}$$

$$\frac{ds(t)}{dt} = s(t)\zeta(t). \tag{4.15}$$

Here, M_s is a thermal "mass" parameter that controls the rate of heat exchange with the exterior of the atomic sub-system.[5] When the kinetic energy E_{kin} exceeds $\frac{3}{2}Nk_BT$, i.e. when the instantaneous temperature exceeds the desired target temperature T, ζ will increase. As soon as ζ becomes positive, it exerts a viscous drag force on all atoms, forcing them to reduce their velocities and lowering the instantaneous temperature. This feedback loop makes the instantaneous temperature fluctuate around the desired temperature T. It is possible to show that MD simulations that use the Nose–Hoover thermostat generate Boltzmann's distribution and, hence, correspond to the canonical ensemble. Similar to the Parrinello–Rahman method

---

[5] M_s is analogous to the wall "mass" parameter in the Parrinello–Rahman method for stress control described in Section 4.3.

(Section 4.3), the Nose–Hoover equations of motion are derived from the standard variational principle of classical mechanics. The trajectories conserve the following total energy of the extended system,

$$\tilde{H} = \sum_{i=1}^{N} \frac{1}{2} m_i |v_i|^2 + V(\{R_i\}) + \frac{1}{2} M_s \zeta^2 + 3 N k_B T \ln s. \qquad (4.16)$$

The first two terms on the right-hand side represent the total energy of the atomistic sub-system and the last two terms can be regarded as the total energy of the heat bath. Conservation of $\tilde{H}$ provides a useful check for bugs in the numerical implementation of the Nose–Hoover thermostat. Thus, even if eq. (4.15) is not needed to compute the trajectory of atoms, it is still useful to keep it in the simulation in order to compute the total energy $\tilde{H}$ of the extended system. In our case study, we use the Nose–Hoover thermostat with the following set of parameters: $3 N k_B T^* / M_s = 10^{-2}$ fs^{-2}, $s(0) = 1$ and $\dot{s}(0) = 0$.

Numerical implementation of the Nose–Hoover equation of motion In the Nose–Hoover dynamics acceleration depends on velocity, as in eq. (4.13). Therefore, the original Verlet algorithm given in eq. (2.41) has to be modified. In the following, we describe a time-reversible centered-difference variant of the Verlet algorithm appropriate for simulations with the Nose–Hoover thermostat [54]. First, let us discretize eq. (4.13):

$$\frac{r_i(t + \Delta t) - 2 r_i(t) + r_i(t - \Delta t)}{\Delta t^2} = \frac{f_i(t)}{m} - \zeta(t) v_i(t), \qquad (4.17)$$

where velocity $v_i(t)$ is expressed as

$$v_i(t) = \frac{r_i(t + \Delta t) - r_i(t - \Delta t)}{2 \Delta t}. \qquad (4.18)$$

Solving these two equations for the atomic positions at time $t + \Delta t$ leads to

$$r_i(t + \Delta t) = \frac{2 r_i(t) - r_i(t - \Delta t) + \Delta t^2 \cdot f_i(t)/m + \Delta t \cdot \zeta(t) r_i(t - \Delta t)/2}{1 + \Delta t \cdot \zeta(t)/2}. \qquad (4.19)$$

The equation for the fictitious parameter ζ is

$$\zeta(t + \Delta t) = \zeta(t) + \frac{\Delta t}{M_s} \left[\sum_{i=1}^{N} m |v_i(t + \Delta t/2)|^2 - 3 N k_B T \right], \qquad (4.20)$$

where

$$v_i(t + \Delta t/2) = \frac{r_i(t + \Delta t) - r_i(t)}{\Delta t}. \qquad (4.21)$$

For more discussions on symplectic integrators for Nose-Hoover dynamics see Ref [131].

Summary

- The work expended by external stress on dislocation motion is dissipated into heat, which can make the temperature rise.

- The Nose–Hoover thermostat is an effective method for controlling the temperature in MD simulations.

Problem

4.4.1. Estimate the rate at which the temperature would increase if no thermostat were used in an MD simulation of dislocation motion. Assume that all of the work of applied stress expended on moving the dislocation is dissipated as heat. The force per unit length on the dislocation is given by the Peach–Koehler formula: $f = \tau b$, where τ is the applied driving stress and b is the Burgers vector. Use the following parameters: $\tau = 300$ MPa, $b = 2.73$ Å, total number of atoms in the simulation volume $N = 24720$, dislocation length $L_z = 65$ Å, and dislocation velocity $v \approx 1000$ m s^{-1}.

4.5 Extracting Dislocation Velocity

By now we have collected all the tools necessary to run an MD simulation of dislocation motion. As an example, let us perform an MD simulation under stress $\sigma_{xy} = 300$ MPa and temperature $T = 300$ K, with the initial and boundary conditions as described in the preceding sections. The major result of this case study is plotted in Fig. 4.4, showing the instantaneous dislocation position as a function of time. In this section, we describe how this information was extracted from the "raw" simulation data.

A simple definition of the dislocation position would be the center of mass of the core atoms. Here, the core atoms are defined as atoms whose CSD parameter (Section 3.3) falls in the range between 1.5 Å^2 and 10 Å^2. As the dislocation moves, atoms may join or leave the set of core atoms. Every time this happens, the dislocation position will jump instantaneously. Fortunately, such discrete jumps are hardly noticeable on the scale of hundreds of atomic spacings, as shown in Fig. 4.4.

Because we use periodic boundary conditions along the direction of dislocation motion, it is necessary to make sure that the positions of all core atoms belong to the same periodic image of the moving dislocation. This can be ensured, for

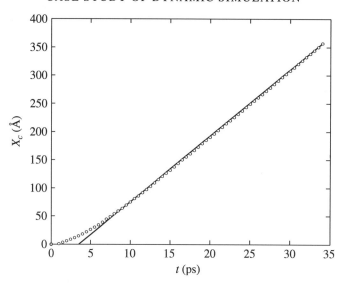

FIG. 4.4. Dislocation position as a function of time (circles), obtained in an MD simulation at $T = 300$ K and stress $\sigma_{xy} = 300$ MPa. The straight line is a linear fit of the data for $t > 10$ ps, corresponding to an average dislocation velocity of $v = 1170$ m s^{-1}.

example, by always selecting among the multiple images of the core atoms those that are nearest to the core atom with the lowest index. Furthermore, when the dislocation leaves the simulation cell and one of its images enters from the other side, the computed dislocation position could jump by L_x over one time step, which, needless to say, would be very unfortunate. A simple solution to this problem is to always select, among the multiple periodic images of the moving dislocation, the one whose position is closest to the position computed at the previous time step.[6] Defined in this manner, the coordinate of the dislocation center continues to increase even after the dislocation and its images have crossed the simulation cell multiple times. The combination of these two techniques allowed us to obtain the smooth curve for the dislocation position as a function of time shown in Fig. 4.4.

Dislocation motion depicted in Fig. 4.4 is similar to the motion of a particle subjected to a constant force in a viscous medium. Initially, there is a transient period over which the dislocation experiences acceleration. Rather quickly, within about 10 ps, steady-state motion sets in with an average velocity $v \approx 1170$ m s^{-1}. This value is extracted from a linear fit to the displacement-time data for $t > 10$ ps, illustrated by a straight line in Fig. 4.4. Beyond the initial transient, dislocation motion remains steady, showing only small velocity fluctuations around its average

---

[6] This assumes that the dislocation cannot move by more than the length of the simulation box in one time step, which is certainly a reasonable assumption.

value. MD simulations like this can produce valuable data on dislocation velocity and its dependence on stress and temperature. Under conditions when the dislocation velocity is proportional to the stress, the viscous drag coefficient $B \equiv \sigma b/v$ can be obtained by measuring dislocation velocities under different values of external stress [55, 56]. Furthermore, by running a series of simulations at different temperatures, dependence of the drag coefficient on temperature can also be obtained (Problem 4.5.1). This information can be used to construct mobility functions for dislocation dynamics simulations to be discussed in Chapter 10. The initial transient period of dislocation motion (the first 10 ps in Fig. 4.4) is also informative. The duration of the transient period is a measure of dislocation's effective mass.[7] In addition to quantifying dislocation mobility and inertia parameters, MD simulations provide valuable insights into various aspects of dislocation motion, such as dislocation–sound-wave interactions and relativistic effects at high dislocation velocities [57, 58].

Summary

- Dislocation position can be defined as the center of mass of the core atoms.

- Dislocation velocity is obtained by measuring the slope of the dislocation position plotted as a function of time.

Problem

4.5.1 Perform a series of MD simulations of dislocation glide motion at a single temperature, say $T = 300$ K, but for different values of stress: 200, 400, 600, 800, 1000 MPa. Plot the dislocation velocity as a function of stress. In the region where the plot is linear, compute the drag coefficient $B = \sigma b/v$. Repeat simulations at $T = 100$ K and $T = 600$ K and examine how the drag coefficient B changes with the increasing temperature.

---

[7] For this estimate to be more accurate, it is necessary to subtract the time for the sound wave generated by the surface force to reach the dislocation. This time lag is around 1ps in this case study.

5

MORE ABOUT PERIODIC BOUNDARY CONDITIONS

We have already used periodic boundary conditions (PBC) for the static and dynamic simulations described in Chapters 2 and 3. There, PBC were applied along one or two directions of the simulation cell. Application of PBC in all three directions holds an important advantage when one's goal is to examine the behavior in the bulk: under fully three-dimensional (3D) PBC, the simulated solid can be free of any surfaces. By comparison, the simulations discussed in the previous chapters all contained free surfaces or artificial interfaces in the directions where PBC were not applied.

Full 3D PBC are easy to implement in an atomistic simulation through the use of scaled coordinates (see Section 3.2). However, there are important technical issues specific to simulations of lattice dislocations. First, a fully periodic simulation cell can accommodate only such dislocation arrangements whose net Burgers vector is zero. Thus, the minimal number of dislocations that can be introduced in a periodic supercell is two, i.e. a dislocation dipole.[1] Two dislocations forming a dipole are bound to interact with each other, as well as with their periodic images. Associated with these interactions are additional strain, energy, and forces whose effects can "pollute" the calculated results. The good news is that, in most cases, the artifacts of PBC can be quantified through the use of linear elasticity theory so that physical properties of dislocations can be accurately extracted. Given the simplicity and robustness of PBC, the extra work required to extract physical results is well worth it.

This chapter describes how to evaluate and eliminate the artifacts that inevitably appear when 3D PBC are used for atomistic simulations of dislocations. In the following three sections, we show how to take full advantage of PBC when one wants to calculate the displacement field induced by a dislocation (Section 5.1), the dislocation's core energy (Section 5.2) and Peierls stress (Section 5.3). The common theme for all three case studies is an attempt to construct a solution of the elasticity equations in a periodic domain by superimposing a periodic array of solutions of an infinite domain.

---

[1] The simulation cell in Chapter 4 contains only one dislocation because it has free surfaces along the y direction.

5.1   Setting up an Initial Configuration

The task of setting up initial conditions for a dislocation dipole under PBC is not as trivial as it may appear at a first glance. Similar to what was done in Section 3.1, we would like to introduce a dislocation by displacing the atoms according to an appropriate solution for the displacement field of a dislocation dipole in the periodic domain, $\mathbf{u}^{\text{PBC}}(\mathbf{r})$. However, such solution is not available in an analytic form. What is available is a solution for the displacement field of a dislocation dipole in an infinite (non-periodic) domain, $\mathbf{u}^{\infty}(\mathbf{r})$. Given that $\mathbf{u}^{\text{PBC}}(\mathbf{r})$ and $\mathbf{u}^{\infty}(\mathbf{r})$ are two different displacement fields produced by the same object (a dislocation dipole) under different boundary conditions, it makes sense to try to construct the desired solution $\mathbf{u}^{\text{PBC}}(\mathbf{r})$ from the known solution $\mathbf{u}^{\infty}(\mathbf{r})$.

5.1.1   A naive approach

As an example, consider a periodic supercell containing a screw dislocation dipole, as shown in Fig. 5.1(a). Three repeat vectors $\mathbf{c}_1$, $\mathbf{c}_2$ and $\mathbf{c}_3$ are along the Cartesian axes x, y and z, respectively, and the dislocation line is along the z axis. Within isotropic elasticity, the only non-zero component of the displacement field produced by a screw dislocation dipole is, (see eq. (3.1))

$$u_z^{\infty}(\mathbf{r}) = b \, \frac{\theta_1 - \theta_2}{2\pi}, \tag{5.1}$$

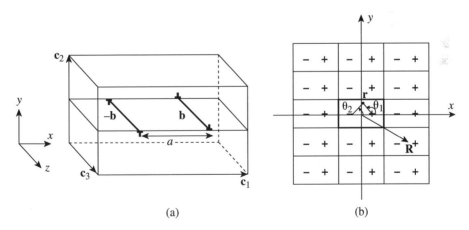

(a)   (b)

FIG. 5.1.   (a) A dislocation dipole in a periodic supercell with mutually orthogonal repeat vectors $\mathbf{c}_1$, $\mathbf{c}_2$, $\mathbf{c}_3$. Here, a is the distance between the two dislocations. (b) The displacement field of a dislocation dipole in PBC is constructed as a superposition of the displacement fields produced by a periodic array of dislocation dipoles.

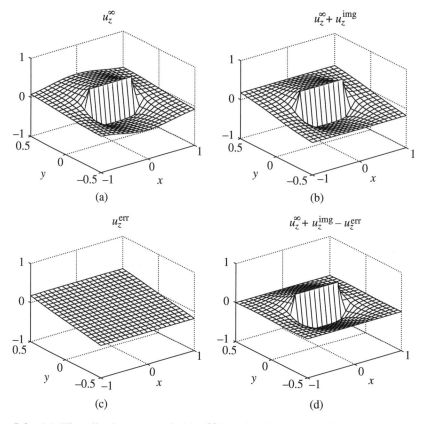

FIG. 5.2. (a) The displacement field $u_z^{\infty}(x, y)$ of a screw dislocation dipole in an infinite medium. (b) The same displacement field after the image contributions are added. (c) The error term resulting from the naive summation of the image contributions. (d) The final result after the error term is subtracted (see text). Reproduced from [59] with the publisher's permission, http://www.tandf.co.uk.

where θ_1 (θ_2) is the angle between the vector connecting the dislocation b ($-b$) to the field point $\mathbf{r}$ and the x axis, as shown in Fig. 5.1(b). The displacement field of eq. (5.1) experiences a discontinuous jump across the area of the cut bounded by the two dislocations (Fig. 5.2(a)). As was discussed earlier, this discontinuity is a normal feature of the dislocation dipole. What is really worrisome is that, as seen from Fig. 5.2(a), $u_z^{\infty}(\mathbf{r})$ is not the same on the two opposite sides of the supercell at $y = -0.5$ and $y = 0.5$.[2] Solution $u_z^{\infty}(\mathbf{r})$ does not satisfy the periodic boundary condition in the y direction and any attempts to fit this displacement

---

[2] Because $u_z^{\infty}(\mathbf{r})$ does not depend on z, the displacement field is trivially periodic along $\mathbf{c}_3$. Therefore, in this chapter it is sufficient to consider its dependence on x and y only.

field into a periodic box will create some mismatch at the box boundaries. This mismatch may relax away during a subsequent energy minimization but this is by no means guaranteed. Part of it can persist in the form of a spurious interface or fault, contaminating the simulation results. To ensure that our simulations are free from such unintended artifacts, it is desirable to initialize atomic positions according to a periodic displacement field.

Because the equations of elasticity are linear, it should be possible to construct a periodic displacement field from the non-periodic solution $u_z^\infty(\mathbf{r})$ using superposition. Specifically, consider the following sum of the otherwise identical (non-periodic) solutions, each offset by a vector in a periodic lattice:

$$u_z^{\text{sum}}(\mathbf{r}) = \sum_{\mathbf{R}} u_z^\infty(\mathbf{r} - \mathbf{R}) = u_z^\infty(\mathbf{r}) + u_z^{\text{img}}(\mathbf{r}), \tag{5.2}$$

$$u_z^{\text{img}}(\mathbf{r}) \equiv \sum_{\mathbf{R}}{}' u_z^\infty(\mathbf{r} - \mathbf{R}). \tag{5.3}$$

Here, the sums $\sum$ and $\sum'$ run over the two-dimensional lattice $\mathbf{R} = n_1\mathbf{c}_1 + n_2\mathbf{c}_2$ (n_1 and n_2 are integers). Each individual term $u_z^\infty(\mathbf{r} - \mathbf{R})$ corresponds to the contribution from an *image* dipole shifted from the *primary* dipole by $\mathbf{R}$, as shown in Fig. 5.1(b). Whereas $u_z^{\text{sum}}(\mathbf{r})$ accounts for the contributions from all dipoles (primary + image), $\sum'$ excludes the contribution from the primary dipole for which $\mathbf{R} = 0$. The sums formally contain an infinite number of terms but, in practice, are evaluated only over a finite number of image dipoles closest to the primary dipole.

Fig. 5.2(b) shows the result for $u_z^\infty(\mathbf{r}) + u_z^{\text{img}}(\mathbf{r})$ obtained with a truncated summation. The mismatch at the boundaries is smaller compared to Fig. 5.2(a), yet this displacement field is still not periodic along the y axis. By inspection of Fig. 5.2(c), the non-periodic part appears to be a linear function of x and y. The desired periodic solution shown in Fig. 5.2(d) is obtained only after this linear field $u_z^{\text{err}}(\mathbf{r})$ is subtracted out. It may seem surprising that the field obtained by superposition of fields due to a periodic array of sources turns out to be non-periodic, but this can be proved [59]. Next we explain why this is the case.

5.1.2 Conditional Convergence and the Linear Error Field

The non-periodic error field $u_z^{\text{err}}(\mathbf{r})$ is an indication that something is wrong with our naive summation procedure. It turns out that, because the individual terms in eq. (5.2) can be either positive or negative, the limit of this sum depends on the ordering of its terms during the summation. This means that the sum depends on exactly how the summation is truncated. When such is the case, the limit is said to depend on a summation *condition*. A sum that behaves in this "strange" way is called *conditionally convergent*, as opposed to an *absolutely convergent* sum, whose limit does not depend on any such condition. For a sum to be absolutely convergent, it is necessary and sufficient that the sum of the absolute values of its terms is convergent.

To see why sum (5.2) is not absolutely convergent, let us examine the sum of the absolute values of its terms. First, notice that, from eq. (5.1),

$$u_z^\infty(\mathbf{r}) \approx \frac{ba \sin \theta}{2\pi r} \tag{5.4}$$

for large r, where $r = |\mathbf{r}|$, a is the distance between the two dislocations in the dipole, and θ is the angle between $\mathbf{r}$ and the x axis. Hence,

$$u_z^\infty(\mathbf{r} - \mathbf{R}) \approx \frac{ba \sin \Theta}{2\pi R} \tag{5.5}$$

for large R, where $R = |\mathbf{R}|$, and Θ is the angle between $\mathbf{R}$ and the x axis. The sum of the absolute values of terms in eq. (5.2) can be approximately evaluated as follows,

$$\sum_{\mathbf{R}} |u_z^\infty(\mathbf{r} - \mathbf{R})| = \sum_{R < R_0} |u_z^\infty(\mathbf{r} - \mathbf{R})| + \sum_{R \geq R_0} |u_z^\infty(\mathbf{r} - \mathbf{R})|$$

$$\approx \sum_{R < R_0} |u_z^\infty(\mathbf{r} - \mathbf{R})| + \int_{R_0}^\infty R \, dR \int_0^{2\pi} d\Theta \, \frac{ba|\sin \Theta|}{2\pi R}$$

$$= \sum_{R < R_0} |u_z^\infty(\mathbf{r} - \mathbf{R})| + \frac{2ba}{\pi} \int_{R_0}^\infty dR, \tag{5.6}$$

where radius R_0 is large enough to replace the sums with the integrals for $R > R_0$. Obviously, the second term in the last equation diverges. This means that the sum in eq. (5.2) is not absolutely convergent and may converge only conditionally, generally to a "wrong" limit. It is no longer surprising that the result of the naive image summation in eq. (5.3) produced a non-periodic field. The good news is that the "right" sum does exist and can be obtained from the "wrong" sum using a simple correction procedure explained below.

Let us first differentiate both sides of eq. (5.2) with respect to the components of $\mathbf{r}$,

$$\partial_i \partial_j u_z^{\text{sum}}(\mathbf{r}) = \sum_{\mathbf{R}} \partial_i \partial_j u_z^\infty(\mathbf{r} - \mathbf{R}), \tag{5.7}$$

where $\partial_i \partial_j \equiv \frac{\partial^2}{\partial x_i \partial x_j}$, $i, j = 1, 2, x_1 \equiv x, x_2 \equiv y$. It is easy to see that

$$\partial_i \partial_j u_z^\infty (\mathbf{r} - \mathbf{R}) \approx \frac{f_{ij}(\Theta)}{R^3}, \tag{5.8}$$

where $f_{ij}(\Theta)$ are functions of Θ whose values are bounded (explicit expressions for $f_{ij}(\Theta)$ are not required for our purposes here). Applying the same analysis as above and using eq. (5.8), it is straightforward to demonstrate that the summation in eq. (5.7) is absolutely convergent (Problem 5.1.1).

The limit of this absolutely convergent sum is unique and can be shown to produce the derivatives of a periodic displacement field $u_z^{\text{PBC}}(\mathbf{r})$. Noting that

$$\partial_i \partial_j u_z^{\text{sum}}(\mathbf{r}) = \partial_i \partial_j u_z^{\text{PBC}}(\mathbf{r}) \tag{5.9}$$

and integrating twice both sides of this equation, we obtain

$$u_z^{\text{sum}}(\mathbf{r}) = u_z^{\text{PBC}}(\mathbf{r}) + \mathbf{s} \cdot \mathbf{r} + \mathbf{u}_0. \tag{5.10}$$

This last result means that, at worst, the "wrong" sum differs from the "right" sum by a constant and a linear term. Because the constant term $\mathbf{u}_0$ does not violate periodicity, it can be ignored here. Hence, in accord with our earlier observation in Fig. 5.2(c), it is the spurious linear term that makes the naive sum non-periodic.

Thus, we arrive at a simple recipe for getting rid of the spurious non-periodic part of the displacement field. First, evaluate a conditionally convergent sum $u_z^{\text{sum}}(\mathbf{r})$ using some (arbitrary) truncation. Second, "measure" the linear spurious part of the resulting field,

$$u_z^{\text{err}}(\mathbf{r}) = \mathbf{s} \cdot \mathbf{r} \tag{5.11}$$

by comparing its values at four points in the periodic supercell (from eq. (5.10)):

$$u_z^{\text{sum}}(\mathbf{r} + \mathbf{c}_i) - u_z^{\text{sum}}(\mathbf{r}) = \mathbf{s} \cdot \mathbf{c}_i, \tag{5.12}$$

where $i = 1, 2$. Finally, subtract the linear term $u_z^{\text{err}}(\mathbf{r})$ from $u_z^{\text{sum}}(\mathbf{r})$ to obtain a corrected solution $u_z^{\text{PBC}}(\mathbf{r})$, as in Fig. 5.2(d). The resulting solution is periodic and independent of the chosen truncation in the limit of large truncation radius. This simple three-step procedure can be safely used for setting up initial atomic positions for dislocation dipoles in the periodic supercells.

5.1.3   Adjusting the Shape of the Supercell

In atomistic simulations of dislocations it is sometimes necessary to use periodic supercells with repeat vectors that are not mutually orthogonal. One reason to

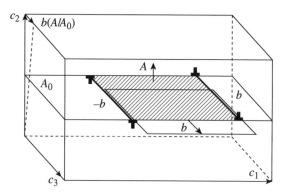

FIG. 5.3. Insertion of a dislocation dipole into the periodic supercell by displacing the atoms on top of cut plane **A** by **b** with respect to the atoms below the cut. Area **A** is part of the horizontal mid-plane of the supercell whose total area is $A_0 = |\mathbf{c}_3 \times \mathbf{c}_1|$. To reset the average stress in the supercell to zero, repeat vector $\mathbf{c}_2$ is shifted by $\mathbf{b}(A/A_0)$.

use non-orthogonal supercells is to compensate for the internal stress produced by dislocations.

Consider again a screw dislocation dipole in an orthogonal supercell, which is replotted in Fig. 5.3. Once a dislocation dipole is inserted through the usual cut-and-shift procedure, the atoms immediately above cut **A** (shaded in Fig. 5.3) are displaced by **b** with respect to the atoms immediately below the cut (here, the magnitude of vector **A** is the area of the cut and its direction is normal to the cut plane). This insertion operation produces the following plastic strain (see also eq. (1.18))

$$\boldsymbol{\varepsilon}^{\mathrm{pl}} = \frac{1}{2\Omega}\,(\mathbf{b} \otimes \mathbf{A} + \mathbf{A} \otimes \mathbf{b}), \tag{5.13}$$

where $\Omega = (\mathbf{c}_1 \times \mathbf{c}_2) \cdot \mathbf{c}_3$ is the supercell volume. In a supercell with fixed periodicity vectors, an increment in plastic strain will be compensated by an oppositely signed increment of elastic strain of the same magnitude, $\boldsymbol{\varepsilon}^{\mathrm{el}} = -\boldsymbol{\varepsilon}^{\mathrm{pl}}$. In response to this elastic strain, there will be an internal back-stress acting to eliminate the source of the strain, i.e. the dislocation dipole. This back-stress may be large enough to push the dislocations back from their intended positions and may even lead to dislocation recombination.

If we were to allow the simulation box to adjust its shape during energy minimization (Section 4.3), it could reach a state of zero average internal stress on its own. Unfortunately, even in this case the dislocations can move significantly during the relaxation. To avoid such unwanted behaviors, we can set the average internal stress to zero by adjusting the shape of the supercell *before* the relaxation.

The idea is to intentionally distort the supercell just enough to accommodate or match the plastic strain produced by the dislocation dipole. In our case study, the cut plane bounded by two dislocations is parallel to two of the repeat vectors, $\mathbf{c}_1$ and $\mathbf{c}_3$. In this case, the internal stress induced by the dipole can be easily removed by adjusting only one repeat vector, $\mathbf{c}_2$,

$$\mathbf{c}_2 \rightarrow \mathbf{c}_2 + \mathbf{b}\, \frac{A}{A_0}, \qquad (5.14)$$

where $A_0 = |\mathbf{c}_3 \times \mathbf{c}_1|$. An easy way to understand the above relation is to consider the case when $A = A_0$. After $\mathbf{c}_2$ is displaced by $\mathbf{b}$, the entire upper half of the crystal would be uniformly displaced by $\mathbf{b}$ with respect to the lower half, so there would be no internal stress.[3]

Adjusting $\mathbf{c}_2$ according to eq. (5.14) amounts to adding an extra term $\mathbf{u}^{\text{tilt}}(\mathbf{r})$ to the solution $\mathbf{u}^{\text{PBC}}(\mathbf{r})$ obtained previously. In this case study,

$$u_z^{\text{tilt}}(\mathbf{r}) = b\, \frac{Ay}{A_0 c_2}, \qquad (5.15)$$

where c_2 is the length of vector $\mathbf{c}_2$ before it is tilted. Even though $u_z^{\text{tilt}}(\mathbf{r})$ and the previously found $u_z^{\text{err}}(\mathbf{r})$ are both linear functions of x and y, their meanings are completely different. $u_z^{\text{err}}(\mathbf{r})$ is a spurious displacement resulting from the conditional convergence of the lattice sum (5.2)—this unwanted term violates PBC and should be subtracted from the sum. On the other hand, $u_z^{\text{tilt}}(\mathbf{r})$ is introduced intentionally in order to remove the internal stress induced by the insertion of a dislocation dipole. The procedure described above is general and works equally well for screw and non-screw dislocations.

Summary

- The displacement field of a dislocation dipole in a periodic supercell can be reconstructed by superposition of the displacement fields of a periodic array of dipoles in an infinite medium.

- The resulting lattice sum contains a spurious non-periodic term that is a linear function of the coordinates in the supercell. To obtain the desired periodic solution, the spurious term can be quantified and subtracted from the sum.

- To eliminate internal stress induced by the insertion of a dislocation dipole, the repeat vectors of the supercell can be adjusted.

---

[3] In this case, there would be no dislocations either.

Problem

    5.1.1. Using a technique similar to eq. (5.6), show that the summation of $\partial_i \partial_j u_z^\infty(\mathbf{r} - \mathbf{R})$ in eq. (5.7) is absolutely convergent. In other words, the summation of the absolute values, $|\partial_i \partial_j u_z^\infty(\mathbf{r} - \mathbf{R})|$, converges.

5.2 Dislocation Core Energy

Here we apply the methods discussed in the preceding section to a specific example of a screw dislocation dipole in silicon. First, let us construct a perfect diamond-cubic crystal (Section 1.1) in a periodic supercell with repeat vectors $\mathbf{c}_1 = 4[11\bar{2}]$, $\mathbf{c}_2 = 3[111]$ and $\mathbf{c}_3 = [1\bar{1}0]$ (in units of lattice constant, 5.4309 Å). Second, let us introduce a screw dislocation dipole with Burgers vector $\mathbf{b} = [1\bar{1}0]/2$, and the separation between two dislocations of the dipole $\mathbf{a} = \mathbf{c}_1/2$. To eliminate average internal stress the supercell is tilted, $\mathbf{c}_2 \to \mathbf{c}_2 + \mathbf{b}/2$. Using the Stillinger–Weber (SW) potential [18] to represent the interaction among the atoms, let us relax the atomistic configuration by the CGR method (Section 2.3). The resulting equilibrated configuration is shown in Fig. 5.4, where the atoms highlighted in grey have a local energy 0.08 eV higher than that in the perfect crystal.

    The resulting total energy E^* can be expressed as a sum of several terms,

$$E^* = E_{\text{atm}} \cdot c_3 + N E_{\text{coh}} + \frac{1}{2} \boldsymbol{\sigma}^* \cdot \mathbf{S} \cdot \boldsymbol{\sigma}^*, \qquad (5.16)$$

where E_{atm} is the excess energy (per unit length) of the dislocation dipole, N is the total number of atoms, E_{coh} is the cohesive energy,[4] $\boldsymbol{\sigma}^*$ is the residual

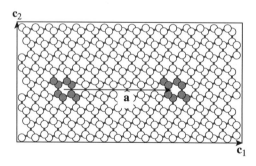

FIG. 5.4. A dipole of screw dislocations in silicon. The separation between two dislocations is $\mathbf{a} = \mathbf{c}_1/2$. Reproduced from [59] with publisher's permission, http://www.tandf.co.uk.

---

[4] The cohesive energy is the energy per atom in the perfect crystal. For the SW potential model used here [18], $E_{\text{coh}} = -4.63$ eV.

stress (Section 4.3), and $\mathbf{S}$ is the elastic compliance tensor.[5] Equation (5.16) can be regarded as the definition of E_{atm}. The third term in eq. (5.16) corresponds to the energy associated with internal stresses. Even though most of internal stress induced by the dipole is taken out by pretilting the vector $\mathbf{c}_2$, some residual stress may persist. The magnitude of this residual stress can be further reduced by allowing the box vectors to change during CGR minimization, as in the Parrinelo–Rahman method. Any residual stress left after tilting and relaxation should be included in the energy balance. The excess energy E_{atm} obtained from eq. (5.16) depends on the supercell geometry. The purpose of this section is to extract a more invariant property of the dislocation—its core energy, which should be independent of the simulation cell geometry.

We can extract the core energy if we can compute the elastic energy E_{el} of a dislocation dipole in a supercell of the same geometry as that in the atomistic simulation. The difference between E_{el} and E_{atm} can be used to define the dislocation core energy E_{core} as follows:

$$E_{atm}(\mathbf{a}, \mathbf{c}_i) = E_{el}(\mathbf{a}, \mathbf{c}_i, r_c) + 2E_{core}(r_c), \tag{5.17}$$

where r_c is a cut-off radius introduced to eliminate the singularity in the linear elasticity theory (both dislocations are assumed to have the same core energy). While both E_{atm} and E_{el} depend on the dipole width ($\mathbf{a}$) and the supercell geometry ($\mathbf{c}_i$, $i = 1, 2, 3$), the core energy should depend only on r_c. Thus, the task of extracting the dislocation core energy reduces to computing the elastic energy E_{el} of a dislocation dipole under PBC.

The analytic expression for $E_{el}(\mathbf{a}, \mathbf{c}_i, r_c)$ is not available. What is available is the elastic energy of a dislocation dipole in an infinite medium. For a screw dislocation dipole in isotropic elasticity, this energy is,

$$E_{el}^{\infty}(\mathbf{a}, r_c) = \frac{\mu b^2}{4\pi} \ln \frac{|\mathbf{a}|}{r_c}. \tag{5.18}$$

The difference between two energies,

$$E_{img} \equiv E_{el}(\mathbf{a}, \mathbf{c}_i, r_c) - E_{el}^{\infty}(\mathbf{a}, r_c), \tag{5.19}$$

is the as yet unknown contribution from the periodic boundary conditions. Similar to the previous section, E_{img} can be thought of as the interaction energy between

---

[5] The rank-4 tensor $\mathbf{S}$ is the inverse of the elastic stiffness tensor $\mathbf{C}$.

the primary dislocation dipole with an infinite array of image dipoles, i.e.

$$E_{\text{img}} = \frac{1}{2} \sum_{\mathbf{R}}{}' E_{\text{dd}}(\mathbf{R}). \tag{5.20}$$

Here, summation in $\sum'$ runs over the two-dimensional lattice of image dipoles $\mathbf{R} = n_1\mathbf{c}_1 + n_2\mathbf{c}_2$, excluding the self-interaction $\mathbf{R} = 0$. Each term $E_{\text{dd}}(\mathbf{R})$ is the interaction energy between two identical dislocation dipoles separated by distance $\mathbf{R}$. The factor of $\frac{1}{2}$ appears because only half of the interaction energy is attributed to the primary dipole while the other half goes to the image dipole. For our example of two screw dislocations in an isotropic crystal,

$$E_{\text{dd}}(\mathbf{R}) = \frac{\mu b^2}{2\pi} \ln \frac{|\mathbf{R}|^2}{|\mathbf{R}+\mathbf{a}| \cdot |\mathbf{R}-\mathbf{a}|}. \tag{5.21}$$

Because $E_{\text{dd}}(\mathbf{R}) \sim 1/R^2$ for large R, the sum in eq. (5.20) is conditionally convergent. Similar to the preceding section, it is necessary to find a way to evaluate the "true" limit of lattice sum (5.20) to be able to extract the dislocation core energy.

A solution to this problem is presented in [59]. We only outline its ideas and describe its findings here. The interaction energy between two dislocation dipoles can be written as the stress field produced by one dipole integrated over the cut plane of the other dipole. Consequently, eq. (5.20) can be rewritten as a sum over the stress fields produced by the image dipoles. Although this sum is still conditionally convergent, its spurious part corresponds to an additional stress, which is constant in the supercell.

For an arbitrary order of summation in eq. (5.20), the error term can be conveniently computed by introducing a set of four test or "ghost" dislocations with Burgers vectors $\alpha\mathbf{b}$, $-\alpha\mathbf{b}$, $\beta\mathbf{b}$, $-\beta\mathbf{b}$, such that $\alpha\mathbf{b}$ is separated from $-\alpha\mathbf{b}$ by $\mathbf{c}_1$, $\beta\mathbf{b}$ is separated from $-\beta\mathbf{b}$ by $\mathbf{c}_2$ and $\alpha\mathbf{c}_1 + \beta\mathbf{c}_2 = \mathbf{a}$. Denoting $E_{\text{dg}}(\mathbf{R})$ as the interaction energy between the dislocation dipole at location $\mathbf{R}$ and the four "ghost" dislocations, then eq. (5.20) should be replaced by

$$E_{\text{img}} = \frac{1}{2} \left(\sum_{\mathbf{R}}{}' E_{\text{dd}}(\mathbf{R}) - \sum_{\mathbf{R}} E_{\text{dg}}(\mathbf{R}) \right), \tag{5.22}$$

where, for consistency, one must use exactly the same set of images in both summations in eq. (5.22). Notice also that the second sum in eq. (5.22) includes the $\mathbf{R} = 0$ term.

A special utility code MadSum (from *Madelung Summation*) implements the above procedure and computes the elastic energy of a dislocation dipole in an

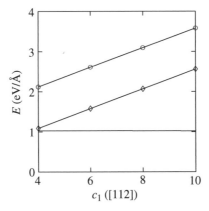

FIG. 5.5. The atomistic and the elastic energies of a dislocation dipole as functions of the supercell shape. E_{atm} is shown by o and E_{el} by ◇. The solid line is the difference, $2E_{core} = E_{atm} - E_{el}$.

arbitrary supercell using full anisotropic elasticity. MadSum is available at the book's web site. The data plotted in Fig. 5.5 is obtained using MadSum.

Figure 5.5 contains the results of a series of atomistic simulations using different cell dimensions along c_1, while keeping $a = c_1/2$. The elasticity predictions for the dipole energies in the same cell geometries are also plotted. The elastic constants used in MadSum calculations are those reported for the same SW potential model, i.e. $C_{11} = 161.6$ GPa, $C_{12} = 81.6$ GPa, $C_{44} = 60.3$ GPa. The data for E_{atm} and E_{el} falls onto two straight lines whose slopes agree to within 0.5 per cent. The resulting core energy is $E_{core} = 0.565 \pm 0.001$ eV Å^{-1}, at $r_c = b = 3.84$ Å. This core energy is manifestly independent of c_1.

Further experimentation with the cell dimensions and geometry showed that atomistic and continuum predictions for the dipole energies agree well for even smaller supercells. This is an important observation making it possible to use accurate DFT models (Section 2.1) for computing dislocation core energies. To retain the accuracy of computationally expensive DFT models, it is important to use full anisotropic elasticity for computing the elastic energies and to correct the resulting sums for conditional convergence. The utility MadSum does both.

Summary

- To extract the dislocation core energy from an atomistic simulation data obtained under PBC, it is necessary to compute the elastic energy of a dislocation dipole in PBC.

- The sum of image interactions is only conditionally convergent, but the associated error term can be computed by introducing a set of "ghost" dislocations.

Problems

5.2.1. Build a perfect BCC crystal with vectors $c_1 = 8[11\bar{2}]$, $c_2 = 19[110]$ and $c_3 = 3[111]/2$ along the x, y and z directions respectively. Introduce a screw dislocation dipole with $b = [111]/2$, where one dislocation, b, is positioned at $s_x = 0$, $s_y = 0.2368$ and the other one, $-b$, is positioned at $s_x = 0$, $s_y = -0.2632$ (in scaled coordinates). Using the Finnis–Sinclair potential for tantalum, relax this configuration using the CGR algorithm. Compute E_{atm} as the excess energy (per unit length) of the dislocation dipole using eq. (5.16). Repeat the calculation with $c_2 = 9, 29, 39[110]$. Notice that positions of two dislocations need to be adjusted slightly each time a different supercell is used to ensure that they are introduced exactly in the centers of triangles formed by three neighbor rows of atoms.

5.2.2. Using the MadSum utility to compute E_{img} for the four cell geometries in Problem 5.2.1, extract dislocation core energies from the difference between atomistic and elastic energies. How much do the core energies depend on the supercell geometry? [59]

5.3 Peierls Stress

The tools developed in the previous two sections are useful for calculations of the Peierls stress under PBC as well. The Peierls stress τ_P is the stress necessary to move a dislocation at zero temperature (Section 1.3). Atomistic calculations of Peierls stress are often complicated by image forces induced by the boundary conditions. When PBC are used, the image artifacts are relatively easy to quantify and reduce with the help of auxiliary elasticity calculations.

For Peierls stress calculations, it is more convenient to position the dislocation dipole perpendicular to the glide plane. If, as before, the supercell is a rectangular parallelepiped with c_1 and c_3 parallel to the glide plane, then positioning two dislocations of the dipole vertically means placing them at $a = c_2/2$ from each other as shown in Fig. 5.6(a). In this orientation, dislocations glide horizontally (in the xz plane) and can avoid recombination that would be inevitable for $a = c_1/2$.

As an example, let us compute the Peierls stress of the same screw dislocation in the SW model of silicon as in the preceding section. Assume that x, y, z axes are parallel to c_1, c_2, c_3, respectively. To induce dislocation motion let us apply a constant stress σ_{yz} using the Parrinello–Rahman method (Section 4.3). This stress exerts equal but opposite forces on the two dislocations in the x and $-x$ directions, respectively. The minimum stress at which dislocations begin to move can be determined by performing a series of calculations in which the applied stress is gradually increased. Each stress increment is now followed by a relaxation using the CGR algorithm. The critical stress τ_c is defined as the lowest value of σ_{yz} at

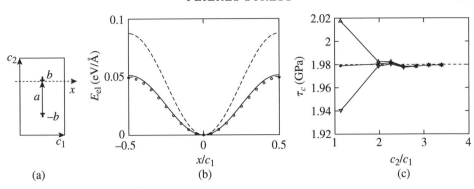

FIG. 5.6. (a) A periodic supercell suitable for Peierls stress calculations—two dislocations gliding on two parallel planes are less likely to recombine. (b) Variation of the dislocation dipole energy as a function of the relative position of two dislocations along c_1. Atomistic simulation results are shown by o. The solid line is the anisotropic elasticity prediction and the dashed line is the isotropic elasticity result. (c) Critical stress τ_c computed in supercells with different aspect ratios (c_2/c_1) at fixed $c_1 = 5[11\bar{2}]$. $\triangle$ are the data points for $x = 0$ and $\triangledown$ are those for $x = c_1/2$, where x is the initial offset of the two dislocations along c_1. $\bullet$ are obtained by averaging $\triangle$ and $\triangledown$.

which dislocation motion is observed. Because of possible effects of the boundary image forces, τ_c is not necessarily equal to the Peierls stress τ_P. The latter corresponds to a critical stress in the idealized situation of a single dislocation in an infinite crystal. Next, we examine the nature of image forces in PBC and consider methods for minimizing the difference between τ_c and τ_P.

Fig. 5.6(b) shows the values of excess energy E_{atm} of the dislocation dipole computed under zero stress as a function of x, the distance between the two dislocations along their motion direction c_1. The energy oscillates periodically, obviously as a result of the interaction between two primary dislocations and their images. For comparison, the elastic energies computed by MadSum using anisotropic elasticity are also plotted. The atomistic and the elastic energies agree very well. Another prediction based on the isotropic elasticity is also shown in Fig. 5.6(b). The latter overestimates the magnitude of the oscillations by a factor of 2, indicating its inaccuracy in describing dislocation interactions. This comparison is given here to emphasize the necessity of using anisotropic elasticity for computing dislocation energies in the supercells. Since the methods implemented in MadSum handle both isotropic and anisotropic cases equally well, there is no excuse for not using the more accurate anisotropic solutions [60].

The slope of the $E(x)$ curve in Fig. 5.6(b) is the image force that the dislocations see in addition to the Peach–Koehler force exerted by the applied stress σ_{yz}. This extra force introduces an error in the Peierls stress value computed in PBC.

Considering the shape of the $E(x)$ curve, it should be obvious that this error is minimized for dislocation positions where $dE/dx = 0$, i.e. when $x = 0$ or $x = c_1/2$. A second-order error still exists even in these two special configurations, due to a finite curvature d^2E/dx^2 and the lattice discreteness. To examine the nature of this second-order effect let us define

$$\Delta E \equiv E(x = 0) - E(x = c_1/2). \tag{5.23}$$

When $\Delta E < 0$, as is the case in Fig. 5.6(b), τ_c computed at $x = 0$ overestimates τ_P because the next lattice position has a slightly higher energy. In contrast, τ_c computed at $x = c_1/2$ will underestimate τ_P. For situations when $\Delta E > 0$, the sign of this second-order error inverts.

Having examined the origin of error in Peierls stress calculations in PBC, let us make use of these observations to devise a practical procedure for minimizing the error in Peierls stress calculations in PBC. Fig. 5.6(c) shows the critical stress τ_c computed for the screw dislocation in Si for different c_2 but the same $c_1 = 5[11\bar{2}]$. Values of τ_c computed at $x = 0$ are shown by $\triangle$, while those computed at $x = c_1/2$ are plotted as $\triangledown$. Both sets of data converge to 1.98 GPa with increasing c_2; however, their averaged values (in •) reach this asymptotic value at smaller cell dimensions. This cancellation of the second-order errors is also observed for even smaller supercells.

The study of convergence behavior presented in Fig. 5.6 suggests that considerable care is necessary in computing the Peierls stress, especially when the latter is expected to be low. However, using the error cancellation procedures described above, it is possible to accurately determine the Peierls stress using relatively small supercells. This opens the possibility of using accurate DFT methods to compute the Peierls stress.

Summary

- It is possible to cancel first- and second-order errors in the Peierls stress calculations in periodic supercells.

Problems

5.3.1. For the supercells described in Problem 5.2.1, calculate the Peierls stress by applying σ_{yz} to drive the dislocations along x. For each supercell, perform the calculations in two "optimal" configurations that produce the smallest errors, i.e. $x = 0$ and $x = c_1/2$. Plot the critical stress computed for two different dislocation positions and observe its convergence with increasing c_2. Calculate and plot the average of two values for each supercell and observe its convergence behavior.

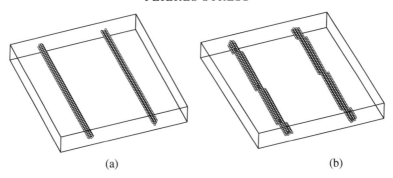

(a)   (b)

FIG. 5.7. (a) A supercell containing two straight dislocations parallel to c_3.
(b) Three kinks develop on each of the two dislocations after c_3 is changed
from $60[111]/2$ to $60[111]/2 + [11\bar{2}]$, while keeping the real coordinates $(\mathbf{r})$ of
all atoms fixed.

5.3.2. Peierls stress of a near-screw dislocation. Build a perfect BCC crystal
in PBC with repeat vectors $\mathbf{c}_1 = 19[11\bar{2}]$, $\mathbf{c}_2 = 5[1\bar{1}0]$, $\mathbf{c}_3 = 3[111]/2$.
Introduce a screw dislocation dipole with $\mathbf{b} = [111]/2$ at $s_x = 0.2456$,
$s_y = -0.05$ and $-\mathbf{b}$ at $s_x = -0.2544$, $s_y = -0.05$. Relax the configuration to zero stress. Replicate the configuration along the $\mathbf{c}_3$ axis 20
times and splice all 20 slabs together and plot the resulting configuration using the energy filtering. Compare your results with Fig. 5.7(a).
Keeping real coordinates of the atoms $\mathbf{r}$ fixed, change $\mathbf{c}_3$ to $\mathbf{c}_3 + \mathbf{c}_1/19$.
Relax the structure again and observe that three kinks appear on each
screw dislocation. Compare this with Fig. 5.7(b). Apply stress σ_{yz} in
increments of 100 MPa and measure the critical stress for dislocation
motion. Compare this stress with the Peierls stress of the straight dislocations obtained in Problem 5.3.1. What is the mechanism of dislocation
motion?

5.3.3. Peierls stress of a dislocation network [61]. Build a perfect BCC crystal of molybdenum (lattice constant 3.1472 Å), with repeat vectors
$16[11\bar{2}]$, $19[\bar{1}10]$ and $16[112]$. Notice that $\mathbf{c}_1$ and $\mathbf{c}_3$ are not orthogonal to each other. Create two dislocation dipoles. In dipole
$\mathbf{b}_1 = [111]/2$, two dislocations are parallel to $\mathbf{c}_3$ and separated by $\mathbf{c}_2/2$.
In dipole $\mathbf{b}_2 = [\bar{1}\bar{1}1]/2$, two dislocation are parallel to $\mathbf{c}_1$ and separated
by $\mathbf{c}_2/2$. To introduce the dipoles, first compute the displacement fields
produced by two dipoles, then superimpose them and apply to all the
atoms at once. Because the dislocations are not purely screw, atoms need
to be removed or inserted at the cut planes of both dipoles. It is recom-
mended that the two cuts are positioned so that they do not intersect each
other, as shown in Fig. 5.8(a). After relaxation, the dislocations form
two junctions, each containing three screw dislocations according to the

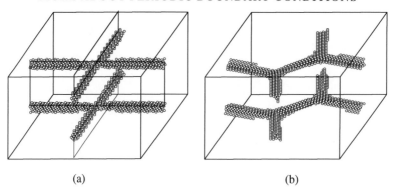

(a)   (b)

FIG. 5.8. (a) Two intersecting dislocation dipoles in one supercell. (b) A pair of
dislocation junctions form after relaxation.

reaction $[111]/2 + [\bar{1}\bar{1}1]/2 = [001]$, as shown in Fig. 5.8(b). Apply stress
on the y plane, with various combinations of σ_{xy} and σ_{yz}, and measure the
critical stress at which the network begins to move. Compare this stress
to the Peierls stress of straight dislocations and observe the mechanism
of dislocation motion.

6

FREE-ENERGY CALCULATIONS

6.1 Introduction to Free Energy

Free energy is of central importance for understanding the properties of physical systems at finite temperatures. While in the zero temperature limit the system should evolve to a state of minimum energy (Section 2.3), this is not necessarily the case at a finite temperature. When an *open* system exchanges energy with the outside world (a *thermostat*) and maintains a constant temperature, its evolution proceeds towards minimizing its *free energy*. For example, a crystal turns into a liquid when the temperature exceeds its melting temperature precisely because the free energy of the liquid state becomes lower than that of the crystalline state.[1] In the context of dislocation simulations, free energy is all important when one has to decide which of the possible core configurations the dislocation is likely to adopt at a given temperature.

The free energy F of a system of N particles with Hamiltonian $H(\{\mathbf{r}_i, \mathbf{p}_i\})$ is defined by [26, 27]

$$e^{-\beta F} = Q = \frac{1}{(2\pi\hbar)^{3N}} \int \prod_{i=1}^{N} d\mathbf{r}_i \, d\mathbf{p}_i \, e^{-\beta H(\{\mathbf{r}_i, \mathbf{p}_i\})}, \qquad (6.1)$$

where $\beta = 1/(k_B T)$, $\hbar = 1.05457148 \times 10^{-34} \, \text{m}^2 \, \text{kg s}^{-1}$ is Planck's constant, and $\mathbf{r}_i$ and $\mathbf{p}_i$ are the position and momentum of each atom i. Q is the *statistical integral* related to the partition function Z in eq. (2.14) by $Q = Z/(2\pi\hbar)^{3N}$. The free energy F, the average energy E and the *entropy* S of the same system are related through

$$F = E - TS, \qquad (6.2)$$

---

[1] Melting is usually observed at constant pressure. A system maintained at constant temperature *and* pressure evolves to a state of minimum *Gibbs* free energy. If, instead of pressure, the system's volume remains constant, the system evolves to a state minimizing its *Helmholtz* free energy. In the following, we will not distinguish Gibbs and Helmholtz free energies because the difference between them is typically small in solids under ambient conditions.

where the average energy E of the system is (Section 2.2)

$$E = \frac{1}{Z} \int \prod_{i=1}^{N} \mathrm{d}\mathbf{r}_i \, \mathrm{d}\mathbf{p}_i \, H(\{\mathbf{r}_i, \mathbf{p}_i\}) \, e^{-\beta H(\{\mathbf{r}_i, \mathbf{p}_i\})}. \tag{6.3}$$

6.1.1   Free Energies of Configurational States

The potential energy function of a typical many-body system has many local minima, each one corresponding to a particular set of atomic coordinates $\mathbf{R}_a = \{\mathbf{r}_i\}$ called a *configurational microstate*. Although a tiny subset of all possible configurations, these special microstates are important because (almost) every other microstate of the system can be associated with a local energy minimum $\mathbf{R}_a$ by a steepest-descent relaxation (Section 2.3). In this way, the entire configurational space can be divided into of a set of non-overlapping energy basins Ω_a, each corresponding to a local energy minimum $\mathbf{R}_a$ and separated from the neighboring basins by the dividing surfaces, as illustrated in Fig. 6.1. When the thermal energy $k_B T$ is much lower than the energy barriers separating the basins, the system's dynamics can be viewed as a sequence of jumps from one basin to the next, separated by long residence periods τ_a.[2] For example, under ambient conditions, atoms in crystals spend most of their time vibrating around a minimum energy configuration except for rare occasions when large atomic displacements occur. When such is the case, it makes sense to regard each energy basin as a distinct *configurational state*.

During the residence periods τ_a, the system continues to interact with the thermostat and may have enough time to achieve *partial* thermal equilibrium. This

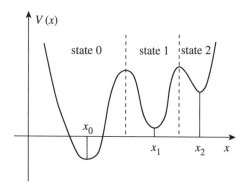

FIG. 6.1. A one-dimensional illustration of the many-body potential energy landscape. Each configurational state or energy basin (0, 1 or 2) includes all microstates that map to the same underlying energy minimum (x_0, x_1 or x_2) by steepest-descent relaxation.

---

[2] The rare transitions from one energy basin to another will be discussed in Chapter 7.

means that the relative probabilities of various microstates within the basin obey Boltzmann's distribution. In this context, it makes sense to define the free energy of a given configurational state (basin) a as $F_a = -k_B T \log Q_a$, where the partial statistical integral Q_a is obtained by integration over the volume Ω_a of the energy basin a,

$$Q_a = \frac{1}{(2\pi\hbar)^{3N}} \int_{\Omega_a} \prod_{i=1}^{N} d\mathbf{r}_i \, d\mathbf{p}_i e^{-\beta H(\{\mathbf{r}_i, \mathbf{p}_i\})}. \tag{6.4}$$

Partial free energies of configurational states a are related to the total free energy through the sum of partial statistical integrals,

$$e^{-\beta F} = \sum_a e^{-\beta F_a}. \tag{6.5}$$

Because the total number of distinct configurational states in a many-body system is very large, while the transitions between them can be rare, the total free energy F is often impossible to compute. At the same time, calculations of the free energy of a given configurational state are generally more manageable.

When the system reaches *complete* thermal equilibrium, the distribution of micro-states in the entire phase space satisfies Boltzmann's distribution. In this case, the probability that the system can be found in a configurational state a is simply

$$P_a = \frac{e^{-\beta F_a}}{e^{-\beta F}}. \tag{6.6}$$

The relative probability of finding the system in two different configurational states is defined by their free energy differences,

$$\frac{P_1}{P_2} = \frac{e^{-\beta F_1}}{e^{-\beta F_2}} = e^{-\beta(F_1 - F_2)}. \tag{6.7}$$

Even if the system is initially in configurational state 1, if $F_1 > F_2$, then after a long equilibration time it is more likely to be found in state 2, simply because $P_1 < P_2$ at equilibrium. This explains why the system evolves towards configurational states with the lower free energies.

From its very definition in eq. (6.1) it should be clear that free energy must be difficult to calculate. While many other quantities, such as internal energy E, can be expressed as averages over Boltzmann's distribution, the free energy cannot be written as a statistical average. Instead, it requires evaluation of the statistical integral Q, which is an integral over the entire phase space weighted with the Boltzmann's factor $e^{-\beta H}$. In the next section, we describe a method for computing the free energy of a simple system of N atoms whose potential energy is a quadratic (harmonic) function of the atomic coordinates. Section 3 presents more sophisticated methods for computing the free energies of more realistic systems and discusses their application for obtaining the free energies of crystal defects.

Summary

- The free energy depends on the volume of phase space accessible to the system at a given temperature. To compute the free energy, extensive sampling of the phase space is necessary.

- The free energy of a configurational state is defined through the partial statistical integral evaluated over a volume of phase space associated with this state. In thermal equilibrium, the relative probabilities of two configurational states are determined by the difference of their free energies.

6.2   Harmonic Approximation

For several particularly simple systems, their statistical integral Q can be obtained analytically. An important example is a one-dimensional harmonic oscillator described by the following Hamiltonian:

$$H(x, p) = \frac{k}{2} x^2 + \frac{p^2}{2m},$$ (6.8)

where k is the spring constant and m is the mass. By direct integration over x and p we obtain

$$Q = \frac{k_B T}{\hbar} \sqrt{\frac{m}{k}} = \frac{k_B T}{\hbar \omega},$$ (6.9)

$$F = -k_B T \ln Q = -k_B T \ln \frac{k_B T}{\hbar \omega},$$ (6.10)

where $\omega \equiv \sqrt{k/m}$ is the vibrational frequency of the oscillator. Obviously, the free energy of a collection of N independent harmonic oscillators, each with frequency ω_i, is simply the sum,

$$F = -k_B T \sum_{i=1}^{N} \ln \frac{k_B T}{\hbar \omega_i}.$$ (6.11)

This is an important result used below to compute the free energy of a solid at a low temperature.

Consider a solid containing N atoms and let q_i be the coordinates and p_i be the momentum of atom i. The Hamiltonian of this solid can be written as

$$H(\{q_i, p_i\}) = \sum_{i=1}^{3N} \frac{p_i^2}{2m} + V(\{q_i\}). \tag{6.12}$$

Assuming that $V_0 = V(\{q_i^{(0)}\})$ is a local minimum of the potential function V, its Taylor expansion around $\{q_i^{(0)}\}$ is

$$H(\{q_i, p_i\}) = V_0 + \frac{1}{2} \sum_{i,j=1}^{3N} K_{ij}(q_i - q_i^{(0)})(q_j - q_j^{(0)})$$

$$+ \mathcal{O}((q_i - q_i^{(0)})^3) + \sum_{i}^{3N} \frac{p_i^2}{2m}, \tag{6.13}$$

where $K_{ij} \equiv \partial^2 V / \partial q_i \partial q_j$ are the components of the $3N \times 3N$ Hessian matrix $\mathbf{K}$.[3] Because the Hessian is a symmetric matrix, i.e. $K_{ij} = K_{ji}$, it can be diagonalized, or transformed to a diagonal form, by an appropriate orthogonal matrix $\mathbf{U}$,[4]

$$\mathbf{K} = \mathbf{U}^{\mathrm{T}} \cdot \mathbf{\Lambda} \cdot \mathbf{U}, \tag{6.14}$$

where the off-diagonal elements of the transformed Hessian $\mathbf{\Lambda}$ are zero. In terms of the new transformed variables,

$$\tilde{q}_i \equiv \sum_{j=1}^{3N} U_{ij}(q_j - q_j^{(0)})$$

$$\tilde{p}_i \equiv \sum_{j=1}^{3N} U_{ij} p_j$$

the Hamiltonian can be rewritten as,

$$H(\{\tilde{q}_i, \tilde{p}_i\}) = V_0 + \sum_{i=1}^{3N} \left(\frac{1}{2} \Lambda_{ii} \tilde{q}_i^2 + \frac{\tilde{p}_i^2}{2m} \right) + \mathcal{O}(\tilde{q}_i^3), \tag{6.15}$$

where Λ_{ii} $(i = 1, \ldots, 3N)$ are the diagonal elements of $\mathbf{\Lambda}$.

In a crystal at temperatures much lower than its melting point, the amplitude of atomic vibrations is small so that the anharmonic terms $\mathcal{O}(\tilde{q}_i^3)$ can be

---

[3] In various contexts $\mathbf{K}$ is also called the dynamical matrix or the stiffness matrix.
[4] An orthogonal matrix satisfies the orthogonality condition $\mathbf{U}^{\mathrm{T}} = \mathbf{U}^{-1}$.

ignored. This is when the representation of the Hamiltonian by its harmonic approximation (HA) is reasonably accurate. Retaining only the quadratic terms, the transformed Hamiltonian is equivalent to that of $3N$ independent harmonic oscillators, each with frequency $\omega_i = \sqrt{\Lambda_{ii}/m}$. The free energy of this system is given in eq. (6.11).

6.2.1   Free Energy of a Vacancy in the Dislocation Core

As an example, let us use the harmonic approximation to compute the free energy of a vacancy in the dislocation core [62], through the following steps.

1. Build a 192-atom fragment of a perfect diamond-cubic crystal in a periodic supercell with dimensions $2[11\bar{2}] \times [111] \times 2[1\bar{1}0]$.

2. Following the procedure given in Section 5.1, make a cut-plane along one of the (111) planes and insert a dipole of 30° partial dislocations with Burgers vector $\mathbf{b} = \frac{1}{6}[2\bar{1}1]$ and line direction along $[1\bar{1}0]$, as shown in Fig. 6.2.[5]

3. Relax the configuration to a local energy minimum using the CGR algorithm and the SW interatomic potential function for silicon.

4. Remove the atom from site A near the core (see Fig. 6.2) and relax the resulting configuration to its energy minimum E_1^0.[6] The superscript 0 emphasizes

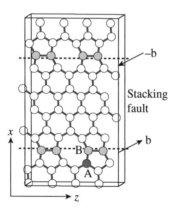

FIG. 6.2. Atomic core structure of two 30° partial dislocations in silicon. Only atoms immediately above and below the glide plane are shown. The core atoms are shown in light gray and the atom to be removed in dark gray.

---

[5] Because the Burgers vector of a partial dislocation is smaller than the full repeat vector of the crystal lattice, the atomic arrangement across the planar area of the cut is not the same as in the perfect crystal: this area forms a specific planar defect called a stacking fault. (see also Section 8.3)

[6] It may seem that a more favorable, lower energy position for vacancy is at site B, i.e. in the center of the dislocation core. However, we observed that if, instead of atom A, atom B is removed,

that E_1^0 is the energy of system 1 at zero temperature. The average energy of system 1 at finite temperature is higher than E_1^0.

5. Compute the Hessian matrix **K**. This can be done numerically by adding a small displacement Δ to each atomic coordinate q_i, one at a time, and computing and recording the resulting forces on all other coordinates $f_j \equiv -\partial V / \partial q_j$. The elements of the Hessian are then computed as

$$K_{ij} = \frac{\partial^2 V}{\partial q_i \partial q_j} = -\frac{\partial f_j}{\partial q_i} \approx -\frac{f_j(q_i + \Delta) - f_j(q_i - \Delta)}{2\Delta} + \mathcal{O}(\Delta^2). \quad (6.16)$$

6. Diagonalize the resulting (573×573) matrix **K** and obtain its 573 eigenfrequencies ω_i. The three lowest eigenfrequencies should be zero, corresponding to translations of the crystal as a whole in three directions. These three translational modes should be excluded from the free-energy calculations; this is equivalent to fixing the position of the system's center of mass.

Now it is straightforward to compute the free energy of this system by summing over 570 harmonic oscillators with non-zero frequencies as in eq. (6.11). For example, at $T = 600$ K, $F_1^{(HA)} = E_1^0 - 6.0509$ eV. Obtained under the harmonic approximation, $F_1^{(HA)}$ is the free energy of the 191-atom system containing a dislocation dipole and a vacancy at site A near the core.

6.2.2 Bookkeeping of the Free Energies

By itself, the magnitude of the free energy of a given configuration has little significance. It becomes useful only when compared to the free energy of another configuration of the same system. In the case study discussed above, the important quantity we would like to extract from the calculations is the formation free energy of a vacancy in the dislocation core. This is defined as the change in the free energy of the system containing the core vacancy relative to the free energy of the same system without the core vacancy. It may appear that to enable the necessary comparison it is sufficient to compute the free energy of system 2 with atom A still in it. However, the two configurations of interest here are not two states of the same system, because the number of atoms is 191 in one case and 192 in the other. One possible way to remedy this difficulty is to assume that the atom removed from system 1 to make a vacancy is never destroyed but somehow added to a large mass reservoir. For a more physical interpretation of this construction, just imagine that the atom removed from the vacancy site is transported to a surface ledge of a large

then atom A jumps spontaneously into the vacant site B during an MD simulation at $T = 600$ K. This preference for a vacancy to form at site A may be an artifact of the SW potential.

crystal. This operation can be mimicked by adding to the free energy of system 1 the free energy $F_0^{(HA)}$ per atom in a perfect crystal (system 0). After this, it becomes meaningful to compare the free energy of system 2 with that of the "augmented" system 1; both systems now have 192 atoms.

Repeating the above sequence of steps (skipping step 4), we find the energy minimum E_2^0 and the free energy $F_2^{(HA)}$ of system 2 (with 192 atoms): $E_2^0 = E_1^0 - 6.7845\,\mathrm{eV}$ and $F_2^{(HA)} = E_2^0 - 5.8121\,\mathrm{eV}$. Following the prescription in Problem 6.2.1, we obtain the *cohesive energy* of the perfect crystal $E_0^0 = -4.63\,\mathrm{eV}$ and its per-atom free energy $F_0^{(HA)} = E_0 - 0.0232\,\mathrm{eV}$ at $T = 600\,\mathrm{K}$. Finally, the energy of vacancy formation in the dislocation core is

$$E_{VC} = E_1^0 - E_2^0 + E_0^0 = 2.1545\,\mathrm{eV} \tag{6.17}$$

and the corresponding free energy is

$$F_{VC}^{(HA)} = F_1^{(HA)} - F_2^{(HA)} + F_0^{(HA)} = 1.8925\,\mathrm{eV}. \tag{6.18}$$

Another important parameter of the core vacancy is its excess entropy. According to eq. (6.2) this entropy is

$$S_{VC}^{(HA)} = \left(F_{VC}^{(HA)} - E_{VC} \right) \Big/ T = 5.067 k_B, \tag{6.19}$$

where E_{VC} is the energy difference between the same two systems with 192 atoms. We intentionally omit the superscript 0 on E_{VC} in eq. (6.17) even though it is computed based on zero-temperature quantities. This is because E_{VC} does not depend on temperature within the harmonic approximation, where the thermal energy of any configurational state a, defined as the average energy minus the potential energy at the minimum $\mathbf{R}_a$, is simply $3Nk_BT$. Therefore, E_{VC}, which is the energy difference between two states with the same number of atoms, does not change with temperature.

It is of interest to compare the above results with the vacancy formation energy in the bulk, $E_V = 3.0272\,\mathrm{eV}$, its free energy $F_V^{(HA)} = 2.7197\,\mathrm{eV}$ and entropy $S_V^{(HA)} = 5.9472 k_B$ (Problem 6.2.2). Since $F_{VC}^{(HA)} < F_V^{(HA)}$, vacancies have a tendency to segregate at the dislocation core.

Summary

- For a solid at a temperature much lower than its melting point, the potential energy can be approximated by a harmonic (quadratic) function of the coordinates. In the same approximation, the free energies of the configurational states can be obtained from the eigen-frequencies of the vibrational modes.

- In comparing the free energies, it is important to make sure that the states have the same number of atoms. When this is not the case, it is possible to account for the difference using the per-atom free energy computed in a large mass reservoir.

Problems

6.2.1. Compute the per-atom free energy of a perfect crystal of silicon in the harmonic approximation. Build a small diamond-cubic crystal in a periodic box with edges $2[100] \times 2[010] \times 2[001]$ (64 atoms) and with the lattice constant $a = 5.4310\,\text{Å}$. Use the SW interatomic potential to compute the Hessian matrix numerically. Diagonalize the resulting (192×192) matrix to obtain its 192 eigenfrequencies ω_i. The three lowest ω_i are zero, corresponding to uniform translations of the crystal in three directions. These should be excluded from the free-energy calculation. Compute and plot the per-atom free energy as a function of temperature, from 0 to 600 K. Compute its per-atom entropy $S = (E - F)/T$.

6.2.2. Compute the free energy of a silicon crystal containing a vacancy at $T = 600\,\text{K}$. Starting from the same supercell with 64 atoms as in Problem 6.2.1, remove one atom. Compute the free energy of the 63-atom system under the harmonic approximation. Compute the free energy and the entropy of vacancy formation in the bulk at $T = 600\,\text{K}$.

6.3   Beyond Harmonic Approximation

While it is relatively simple to compute free energies under the harmonic approximation, there are many situations (even in solids) where HA is inaccurate. When such is the case, it becomes necessary to use other, generally more expensive methods that do not rely on HA. In Section 6.1, we noted that the free energy of a system cannot be expressed as a statistical average, making it difficult to compute by simulation. In turns out that the *free-energy difference* between two systems *can* be written in the form of a statistical average. In this section we discuss why this is indeed the case. Furthermore, we take advantage of this fact by noting that if the free energy of one system is known exactly, such as for a set of harmonic oscillators, then the free energy of the other system can be obtained by computing the free-energy difference between the two.

Consider systems 1 and 2 with the same number of atoms N but described by two different Hamiltonians,

$$H_1(\{\mathbf{r}_i, \mathbf{p}_i\}) = \sum_{i=1}^{N} \frac{|\mathbf{p}_i|^2}{2m} + V_1(\{\mathbf{r}_i\}), \tag{6.20}$$

$$H_2(\{\mathbf{r}_i, \mathbf{p}_i\}) = \sum_{i=1}^{N} \frac{|\mathbf{p}_i|^2}{2m} + V_2(\{\mathbf{r}_i\}). \tag{6.21}$$

From the definition of free energy, eq. (6.1), it follows that

$$e^{-\beta(F_1-F_2)} = \frac{Q_1}{Q_2} = \frac{\int \prod_{i=1}^{N} d\mathbf{r}_i\, d\mathbf{p}_i\, e^{-\beta H_1(\{\mathbf{r}_i, \mathbf{p}_i\})}}{\int \prod_{i=1}^{N} d\mathbf{r}_i\, d\mathbf{p}_i e^{-\beta H_2(\{\mathbf{r}_i, \mathbf{p}_i\})}}$$

$$= \frac{\int \prod_{i=1}^{N} d\mathbf{r}_i\, d\mathbf{p}_i e^{\beta[H_2(\{\mathbf{r}_i, \mathbf{p}_i\}) - H_1(\{\mathbf{r}_i, \mathbf{p}_i\})]} e^{-\beta H_2(\{\mathbf{r}_i, \mathbf{p}_i\})}}{\int \prod_{i=1}^{N} d\mathbf{r}_i\, d\mathbf{p}_i e^{-\beta H_2(\{\mathbf{r}_i, \mathbf{p}_i\})}}$$

$$\equiv \langle e^{\beta(H_2-H_1)} \rangle_2$$

$$F_2 - F_1 = k_B T \ln\langle e^{\beta(H_2-H_1)} \rangle_2. \tag{6.22}$$

Therefore, the difference between the absolute free energies of systems 1 and 2 can be formally expressed in terms of an average of the function $e^{\beta(H_2-H_1)}$ over the canonical ensemble generated by Hamiltonian 2. However, unless Hamiltonians H_1 and H_2 are very close to each other, the fluctuations of $e^{\beta(H_2-H_1)}$ can be very large,[7] leading to large statistical errors when computing the average. Thus, eq. (6.22) is not well suited for computing free-energy differences.

To construct a more efficient method for computing $F_2 - F_1$, consider a family of Hamiltonians $H(\{\mathbf{r}_i, \mathbf{p}_i\}; \lambda)$ where λ is a numerical parameter [63, 36]. Assume further that, when $\lambda = 0$, $H(\lambda) = H_1$ and, when $\lambda = 1$, $H(\lambda) = H_2$ (here, we have dropped the dependence on $\{\mathbf{r}_i, \mathbf{p}_i\}$ for brevity). The following is the simplest example of such a family:

$$H(\lambda) = (1 - \lambda)H_1 + \lambda H_2. \tag{6.23}$$

Let $F(\lambda)$ be the free energy of the system described by the Hamiltonian $H(\lambda)$, i.e.

$$F(\lambda) = -k_B T \ln \left[\frac{1}{(2\pi\hbar)^{3N}} \int \prod_{i=1}^{N} d\mathbf{r}_i\, d\mathbf{p}_i e^{-\beta H(\{\mathbf{r}_i, \mathbf{p}_i\}; \lambda)} \right]. \tag{6.24}$$

---

[7] This is because many of the states from the statistical ensemble generated by Hamiltonian H_2 are statistically improbable for the system described by Hamiltonian H_1.

By taking a derivative of the above equation with respect to λ, we arrive at,

$$\frac{dF(\lambda)}{d\lambda} = \frac{\int \prod_{i=1}^{N} d\mathbf{r}_i \, d\mathbf{p}_i e^{-\beta H(\{\mathbf{r}_i, \mathbf{p}_i\}; \lambda)} \frac{\partial H(\lambda)}{\partial \lambda}}{\int \prod_{i=1}^{N} d\mathbf{r}_i \, d\mathbf{p}_i e^{-\beta H(\{\mathbf{r}_i, \mathbf{p}_i\}; \lambda)}} \equiv \left\langle \frac{\partial H(\lambda)}{\partial \lambda} \right\rangle_\lambda. \tag{6.25}$$

To put this result in words, the derivative of the absolute free energy with respect to λ is equal to the average of function $\partial H(\lambda)/\partial \lambda$ over the canonical ensemble generated by the "mixed" Hamiltonian $H(\lambda)$. Function $\partial H(\lambda)/\partial \lambda$ is much better suited for numerical averaging over the statistical ensemble generated by the mixed Hamiltonian.

From the above, the free-energy difference between systems 1 and 2 is obtained by integration,

$$F_2 - F_1 = \int_0^1 \frac{dF(\lambda)}{d\lambda} \, d\lambda = \int_0^1 \left\langle \frac{\partial H(\lambda)}{\partial \lambda} \right\rangle_\lambda d\lambda. \tag{6.26}$$

When $H(\lambda)$ is given by eq. (6.23), this reduces to

$$F_2 - F_1 = \int_0^1 \langle H_2 - H_1 \rangle_\lambda \, d\lambda. \tag{6.27}$$

Equations (6.26) and (6.27) form the basis of the method of *thermodynamic integration* (TI). In this method, one computes averages of $\partial H(\lambda)/\partial \lambda$ over a set of equilibrium simulations for systems described by Hamiltonian $H(\lambda)$ with different values of λ, from 0 to 1. The averaging produces the values of $dF(\lambda)/d\lambda$ for a selected set of λ and, finally, $F_2 - F_1$ is obtained by numerical integration of curve $dF(\lambda)/d\lambda$ from $\lambda = 0$ to $\lambda = 1$. While the TI method is exact in principle, it requires a lot of equilibrium simulations (at least one for each value of λ) to obtain accurate results. The method of *adiabatic switching* (AS) derives from the TI but is often more efficient for computing the free-energy differences.

The AS method is based on the following physical interpretation of eq. (6.26). Let us define the generalized force conjugate to variable λ as follows:

$$f_\lambda \equiv \left\langle -\frac{\partial H(\lambda)}{\partial \lambda} \right\rangle_\lambda. \tag{6.28}$$

Then $\Delta W = -\int_0^1 f_\lambda \, d\lambda$ is the thermodynamic work required to drive (or switch) the system from $\lambda = 0$ to $\lambda = 1$, against the generalized force f_λ. Equation (6.26) is a statement that this work is stored into the free energy of the transformed system.

Given this interpretation, it becomes possible to compute the free-energy difference $F_2 - F_1$ using a single *non-equilibrium* (MD or MC) simulation, in which the Hamiltonian gradually changes from H_1 to H_2. For every increment $d\lambda$, the instantaneous work of switching is $-f_\lambda \, d\lambda$. The total work accumulated in one such simulation gives an estimate of the free-energy difference $F_2 - F_1$. In order to use MD simulations to sample statistical ensembles of the mixed Hamiltonians, it is necessary to define a switching function $\lambda(t)$ such that, at the beginning of the simulation $\lambda(0) = 0$ and at the end $\lambda(t_s) = 1$ (here t_s is the total duration of the MD simulation). The estimate of the free-energy difference in an MD AS simulation is

$$\Delta F \equiv F_2 - F_1 = \int_0^{t_s} dt \frac{d\lambda(t)}{dt} \left[H_2(p_i(t), q_i(t)) - H_1(p_i(t), q_i(t)) \right]. \quad (6.29)$$

Strictly speaking, this equation is valid only in the limit of $t_s \to \infty$. For any finite switching time t_s, the right-hand side provides an estimate for ΔF subject to statistical and systematic errors.

An effective way to compute the absolute free energy of a given system H_2 is to run a switching simulation from a reference system H_1 whose free energy is known exactly. One widely used reference system is the Einstein crystal equivalent to $3N$ harmonic oscillators,

$$H_1(\{\mathbf{r}_i, \mathbf{p}_i\}) = \sum_{i=1}^{N} \frac{k}{2} (\mathbf{r}_i - \mathbf{r}_{0,i})^2 + \sum_{i}^{N} \frac{|\mathbf{p}_i|^2}{2m} + V_0. \quad (6.30)$$

Given the freedom to choose parameters k, m, $\mathbf{r}_{0,i}$ and V_0, it is useful to exercise this freedom to make the reference system as close as possible, *structurally* and *dynamically*, to the system of interest. A good choice reduces dissipation (see below) and makes the sampling more efficient, allowing the use of shorter switching times. One particularly good choice for the reference system is the harmonic approximation of the target Hamiltonian itself (see Problem 6.3.3). Even then, it is important to switch slowly to allow an adequate sampling of the phase space.

An additional complication arises in MD simulations if H_1 or H_2 is exactly or nearly harmonic. When this is the case, MD sampling becomes *non-ergodic*, meaning that the MD trajectory samples a small region of the phase space over and over again, instead of sampling the entire phase space according to Boltzmann's distribution. To overcome this problem, a more sophisticated version of the Nose–Hoover thermostat (Section 4.4), namely a Nose–Hoover chain [64], can be used.

Such a complication does not arise in Monte Carlo simulations because Monte Carlo is designed to sample from the canonical (isothermal) ensemble, requiring no thermostat at all. It is mostly for this reason that we use MC in this case study. Because physical "time" does not exist in MC simulations, the switching parameter λ is simply a function of the Monte Carlo step n.

In the standard Metropolis MC algorithm (Section 2.4), the acceptance probability at step n should be computed using the mixed potential energy $V(\lambda(n)) = [1 - \lambda(n)]V_1 + \lambda V_2$. Since the kinetic energies of the two Hamiltonians H_1 and H_2 are exactly the same, only the difference between two potential energies $V_2 - V_1$ is computed and added to the accumulated work of switching:

$$\Delta W \equiv \sum_{n=0}^{n_{max}} \frac{d\lambda(n)}{dn} [V_2(\{q_i(n)\}) - V_1(\{q_i(n)\})]. \tag{6.31}$$

Again, ΔW approaches ΔF only in the limit of the total number of MC steps n_{max} approaching infinity. In practice, because n_{max} is finite, the accumulated work ΔW overestimates the free-energy difference ΔF since some part of it is dissipated as heat during the switching. So, in addition to the random statistical error, a systematic error I exists, $\Delta W = I + \Delta F$. From its physical meaning, the dissipated work I should be positive and is expected to decrease with increasing n_{max}. Switching along the same trajectory $\lambda(n)$ but in the opposite direction, from V_2 to V_1, should produce a different amount of work, $\Delta W' = I - \Delta F$. Even if the sign of the free-energy difference reverses, the dissipation remains positive [63]. Therefore, a more accurate value of the free-energy difference is obtained by averaging: $\Delta F = (\Delta W - \Delta W')/2$.

6.3.1 Free Energy of a Core Vacancy Revisited

Let us compute once again the formation free energy of a core vacancy in the same setting as in Section 6.2. However, this time we shall not rely on the harmonic approximation but instead use the Monte Carlo adiabatic switching simulation. There are multiple ways to compute the same free energy depending on the direction of switching between the two systems. One option is to run a switching simulation from system 1 (191 atoms with vacancy at site A) to an Einstein crystal with the same number of atoms, which will produce F_1. Then, similar to the steps taken in Section 6.2, we can compute the free energy F_2 of the 192-atom system with atom A still present and, finally, the free energy per atom F_0 in a perfect crystal. As before, the formation free energy of a core vacancy is $F_{VC} = F_1 - F_2 + F_0$ except that all three free energies on the right-hand side are computed beyond the harmonic approximation. Unfortunately, this approach does not work very well in combination with the TI and AS methods because it is necessary to compute the difference between F_1 and F_2, both of which have statistical and systematic errors that may be larger than their difference.

A more intelligent approach is to obtain the needed free-energy difference by direct switching from system 1 to system 2. This idea may appear a bit strange because systems 1 and 2 do not have the same number of atoms. Yet, this issue is easy to resolve by augmenting system 1 with an extra atom in a harmonic potential (so that its free-energy contribution is known) that does not interact with the other $N - 1$ atoms. Let $\mathbf{r}_1$ be the extra atom introduced in place of atom A to augment system 1, while $\mathbf{r}_2, \ldots, \mathbf{r}_N$ are the remaining $N - 1$ atoms. The potential function for the modified system $1'$ becomes

$$\tilde{V}_{1'}(\mathbf{r}_1, \ldots, \mathbf{r}_N) = \frac{1}{2} k |\mathbf{r}_1|^2 + V_1(\mathbf{r}_2, \ldots, \mathbf{r}_N). \tag{6.32}$$

To reduce dissipation while switching between systems $1'$ and 2, let us slightly modify the former to make it structurally similar to the latter. Let us assume that $\mathbf{r}_i^{(0)1}$ and $\mathbf{r}_i^{(0)2}$ are the minimum-energy configurational states of $\tilde{V}_{1'}$ and V_2, respectively. Then the system described by the following potential energy function,

$$V_{1'}(\{\mathbf{r}_i\}) = \tilde{V}_{1'}(\{\mathbf{r}_i - \mathbf{r}_i^{(0)2} + \mathbf{r}_i^{(0)1}\}), \tag{6.33}$$

will have the same minimum-energy structure as V_2 and, at the same time, the same free energy as $\tilde{V}_{1'}$. Next, we will perform switching simulations to compute the free-energy difference between systems $V_{1'}$ and V_2.

For a fixed total number of Monte Carlo steps, the switching trajectory $\lambda(n)$ is somewhat arbitrary. The simplest form $\lambda = n/n_{max}$ is usually not a good choice (see Problem 6.3.2). Other switching trajectories of the same total length n_{max} exist that can improve the simulation efficiency by reducing both statistical errors and heat dissipation. A reasonable choice is shown in Fig. 6.3(a), where $\lambda(s) = s^5(70s^4 - 315s^3 + 540s^2 - 420s + 126)$ and $s \equiv n/n_{max}$ [63]. This switching function is constructed specifically to make the increase rate of λ very

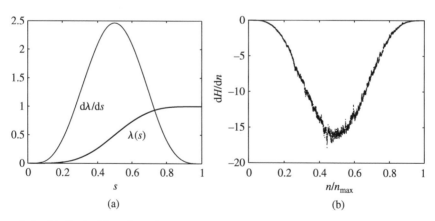

FIG. 6.3. (a) Switching function $\lambda(s)$, where $s \equiv n/n_{max}$. (b) Instantaneous value of $(V_2 - V_{1'}) \, d\lambda/dn$ as a function of MC step n for a switching simulation at $T = 600 \, \text{K}$ and $n_{max} = 5 \times 10^5$.

low both at the beginning and at the end of the switching trajectory where the
fluctuations of $V_2 - V_{1'}$ tend to be largest.

Before switching begins, the system should be equilibrated at the temperature
where the free energy is to be computed. This is done here by Monte Carlo
simulation using the Metropolis algorithm (Section 2.4). Starting from the minimum energy configuration, the system $V_{1'}$ is equilibrated at $T = 600$ K for 10^6
MC steps, with the parameter k chosen to be 20 eV/Å. The equilibrated configuration serves as the initial state for a subsequent switching simulation with
$n_{max} = 5 \times 10^5$ MC steps. The instantaneous values of $[V_2(n) - V_{1'}(n)] \, d\lambda(n)/dn$
obtained during one such simulation are plotted in Fig. 6.3(b). The total area under
this curve is the work ΔW done by the switching force. Because of statistical
fluctuations, another switching simulation under the same conditions would yield
a slightly different value for ΔW. By averaging over 50 such switching runs,
we obtain $\Delta W = -6.494 \pm 0.045$ eV. In addition to the desired free-energy difference, ΔW includes some amount of heat dissipation I due to the finite rate of
switching.

To eliminate the systematic error due to dissipation, let us now switch in the
opposite direction, i.e. from system 2 back to system $1'$. Again, system 2 is first
equilibrated for 10^6 MC steps and switched to $1'$ over 5×10^5 MC steps. Based on
50 such simulations, the work of reversed switching is $\Delta W' = 6.519 \pm 0.023$ eV.
It is gratifying to observe that, $\Delta W + \Delta W' > 0$, consistent with positive heat
dissipation. Finally, an improved estimate for the free-energy difference between
the two systems at $T = 600$ K is $F_2 - F_{1'} = (\Delta W - \Delta W')/2 = -6.506$ eV.

To arrive at the final answer, we need to obtain two more quantities. First, the
free-energy difference between systems $1'$ and 1 is the free energy of the auxiliary
atom in a three-dimensional harmonic potential, i.e.

$$F_e \equiv F_{1'} - F_1 = -3k_B T \ln \frac{k_B T}{\hbar \sqrt{k/m}} = 0.00843 \text{ eV}. \qquad (6.34)$$

Second, the free energy per atom in a perfect lattice can be obtained from separate
simulations and in our example we find $F_0 = E_0 - 0.0245$ eV (Problem 6.3.1).
Putting all the required ingredients together, the resulting formation free energy of
the core vacancy is

$$F_{VC} = F_1 - F_2 + F_0 = F_{1'} - F_2 - F_e + F_0 = 1.843 \text{ eV} \qquad (6.35)$$

and the entropy of vacancy formation at $T = 600$ K is $S_{Vc} = (E_{Vc} - F_{Vc})/$
$T = 6.025 k_B.$[8] Notice that the free energy is lower whereas the entropy is higher

---

[8] The zero temperature value for E_{Vc} quoted here is the one computed previously, in the harmonic
approximation. A more accurate result can be obtained by computing the average internal energies
of systems 0, 1 and 2 at temperature T. One can show, using a Monte Carlo simulation, that the
anharmonic correction to E_{Vc} is very small in this case.

than those predicted earlier from the harmonic approximation. Such anharmonic corrections are typical for solids and become even greater at higher temperatures.

Summary

- The free-energy difference between two systems can be written in terms of an ensemble average.

- Methods of thermodynamic integration and adiabatic switching are used to compute the free-energy difference between two systems. The absolute free energy of a system can be obtained by computing its difference from that of a reference system whose free energy is known.

Problems

6.3.1. For the same model of single-crystal silicon as described in Problem 6.2.1, compute its per-atom free energy and entropy by MC switching to and from an Einstein crystal at $T = 600$ K. Use the same switching function and parameters as in the case study discussed in the text. Estimate computational cost and feasibility of such calculations if first-principles methods (Section 2.1) are used.

6.3.2. Repeat the MC switching calculations for a perfect crystal of silicon using a linear switching function, $\lambda(n) = n/n_{max}$. Observe the amount of dissipation and statistical fluctuation during the switching. Is this switching function better or worse than the one plotted in Fig. 6.3(a)?

6.3.3. The most direct way to compute the anharmonic contributions to the free energy and the entropy is to switch from a "harmonic version" of the potential to the full anharmonic potential. The accumulated work of switching is a direct measure of the anharmonic part of the free energy. Starting from the same supercell with 64 atoms as in Problem 6.2.1, remove one of the atoms. Perform MC switching simulation between the true Hamiltonian and its harmonic approximation and obtain the free-energy difference at $T = 600$ K. Based on the result of Problem 6.2.2, obtain a more accurate prediction of the free energy and the entropy of vacancy formation at $T = 600$ K, beyond the harmonic approximation.

6.3.4. Compute the anharmonic contribution to the free energy of the 191-atom system containing a dislocation core vacancy in Fig. 6.2 by switching from its full potential function to its HA counterpart.

7

FINDING TRANSITION PATHWAYS

7.1 The Rare Event Problem

As was discussed in Chapter 2, stable and accurate numerical integration of the MD equations of motion demands a small time step. In MD simulations of solids, the integration step is usually of the order of one femtosecond (10^{-15} s). For this reason, the time horizon of MD simulations of solids rarely exceeds one nanosecond (10^{-9} s). On the other hand, dislocation behaviors of interest typically occur on time scales of milliseconds (10^{-3} s) or longer. Such behaviors remain out of reach for direct MD simulations. Time-scale limits of a similar nature also exist in MC simulations. For instance, the magnitude of the atomic displacements in the Metropolis algorithm has to be sufficiently small to ensure a reasonable acceptance ratio, which results in a slow exploration of the configurational space.

This disparity of time scales can be traced to certain topographical features of the potential-energy function of the many-body system, typically consisting of deep energy basins separated by high energy barriers. The system spends most of its time wandering around within the energy basins (metastable states) only rarely interrupted by transitions from one basin to another. Whereas the long-term evolution of a solid results from transitions between the metastable states, direct MD and MC simulations spend most of the time faithfully tracing the unimportant fluctuations within the energy basins. In this sense, most of the computing cycles are wasted, leading to very low simulation efficiency. Because the transition rates decrease exponentially with the increasing barrier heights and decreasing temperature, this problem of time-scale disparity can be severe.

As an illustration, consider a simple potential energy landscape (in arbitrary units):

$$U_1(x, y) = \frac{1}{6}\{4(1 - x^2 - y^2)^2 + 2(x^2 - 2)^2$$

$$+ [(x + y)^2 - 1]^2 + [(x - y)^2 - 1]^2 - 2\}. \tag{7.1}$$

Shown in Fig. 7.1(a), this two-dimensional energy function has two local minima, $\mathbf{r}_A = (-\sqrt{5}/2, 0)$ and $\mathbf{r}_B = (\sqrt{5}/2, 0)$. Starting from an arbitrary point (x, y), the steepest-descent path will eventually slide down either to $\mathbf{r}_A$ (when $x < 0$) or to $\mathbf{r}_B$ (when $x > 0$), except for the points on the *ridge* $x = 0$ separating basins A and B.

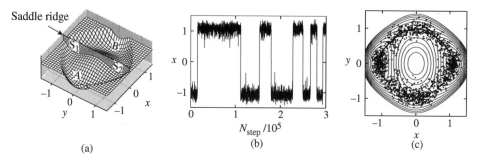

FIG. 7.1. (a) A two-dimensional potential energy function $U_1(x, y)$. The two meta-stable states, A for $x < 0$ and B for $x > 0$, are separated by a ridge at $x = 0$. S_1 and S_2 mark the locations of two saddles on the ridge. (b) Instantaneous values of coordinate x in an MC simulation at $k_B T = 0.15$. (c) The microstates (dots) visited during transitions between A and B plotted on top of the contour lines of $U(x, y)$. The solid (dashed) line connecting two energy minima is the minimum energy path obtained by steepest descent from saddle S_1 (S_2).

Any point on the ridge is a maximum along the directions normal to the ridge. Points S_1 and S_2 on the ridge are even more special because their energies are locally minimal among all the points on the ridge; they are called *saddle points*. Saddles are microstates of unstable equilibrium such that an arbitrarily small displacement from a saddle results in a descent to one of the underlying minima. A small displacement dx from saddle S_1 leads to a steepest descent to A when $dx < 0$ but a small positive dx results in a descent to B. Stitched together, these two descent trajectories form a path connecting A and B, as shown in Fig. 7.1(c), called the *minimum energy path* (MEP). The saddle is the highest energy point along the MEP.

Figure 7.1(b) and (c) show the results of a standard Metropolis MC simulation on this energy landscape at $k_B T = 0.15$ using the step size $\Delta = 0.25$. Figure 7.1(b) shows the evolution of coordinate x that is seen to stay close to -1 or $+1$ for many MC steps interrupted by occasional transitions. Although infrequent, the transitions themselves are fast, each taking just a few tens of MC steps. Focusing just on the transition events themselves, Fig. 7.1(c) shows the states visited during the transitions. These states cluster along two well-defined channels, closely following two MEPs passing through saddles S_1 and S_2. The reason for this is quite obvious. In crossing from one basin to another, the system is more likely to pass through S_1 or S_2, since crossing the ridge anywhere else requires higher energy and is therefore much less probable. The preference to crossing the ridge very near the saddles becomes even stronger at lower temperatures. This is why saddles are important for understanding the mechanisms of a system's evolution in the long time scale, just as the local minima are important for understanding its equilibrium properties. Quantitatively, the free energy of the dividing surface (ridge) is related to the rate of the corresponding transition (see next section). When multiple MEPs exist, the most likely escape route will be the one that crosses the saddle point with the lowest free energy. In the limit of low temperatures, the free energy of the dividing surface reduces to the energy of the lowest saddle point.

We would like to point out that our example problem here is sufficiently simple and the temperature is sufficiently high so that the simulation has been able to produce a few transition events over 10^5 MC steps. However, at lower temperatures and/or for more complex higher-dimensional systems, it is possible that no single transition event will be observed during a long simulation. In such cases, it becomes necessary to invoke special numerical methods geared to the studies of long-term evolution of infrequent events.

Summary

- The potential-energy landscape of many-body systems usually consists of energy basins (metastable states) separated by energy barriers. In solids, transitions between the neighboring basins are rare because the energy barriers are much higher than the thermal energy k_BT.

- In the low-temperature limit, transitions between the neighboring metastable states tend to follow the minimum-energy paths and pass near the saddle points.

7.2   Transition State Theory

Transition state theory (TST) provides an alternative approach to computing transition rates [65, 66], which is applicable even when the intervals between transitions exceed the total time duration of the atomistic simulation. To appreciate the basic ideas of TST, consider a hypothetical system of N atoms such that its potential energy function contains only two metastable states A and B. The $3N$-dimensional configurational space Ω of this system consists of two basins Ω_A (for metastable state A) and Ω_B (for metastable state B) divided by a $(3N-1)$-dimensional hyper-surface Ω^*. For simplicity, let the dividing surface be a hyper-plane $s=0$, where s is one of the $3N$ spatial variables, as shown in Fig. 7.2. In such a case, all configurational states with $s<0$ belong to Ω_A while all states with $s>0$ belong to Ω_B.

Consider the probability $P(\delta t)$ that a system, initially in basin Ω_A, will cross the dividing surface Ω^* towards the neighboring basin Ω_B during a short time interval δt. Then the transition rate from A to B is

$$k_{TST} = \lim_{\delta t \to 0} \frac{P(\delta t)}{\delta t}. \tag{7.2}$$

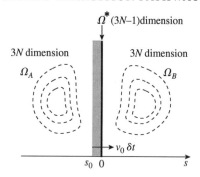

FIG. 7.2. Computing transition rates in the transition state theory. Two energy basins A and B are separated by ridge hyper-surface Ω^*. The microstates in the thin shaded layer on the left of the ridge plane will cross the ridge provided the velocity component v_s is positive and sufficiently high.

Suppose the system has reached a state of partial equilibrium in basin Ω_A; then the fraction of microstates in Ω_A that will cross the dividing plane over time is

$$P(\delta t) = \frac{Q_S(\delta t)}{Q_A}. \tag{7.3}$$

Here, Q_A is the statistical integral over basin Ω_A and $Q_S(\delta t)$ is the statistical integral over the sub-region of microstates that cross Ω^* from A to B within time δt. The following derivation takes into account that, in the limit $\delta t \to 0$, the major contribution to $Q_S(\delta t)$ comes from the region immediately adjacent to the dividing surface Ω^*, as shown in Fig. 7.2.

First, let us rewrite the statistical integrals using the step function $\Theta(\cdot)$ defined such that $\Theta(x) = 1$ if $x > 0$ and $\Theta(x) = 0$ otherwise. Then

$$Q_S(\delta t) = \frac{1}{h^{3N}} \int d^{3N}\mathbf{r}\, d^{3N}\mathbf{p}\, \Theta(-s)\Theta(p_s + ms/\delta t)\, e^{-\beta H(\mathbf{r},\mathbf{p})}, \tag{7.4}$$

$$Q_A = \frac{1}{h^{3N}} \int d^{3N}\mathbf{r}\, d^{3N}\mathbf{p}\, \Theta(-s)\, e^{-\beta H(\mathbf{r},\mathbf{p})}, \tag{7.5}$$

where both integrals are now over the entire $6N$-dimensional phase space. The meaning of $\Theta(p_s + ms/\delta t)$ in eq. (7.4) is that, for a microstate at point $s < 0$ to cross Ω^* within time δt, its velocity in the s direction has to be greater than $-s/\delta t$. The integral over s and p_s in eq. (7.4) can be carried out explicitly. In the limit of $\delta t \to 0$, this leads to

$$Q_S(\delta t) = \frac{k_B T \delta t}{h}\, \frac{1}{h^{3N-1}} \int_{\Omega^*} d^{3N-1}\mathbf{r}\, d^{3N-1}\mathbf{p}\, e^{-\beta H(\mathbf{r},\mathbf{p})}, \tag{7.6}$$

where the integral is now over the $(6N - 2)$-dimensional phase space of the dividing surface (excluding s and p_s). Define the statistical integral over the dividing surface Ω^* as

$$Q^* = \frac{1}{h^{3N-1}} \int_{\Omega^*} d^{3N-1}\mathbf{r}\, d^{3N-1}\mathbf{p}\, e^{-\beta H(\mathbf{r},\mathbf{p})}. \tag{7.7}$$

Then the rate of crossing from A to B is

$$k_{\text{TST}} = \frac{k_B T}{h} \frac{Q^*}{Q_A} = \frac{k_B T}{h} e^{-\beta(F^* - F_A)}, \tag{7.8}$$

where $F^* = -k_B T \ln Q^*$ can be interpreted as the free energy of microstates in the dividing surface Ω^* and $F_A = -k_B T \ln Q_A$ is the free energy of metastable state A. As usual, the integrals over momenta in Q^* and Q_A can be carried out explicitly, leading to

$$Q^* = \left(\frac{\sqrt{2\pi m k_B T}}{h} \right)^{3N-1} \int_{\Omega^*} d^{3N-1}\mathbf{r}\, e^{-\beta V(\mathbf{r})} \tag{7.9}$$

$$Q_A = \left(\frac{\sqrt{2\pi m k_B T}}{h} \right)^{3N} \int_{\Omega_A} d^{3N}\mathbf{r}\, e^{-\beta V(\mathbf{r})}. \tag{7.10}$$

Although we assumed, for simplicity, that the dividing surface was a hyper-plane, the results in eqs. (7.8)–(7.10) apply to situations where the dividing surface Ω^* is curved [75].

The transition state theory is exact in describing the "flux" across an arbitrary dividing surface in the configurational space. However, this "flux" is not necessarily the rate of transitions between metastable states in the usual sense. This is because, even if a trajectory initiated from metastable state A crosses the dividing surface Ω^*, it may subsequently recross Ω^* multiple times before eventually settling down into another metastable state. Furthermore, the trajectory can cross Ω^* an even number of times and settle back into metastable state A again. In both cases, the k_{TST} rate overestimates the "true" transition rate, which can be written as

$$k = \kappa \cdot k_{\text{TST}}, \tag{7.11}$$

where $\kappa < 1$ is a correction factor that accounts for the recrossing events. κ can be obtained by counting the number of recrossing events in a series of short MD simulations initiated at different microstates on the dividing surface [65, 66].

Summary

- In the transition state theory (TST), the transition rate from one metastable state to another depends exponentially on the difference between the free energy of the dividing surface and the free energy of the initial metastable state.

- TST overestimates the actual transition rate because the TST flux includes trajectories that recross the dividing surface over short periods of time.

7.3 Local Path Optimization

TST relates the equilibrium "flux" between neighboring configurational states to the free energy F^* of the dividing hyper-surface. In solids at temperatures well below melting, the actual values of F^* are defined mostly by the energies of the saddle points on the dividing surface. This is similar to the dominant role of the energy minima in defining the equilibrium properties of solids. This is why calculations of the transition rates often involve finding the relevant saddle points on the dividing surfaces separating the metastable states.

Development of efficient algorithms for searching and finding low-energy transition paths in many-dimensional systems is an area of active research [67–76]. In the following sections, we describe several algorithms that are not necessarily the most efficient but are representative of the current state of the art. In doing so, we pay more attention to the physical issues commonly encountered in searching for the low-energy transition pathways. To illustrate the progressively more complicated aspects of long-time dynamics, we keep coming back to the same example of dislocation kink migration in silicon. Our hope is that such a discussion will help the reader to develop some appreciation for this difficult task and to better understand both the strengths and weaknesses of the various methods described in the literature. In this section, we discuss two algorithms based on the ideas of constrained minimization. Despite their similar ideologies, the algorithms are quite different in terms of their numerical robustness and computational cost.

7.3.1 Constrained Minimization

This method is based on the assumption that the minimum energy path (MEP) is close to a straight line connecting the energy minima of two metastable states.[1] Let the coordinates of the energy minima of two neighboring basins A and B be $\mathbf{r}_A$ and $\mathbf{r}_B$, respectively. The straight path connecting the two minima is parametrically expressed as

$$\mathbf{r}(s) = \mathbf{r}_A + (\mathbf{r}_B - \mathbf{r}_A) \cdot s, \qquad (7.12)$$

---

[1] This is a strong assumption because, generally, the MEPs are curved and can be *very* different from a straight line, in which case the method described here is not expected to work well.

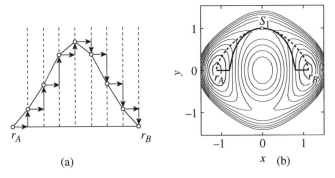

(a) x (b)

FIG. 7.3. (a) A schematic illustration of Algorithm 7.1. The horizontal arrows show
the increments of s. Each vertical line depicts a sub-space in which constrained
relaxation is performed for a given value of path parameter s. The solid line links
the relaxed configurations to show the search trajectory. (b) The solid line is
a transition path obtained by constrained minimization of the two-dimensional
energy function (eq. 7.1). The dashed line is the MEP obtained by steepest
descent from saddle S_1.

where $s \in [0, 1]$ is a path parameter. Obviously, even in the simple situation depicted
in Fig. 7.1, the straight path crosses the ridge at point $(0, 0)$, which is far away from
the saddle points S_1 and S_2.

To allow the candidate path to deviate from the straight line and to better track
an actual MEP, let us search for an MEP among paths of the following form,

$$\mathbf{r}(s) = \mathbf{r}_A + (\mathbf{r}_B - \mathbf{r}_A) \cdot s + \mathbf{r}^\perp(s). \tag{7.13}$$

where $\mathbf{r}^\perp(s)$ is a vector orthogonal to the straight path. The exact shape of $\mathbf{r}^\perp(s)$
can be obtained by minimizing the energy at every point $s \in [0, 1]$ along the path.
This can be done by incrementing s in small steps Δs and finding an optimal value
of $\mathbf{r}^\perp(s)$ for each value of the path parameter s in the hyperplane orthogonal to
the straight path. Illustrated in Fig. 7.3(a), the following algorithm is based on this
simple idea.

Algorithm 7.1

1. Define $\mathbf{r}_{AB} \equiv \mathbf{r}_B - \mathbf{r}_A$ and the unit vector $\hat{\mathbf{r}}_{AB} \equiv \mathbf{r}_{AB}/|\mathbf{r}_{AB}|$.

2. Initialize the path parameter and the starting state $s := 0$, $\mathbf{r}(s) := \mathbf{r}_A$.

3. Increment $s := s + \Delta s$. If $s \geq 1$, return.

4. For the current path parameter s, prepare the initial state for subsequent
   constrained minimization $\mathbf{r}_0 := \mathbf{r}(s - \Delta s) + \mathbf{r}_{AB} \cdot \Delta s$.

5. Find an optimal $\mathbf{r}_0$ by minimizing $V(\mathbf{r}_0)$ under the constraint
   $(\mathbf{r}_0 - \mathbf{r}_A) \cdot \hat{\mathbf{r}}_{AB} = s|\mathbf{r}_{AB}|$.

6. $\mathbf{r}(s) := \mathbf{r}_0$.

7. Go to 3.

The minimization in step 5 can be performed using the standard methods dis-cussed in Section 2.3 (CGR) only modified to enforce the orthogonality constraint $\mathbf{r}^\perp(s) \cdot \hat{\mathbf{r}}_{AB} = 0$. A simple way to prevent the system from moving along the con-straint direction $\hat{\mathbf{r}}_{AB}$ is to remove from the gradient vector $\mathbf{g}$ its component parallel to $\hat{\mathbf{r}}_{AB}$:

$$\mathbf{g}^\perp(\mathbf{r}_0) = \mathbf{g}(\mathbf{r}_0) - \left[\mathbf{g}(\mathbf{r}_0) \cdot \hat{\mathbf{r}}_{AB}\right]\hat{\mathbf{r}}_{AB}. \qquad (7.14)$$

Because modified gradient $\mathbf{g}^\perp(\mathbf{r}_0)$ is orthogonal to $\hat{\mathbf{r}}_{AB}$, the minimization algorithm will never attempt to move $\mathbf{r}_0$ along the $\hat{\mathbf{r}}_{AB}$ direction.

Figure 7.3(b) shows a transition path obtained using Algorithm 7.1 with $\Delta s = 0.02$. Although the path does not trace the actual MEP too well, especially near the energy minima, it passes through one of the saddle points. As a result of the symmetry of the energy function (eq. 7.1), the search trajectory is equally likely to lead to S_1 or S_2. This symmetry is broken here by the numerical noise.

7.3.2   The Chain of States Method

Although Algorithm 7.1 is often successful in identifying a transition path that passes near a saddle point, it is known to fail in certain situations. Depending on the particular shape of the potential energy function, the algorithm may converge to a saddle point that leads to an energy basin different from the intended destination B or, worse still, may fail to find a saddle point at all. These undesirable behaviors may occur when the initial assumption, that the MEP is "close" to the straight-path, turns out to be incorrect. In the case shown in Fig. 7.3(b), the MEPs are not straight but do not have a complex shape either. In fact, at the saddle points, the MEPs are parallel to the initial straight path. This relative closeness to the straight path is why Algorithm 7.1 is successful in this simple example. In other situations the MEP can be so curved that somewhere along the path (especially near the saddle point), it may become nearly orthogonal to the straight-path direction. In such cases, Algorithm 7.1 or other similar method is unlikely to find the relevant saddle point.

More robust methods exist that often find saddle points even when Algorithm 7.1 fails, at the expense of an increased computational cost. The nudged elastic band (NEB) method assumes that the MEP is a smooth curve that can be approximated by a *chain of states*, each state representing a copy of the whole system: $\mathbf{r}(s_i)$, $s_i = i/N_c$, $i = 0, 1, \ldots, N_c$. The end states of the chain are fixed at the energy minima of interest, i.e. $\mathbf{r}(0) \equiv \mathbf{r}_A$ and $\mathbf{r}(1) \equiv \mathbf{r}_B$. The NEB method follows from the observation that each point $\mathbf{r}(s)$ along the MEP should be a local energy minimum in the hyperplane normal to the local tangent direction $\mathbf{t}(s)$, as shown in Fig. 7.4(a).

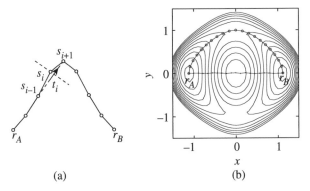

(a)   (b)

FIG. 7.4. (a) A schematic illustration of Algorithm 7.2. The trial path is a chain of $N+1$ replicas of the system. Path optimization includes steepest-descent relaxation of all states in the hyperplanes orthogonal to the local tangents $\mathbf{t}(s_i)$, followed by redistribution of states along the path. (b) Initially, the chain follows the straight path from $\mathbf{r}_A$ to $\mathbf{r}_B$ (thin line). The fully optimized chain (circles) closely follows the MEP (solid line).

The MEP can be found by iterative relaxation of the trial chain, assuming that the local tangent to the path at each state s_i, $i = 1, \ldots, N_c - 1$, is approximated by

$$\mathbf{t}(s_i) \equiv [\mathbf{r}(s_{i+1}) - \mathbf{r}(s_{i-1})]/|\mathbf{r}(s_{i+1}) - \mathbf{r}(s_{i-1})|. \qquad (7.15)$$

Starting from an initial trial chain (usually taken as the straight path from $\mathbf{r}_A$ to $\mathbf{r}_B$), the following algorithm finds an MEP by an iterative steepest-descent relaxation with a step size Δ.

Algorithm 7.2

1. Find the local tangent vectors, $\mathbf{t}(s_i) := [\mathbf{r}(s_{i+1}) - \mathbf{r}(s_{i-1})]/|\mathbf{r}(s_{i+1}) - \mathbf{r}(s_{i-1})|$, for all $i = 1, \ldots, N_c - 1$.

2. Compute the energy gradients, $\mathbf{g}(s_i) := \partial V(\mathbf{r}(s_i))/\partial \mathbf{r}$, for all $i = 1, \ldots, N_c - 1$.

3. Orthogonalize the gradients to the tangent vectors, $\mathbf{g}^{\perp}(s_i) := \mathbf{g}(s_i) - [\mathbf{g}(s_i) \cdot \mathbf{t}(s_i)]\mathbf{t}(s_i)$.

4. If $\max_i |\mathbf{g}^{\perp}(s_i)| < \epsilon$, return; otherwise move states along the negative gradient direction, $\mathbf{r}(s_i) := \mathbf{r}(s_i) - \mathbf{g}^{\perp}(s_i) \cdot \Delta$.

5. Redistribute states $\mathbf{r}(s_i)$ by shifting them along the tangents $\mathbf{t}(s_i)$, to maintain a uniform spacing along the chain.

6. Go to 1.

Step 5 in this algorithm is optional and can be performed once every few cycles of the relaxation. A good redistribution is achieved by demanding that the distance between the neighboring states remains uniform along the chain [76]. Fig. 7.4(b) shows a path obtained by Algorithm 7.2 with $N = 20$ and step size $\Delta = 0.02$. The resulting path tracks the MEP rather faithfully indicating that the NEB method is robust. At the same time, the NEB method involves simultaneous optimization of many replicas of the system and, consequently, is more computationally demanding than the method of constrained minimization.

7.3.3   Case Study: Kink Migration in Silicon

Let us now apply the two methods introduced in this section to finding transition paths in a system of N atoms. The particular example examined here is the path for migration of a dislocation kink in silicon. Because of the strong covalent bonding between the atoms, dislocations in diamond-cubic semiconductors consist of very long straight segments connected by rather narrowly localized kinks. The atomic structure of one such kink on a 30° partial dislocation in silicon is shown in Fig. 7.5(a). The importance of kink migration is that its rate determines how fast the dislocation moves under given temperature and stress. Figures 7.5(b) and (c) show a sequence through which the kink migrates to the right by one lattice vector. Notice that final state B has a different local atomic arrangement from the original state A [14, 77].

To find the saddle point for this transition, let us first apply the method of constrained minimization. For computational expediency, it is helpful to slightly modify Algorithm 7.1 before applying it to a situation with as many dimensions as in this example. The entire simulation cell in this calculation contains $N = 6912$ atoms, and therefore the total number of spatial degrees of freedom is $3N = 20\,736$. Obviously, not all of the available degrees of freedom will participate equally in the transition. In particular, atoms far away from the kink are expected to move only slightly during the transformation from A to B. It is possible and useful to reduce the dimensionality of the problem at hand by focusing on a smaller number of atoms that are close to the defect, while treating the rest of the atoms in a simplified manner. Let us select $n = 109$ atoms close to the kink and perform constrained minimization on this "subsystem". For any given configuration of the selected subsystem, the reduction in complexity is achieved by relaxing positions of the surrounding $(N - n)$ atoms to their minimum energy with no further constraints. This corresponds to searching for a saddle in a reduced 327-dimensional subspace defined by the condition that the components of the gradient corresponding to the surrounding $(N - n)$ atoms are all zero. In addition to reducing the computational cost and the memory requirements, confining the search to a reduced subspace provides for a better control of the transition process.[2]

---

[2] In our experience, it is sometimes possible to find a saddle point in a reduced subspace even when direct application of Algorithm 7.1 produces various unphysical behaviors, e.g. non-smooth energy variations along the path.

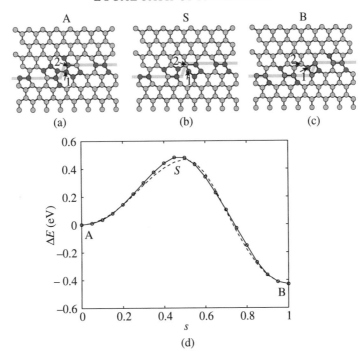

FIG. 7.5. (a) The initial state A of a kink on a $30°$ partial dislocation in silicon. The dark circles show the positions of atoms whose potential energies exceed the ideal bulk energy by 0.19 eV and the thick line traces the dislocation center. (b) The saddle-point configuration S of the kink as it moves to the right towards configuration B. (c) The final state of kink translation. Notice that the atomic configurations of the initial (A) and final (B) states are different: in the nomenclature used in [14, 77] state A corresponds to "left kink" LK', while state B corresponds to "left kink" LK. (d) The variation of the potential energy along the transition path. The dashed line is the result of the constrained minimization (Algorithm 7.1) and the solid line is the result of the chain-of-states relaxation (Algorithm 7.2).

The result of constrained minimization with step size $\Delta s = 0.05$ is plotted as the dashed line in Fig. 7.5(d). The highest energy along the transition path is reached close to $s = 0.5$: the estimate for the saddle-point energy for this transition is 0.47 eV. To verify this prediction, let us now apply the chain of states method, Algorithm 7.2, to the same problem. Again as before, the chain of state optimization is applied only to the reduced subsystem of $n = 109$ atoms.[3] Plotted as the solid line in Fig. 7.5(d),

---

[3] A considerable further reduction in the computational cost can be achieved by making the remaining $(N - n)$ atoms follow the straight path from A to B. This trick is especially useful during the initial relaxation of the transition path. Then, if desired, the path can be refined near the saddle by allowing the $(N - n)$ atoms to relax.

the path obtained by the chain of state optimization almost overlaps with that obtained by the constrained minimization. The highest energy along this path is 0.48 eV, reached at $s = 0.45$. It is satisfying to observe that both algorithms find the same saddle point: the small difference in the predicted saddle-point energy is within the numerical error.

Notice that, while both methods produce essentially the same results, the chain of states method (Algorithm 7.2) takes an order of magnitude longer time to converge compared with the constrained minimization method (Algorithm 7.1). Therefore, the simple method of constrained minimization is worth trying first. If it fails to converge to a saddle point, application of more robust and sophisticated methods becomes justified.

Despite the obvious differences, both methods described in this section have the same fundamental limitations. First, a transition path identified by either of these two methods is not necessarily globally optimal. For example, the paths found by Algorithm 7.2 tend to be very similar to the initial trial path, which is almost invariably assumed to be the straight path from A to B. In many-body systems, it is very possible that multiple MEPs and saddle points connect the same two energy minima. While the most relevant saddle points are those with the lowest free energy, the methods described here may well converge to an MEP whose free energy barrier is significantly higher than the globally optimal MEP. In other words, both algorithms may find "irrelevant" saddle points.

The second limitation is that both methods are applicable only when the destination state B is already known. In many interesting situations, all one knows is that the system is "trapped" inside a metastable state A, while the destination state B is unknown. If such is the case, other methods are needed to identify transition paths and saddle points. These two limitations are serious and will be addressed in the following two sections.

Summary

- The methods of constrained minimization, including the nudged elastic band (NEB) method, are used to search for *local* minimum energy paths between two known metastable states.

7.4   Global Path Optimization

In a way, the task of finding the globally optimal transition path is similar to the problem of finding the global minimum of a potential energy function. Just as we did in Section 2.3, it is possible to use "finite-temperature" methods such as simulated annealing in order to avoid premature "locking" into a locally (but not globally) optimal path. Consider a path represented by a chain of $N_c + 1$ microstates: $\mathbf{r}(s_i)$,

$i = 0, 1, \ldots, N_c$, with the ends fixed at $\mathbf{r}(0) \equiv \mathbf{r}_A$ and $\mathbf{r}(1) \equiv \mathbf{r}_B$. To give this chain enough flexibility to sample curved transition paths, let us make the maximum length of this path $L_{max} = N_c r_0$ substantially longer than the distance between its two end states, $|\mathbf{r}_B - \mathbf{r}_A|$. Here, parameter r_0 caps the distance between any two neighboring states, i.e.

$$|\mathbf{r}(s_i) - \mathbf{r}(s_{i-1})| \leq r_0. \tag{7.16}$$

Let us call the highest energy E_{max} among all states along a given path the *energy barrier* of this path. The goal is to find the optimal path whose energy barrier is the lowest among all paths with the same N_c [78]. Once found, the highest energy state along the optimal path should be close to the saddle state of the MEP whose energy is lowest among all MEPs connecting A to B. The result should not depend on N_c in the limit of large N_c. In the following, we will be searching for the optimal transition paths using the Simulated Annealing method in combination with the Metropolis Monte Carlo sampling (see Section 2.4 and [78–80]).

In the simulations described in this section, the temperature is entirely fictitious and is used only to prevent trapping into a local minimum in the space of transition paths. Therefore, the distribution of paths sampled at an annealing temperature T does not have to satisfy the actual statistical distribution of transition paths between two metastable states at that temperature. At the end of the annealing simulations, the energy of the lowest saddle point defines the rate of the dominant transition in the limit of $T \to 0$. When the actual statistical distribution of transition paths is of interest, the method of transition path sampling (TPS) is designed to sample the paths according to their statistical distribution at a given temperature [67, 81].

In the following algorithm, the annealing temperature varies as $T(n_{step}) = T_0 \exp(-\lambda \cdot n_{step}/N_{step})$, where T_0 is the starting temperature, N_{step} is the total number of MC steps, n_{step} is the current MC step, and λ specifies the rate of temperature reduction.

Algorithm 7.3

1. Initialize a step counter $n_{step} := 0$.

2. Compute the energies of all microstates $E_i := V(\mathbf{r}(s_i))$ along the chain and find the maximum $E_{max} := \max_{i=1}^{N-1}(E_i)$.

3. Adjust temperature $T := T_0 \exp(-\lambda \cdot n_{step}/N_{step})$.

4. Randomly pick state i from $\{1, 2, \ldots, N-1\}$.

5. Add a random displacement to the atomic coordinates of state i. Make sure that the new trial configuration $\mathbf{r}^{new}$ satisfies constraints $|\mathbf{r}^{new} - \mathbf{r}(s_{i-1})| \leq r_0$ and $|\mathbf{r}^{new} - \mathbf{r}(s_{i+1})| \leq r_0$. Set $E^{new} := V(\mathbf{r}^{new})$.

6. Draw a random number ξ uniformly distributed in $[0, 1]$.

7. If $\xi \leq \exp\left[(E_{max} - E^{new})/k_B T\right]$, accept the trial move, set $\mathbf{r}(s_i) := \mathbf{r}^{new}$, $E_i := E^{new}$, $E_{max} := \max_{j=1}^{N-1}(E_j)$. Otherwise, reject the move and do nothing.

8. Increment $n_{step} := n_{step} + 1$. If $n_{step} > N_{step}$ stop; otherwise go to 3.

Notice that the decision to accept or reject the new trial path depends only on its maximum energy. This means that, if altering one state of the path does not change the maximum energy of the path, such a trial move is always accepted. This version of Monte Carlo sampling gives much flexibility to the path, allowing it to sample a wide range of configurations, which is beneficial for finding the global optimum.

7.4.1 Case Study I: The Two-dimensional Potential

Let us first apply Algorithm 7.3 to a slightly modified 2D potential:

$$U_2(x, y) = U_1(x, y) - 0.4 \exp\left(-16y^2\right). \tag{7.17}$$

As shown in Fig. 7.6(a), the new term introduces an additional saddle point at $S_0 = (0, 0)$. The energy of this new saddle S_0 is higher than that of the other two saddle points: $E_{S_0} = 1.6$ versus $E_{S_1} = E_{S_2} = 1.0$. The new saddle S_0 is located right on the straight line connecting $\mathbf{r}_A$ and $\mathbf{r}_B$. If this straight line is chosen as the initial trial path, the algorithms of local path optimization (such as NEB) will be trapped at the high-energy saddle S_0 and will never find S_1 or S_2.

To see how Algorithm 7.3 overcomes this difficulty of local trapping, let us lay the initial chain path along the straight line from A to B, divide it into 100 equally

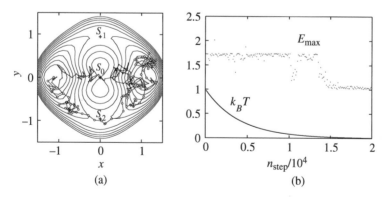

(a)   (b)

FIG. 7.6. (a) Contour plot of $U_2(x, y)$. Two snapshots of the path are shown: one at step 1000 (thin line) and another in the end of the annealing (circles). (b) The evolution of the energy barrier during the annealing simulation. The solid line is the annealing temperature as a function of the Monte Carlo step.

spaced states and set the rest of the simulation parameters as follows: $r_0 = 0.2$, $k_B T_0 = 1.0$, $\lambda = 5$ and $N_{step} = 20\,000$. At the end of the Monte Carlo annealing simulation, the temperature becomes very low, $k_B T = 0.0067$. Fig. 7.6(b) shows how the energy barrier evolves during this simulation. For a while, the energy barrier of the evolving chain stays close to the energy of saddle S_0 (1.6). At the end of this simulation, the energy barrier converges to 1.0. This is a signal that the annealing simulation finds one of the two lower-energy paths. Indeed, Fig. 7.6(b) shows two snapshots of the chain, one at step 1000 and the other at the very end of the simulation. Even though initially the chain passes through the high saddle S_0, in the end it finds a curved path passing through the low saddle S_2. Unlike Algorithms 7.1 and 7.2, Algorithm 7.3 is able to find transition paths that are significantly different from the initial guess. Provided the chain is sufficiently long and the annealing schedule is sufficiently slow, the path annealing algorithm is more likely to find the lowest saddle point between any two metastable states in a complex energy landscape.

At the end of the annealing simulation the chain folds on itself in basins A and B, as shown in Fig. 7.6(a). This happens because Algorithm 7.3 penalizes only the maximum energy on the path during the annealing. There is nothing wrong with this behavior, which simply indicates that the contour length of the chain happens to be considerably longer than the length of the optimal path. If desired, the location of the saddle point can be further refined starting from the maximum energy state on the final path. For this, it should be sufficient to use an algorithm of local path optimization, e.g. the NEB (Algorithm 7.2) or the Sinclair–Fletcher algorithm [82].

7.4.2 Case Study II: Kink Migration in Silicon Revisited

Here we use Algorithm 7.3 to search for an optimal transition path between the same two metastable states of the dislocation kink in silicon introduced in the previous section, Fig. 7.5. To reduce the computational burden, only 39 atoms around the kink are selected for the Monte Carlo moves, whereas positions of the remaining $N - 39$ atoms are linearly interpolated along the chain from state A to state B. The initial path is obtained from a NEB relaxation using a chain with 100 segments. Plotted as the thick line in Fig. 7.7(a), the energy profile of the initial path is very similar to that in Fig. 7.5(d).[4] The rest of the simulation parameters are as follows: the maximum separation between two neighboring states along the chain $r_0 = 0.2$ Å, the initial annealing temperature $k_B T_0 = 2.0$ eV, the annealing rate $\lambda = 10$ and the total number of MC steps $N_{step} = 200\,000$. The temperature reached at the end of this annealing simulation is very close to zero, at $k_B T = 4.54 \times 10^{-5}$ eV.

---

[4] The energy barrier of this path is a bit higher than that in Fig. 7.5(d) because fewer (39 versus 109) atoms are included in the path relaxation.

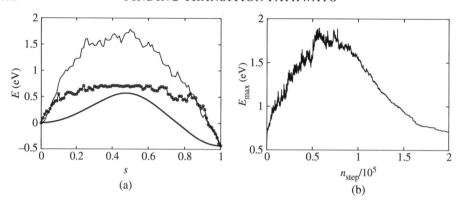

FIG. 7.7. (a) The energy of states along the chain at steps 0 (thick line), 6×10^4 (thin line) and 2×10^5 (circles) of the annealing simulation. (b) Variations of the energy barrier during the annealing simulation.

Fig. 7.7(a) shows three energy profiles along the chain taken at different stages of the annealing simulation. The progress towards a low-energy path during the annealing is shown in Fig. 7.7(b). Initially, E_{max} rises, approaching nearly $2 \, eV$ in the middle of the simulation. Over the second half of the annealing simulation, the barrier gradually decreases to its initial value. Further refinement of the final path by NEB confirms that the saddle along this path is exactly the same as identified in the previous section. Thus, the path annealing simulation confirms that the previously identified saddle point is indeed the global minimum amongst all possible saddle points separating states A and B.[5]

Summary

- The method of path annealing can be used to search for low-energy transitions that are significantly different from the initial guess. The method of path annealing is computationally demanding.

7.5   Temperature-accelerated Sampling

The methods described in the preceding sections of this chapter are designed to search for transition pathways between two known metastable states A and B. However, often only the starting state A is known. When such is the case, it is necessary to first identify the "destination" state B before applying the methods described in the previous sections. The method of temperature-accelerated sampling (TAS) can be used for this purpose.

---

[5] To make this claim with a higher confidence, still longer annealing simulation may be required.

The idea of TAS follows from the observation that at high temperatures the transitions between metastable states occur frequently. Because the potential energy landscape does not depend on temperature, it should be possible to use MD or MC simulations at a fictitiously high temperature (T_2) to explore the energy landscape and identify neighboring metastable states and transition pathways that might be relevant at a lower temperature of interest (T_1). Typically, in a system with many dimensions, there is more than one state that the system can go to after escaping from the origin state A. The relative rates of different escape routes depend on temperature. Some escape routes that appear frequently in the high-temperature simulation may be very improbable at the low temperature, especially when T_2/T_1 is large. To be able to harvest low-energy paths relevant for the low temperature, a reasonable search strategy is to sample multiple escape trajectories out of state A at a high temperature and analyze the relevance of the sampled escape routes for the low temperature. One possible way to do this is by looking at the energies of the destination states and the related saddles.

In the following, we describe a TAS algorithm that identifies possible destination states by Monte Carlo simulations of escapes around the origin state A at a fictitiously high temperature T_2. Once a transition is observed, the same algorithm also returns a path connecting the origin state to the newly sampled destination state. This path can be used as an initial condition for subsequent local path relaxations or path annealing simulations, as described previously.[6]

The algorithm described below is nothing but the standard Metropolis Monte Carlo (Algorithm 2.3) augmented with an additional procedure for checking, periodically, if the system has left the origin state A. This check is done by occasionally relaxing the current configuration to its underlying energy minimum $\mathbf{r}_{new}$ and comparing the resulting structure to the origin state $\mathbf{r}_A$. If the distance between $\mathbf{r}_{new}$ and $\mathbf{r}_A$ exceeds a threshold value r_c, a transition has just taken place. Otherwise, the simulation marches on until a transition is detected.

Algorithm 7.4

1. Allocate a chain of states $\mathbf{r}_i$, $i = 0, \ldots, N_c$ that can hold $N_c + 1$ copies of the system.

---

[6] In a popular version of TAS called temperature-accelerated dynamics (TAD) [83–85], these two steps, i.e. detection of transitions at a high-temperature and search for the saddle points, are combined together. TAD is an automated algorithm that relies on high-temperature MD simulations to evolve the system as if it were at a lower temperature. Here we use the temperature-accelerated sampling solely for identification of the candidate escape routes. Every new escape path is then subjected to additional analysis, using the methods discussed in Sections 7.3 and 7.4, to judge its relevance for the low-temperature dynamics.

2. Copy the initial configuration to the head of the chain $\mathbf{r}_0$. Set $i_c := 0$.

3. Run a Metropolis Monte Carlo simulation at temperature T_2 for n_{save} steps (Algorithm 2.3).

4. Set $i_c := i_c + 1$. Copy the current configuration to state $\mathbf{r}_{i_c}$. If $i_c < N_c$, go to step 3.

5. Starting from $\mathbf{r}_{N_c}$ as the initial state, find the underlying energy minimum $\mathbf{r}_{new}$ using the conjugate gradient relaxation method (Algorithm 2.2).

6. If $|\mathbf{r}_{new} - \mathbf{r}_A| > r_c$, save the current chain of states $\mathbf{r}_i$, $i = 0, \ldots, N_c$ and exit. Otherwise, copy the end state of the chain to the head of the chain, $\mathbf{r}_0 := \mathbf{r}_{N_c}$ and go to step 3.

At the end of a TAS simulation, a chain of states is generated connecting the origin state A to a new metastable state B.[7] Successful applications of this algorithm depend on the choice of the sampling temperature T_2. On the one hand, if the sampling temperature T_2 is too high, the accelerated sampling may produce many transition paths with very high saddle-point energies, none of which are relevant for the low-temperature behavior. On the other hand, if the sampling temperature is too low, the algorithm may not capture any transition events over a very long simulation. One possible strategy to selecting the sampling temperature is to first use a sampling temperature T_2 that is high enough for the transitions to be observed frequently. Then, the sampling temperature can be reduced until transition events are observed rarely but still with an acceptable frequency.

7.5.1 Case Study: Kink Migration in Silicon Yet Again

Let us now apply Algorithm 7.4 to search for low-energy escape paths starting from the same kink configuration LK' already discussed in Sections 7.3 and 7.4, Fig. 7.5(a). The TAS algorithm is applied here only to $n = 612$ atoms around the kink while the remaining $N - 612$ atoms are held fixed at their respective positions in state A (i.e. LK'). Starting with a sampling temperature $T_2 = 2000$ K and using $r_c = 0.1$ Å as a stop-search criterion, we observe multiple transitions from the origin state A (same as LK') to two other distinct states. Interestingly, neither of these two states is the same as state LK shown in Fig. 7.5(c). Labelled A_1 and A_2, the atomic configurations corresponding to the two newly found states are shown in Fig. 7.8. The new states are nearly equidistant from the origin state $|\mathbf{r}_{A_1} - \mathbf{r}_A| \approx |\mathbf{r}_{A_2} - \mathbf{r}_A| \approx 1.3$ Å and their energies are both higher than that of state A by about 0.2 eV.

---

[7] As long as the sampling temperature is not too high, it is unlikely that the chain will contain more than one transition event.

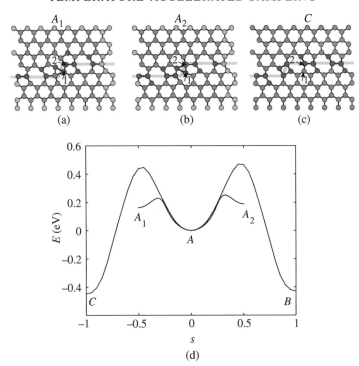

F IG. 7.8. (a) and (b) are two metastable configurations of a dislocation kink that are very close to the standard configuration shown earlier in Fig. 7.5(a). (c) Kink configuration that is equivalent to Fig. 7.5(c), except that it is translated to the left by one lattice vector.

To clarify the connectivity of two new states, we ran further TAS simulations starting from state A_1 (A_2) and observed multiple transitions to states A and A_2 (A_1). Thus, the TAS simulations appear to have discovered a triplet of metastable states A, A_1 and A_2 that should have relatively frequent transitions amongst them.[8]

To explore transitions out of these three states, we ran more TAS simulations starting from state A, but now with a much larger stop-search parameter $r_c = 1.5$ Å. Using the same sampling temperature $T_2 = 2000$ K, a variety of transition events are observed leading to destination states whose energies range from -0.4 eV to 1.7 eV.

To reduce the incidence of high-energy transition events, we ran another series of TAS simulations with the same stop-search parameter but at lower temperature $T_2 = 1800$ K. Among 10 transition events recorded, the energies of the destination states were as follows: -0.4054 eV (5 times), -0.3919 eV (3 times),

---

[8] The existence of states A_1 and A_2 may be an artifact of the SW potential.

0.7956 eV (once), and 1.4670 eV (once). Discarding the two states with the highest energies, the rest of the sampled destination states are familiar. The state with energy -0.3919 eV is state B (LK) already encountered in Sections 7.3 and 7.4, Fig. 7.5(c). The state with energy -0.4054 eV is the same as B but translated to the left by one lattice vector; we label this state C.[9]

Now that several metastable states around state A have been identified, it is possible to apply either of the methods discussed in Section 7.3 to obtain the minimum-energy paths connecting these states. Fig. 7.8(d) shows results obtained by the NEB path relaxation. Consistent with the earlier observation that transitions among A, A_1 and A_2 are much more frequent than between A and B or C, the energy barrier between A and A_1 (or A_2) is lower than that between A and B (or C). A more interesting finding is that the MEP connecting A_1 (or A_2) and B (or C) passes through A as an intermediate state. This means that, at low temperatures, the rate of escapes to state B (or C) from any one of the three A states (A or A_1 or A_2) is controlled by the energy barrier between A and B (or C), at 0.47 eV. Thus, even though two additional metastable states A_1 and A_2 were discovered in the TAS simulations, the overall rate of kink migration is still controlled by the energy barrier between states A and B. This conclusion is fully consistent with the path-annealing simulation described in the previous section.

Summary

- The method of temperature-accelerated sampling (TAS) can be used to search for escape paths from a given energy basin to unknown destination states.

- To ensure that TAS finds escape paths relevant for the low temperature of interest, it is necessary to generate multiple escape trajectories and locate their saddle-points.

---

[9] The small difference between the energies of states B and C is caused by the presence of another kink (RK') on the same dislocation line in our simulation cell.

PART II

CONTINUUM MODELS

8

PEIERLS–NABARRO MODEL OF DISLOCATIONS

Chapter 1 introduced dislocations as dual objects permitting both atomistic and continuum descriptions. The subsequent Chapters 2 through 7 discussed various aspects of atomistic simulations and their application to dislocation modeling. In the rest of the book, from Chapter 8 to Chapter 11, we will be treating dislocations as continuum objects. This is a huge simplification that makes it possible to consider dislocation behavior on length and time scales well beyond reach of the atomistic simulations. The following chapters are organized in the order of increasing length and time scales. This particular chapter deals with the famous Peierls–Nabarro continuum model that is most closely related to the atomistic models discussed earlier.

Fundamentally, dislocations are line defects producing distortions in an otherwise perfect crystal lattice. While this point of view is entirely correct, the atomistic models of dislocations can deal with only relatively small material volumes where every atom is individually resolved. Furthermore, having to keep track of all these atoms all the time limits the time horizon of atomistic simulations (see Chapter 7). On the other hand, when the host crystal is viewed as an elastic continuum, the linear elasticity theory of dislocations offers a variety of useful analytical and numerical solutions that are no longer subject to such constraints. Although quite accurate far away from the dislocation center, where the lattice distortions remain small, continuum theory breaks down near the dislocation center, where lattice discreteness and non-linearity of interatomic interactions become important. To obtain a more efficient description of crystal dislocations, some sort of bridging between the atomistic and continuum models is necessary. For example, it would be very useful to have a hybrid continuum–atomistic approach such that it retains the analytic nature of the continuum theory for the long-range elastic fields but also captures the essential non-linear effects in the atomic core. Bearing the names of Rudolf Peierls [86] and Frank Nabarro [87], the celebrated Peierls–Nabarro (PN) model is one such approach.

Possibly the most attractive feature of the PN model is its simplicity. In some cases, the model can be solved analytically, providing a good starting point for understanding the interaction between the elastic fields of dislocations and the discrete crystal lattice. In other cases, no analytical solutions exist but numerical solutions are easy to obtain. Either way, the PN model holds considerable pedagogical value since it has relatively few parameters and their effect on dislocation behavior is easy to examine and understand.

The PN model has been influential in the development of dislocation theory and continues to affect the way scientists think about dislocations today. Since the 1940s it has been used extensively to study dislocation core properties. Although direct atomistic simulations gradually supplanted the PN model in this role, the PN model is still widely used, referred to and discussed in the literature. In fact, simulations based on the PN approach have experienced a revival in recent years, owing largely to the availability of first principles methods that provide more accurate atomistic inputs to the model.

It is somewhat unconventional to present the PN model after having already discussed the atomistic models. The advantage of this approach is that it should allow us to critically examine the assumptions made or implied in the formulation of the PN model. In Section 8.1, we develop the PN model using a variational formulation, which is also different from the original formulation based on the classic Peierls–Nabarro equation for stress equilibrium. Instead, we arrive at the same equation by taking a variational derivative of the dislocation energy functional. Section 8.2 discusses practical techniques for solving the PN model numerically in 1D. In Section 8.3, the model is extended to the case of two displacement components. Section 8.4 discusses further generalizations of the PN model.

8.1 Model Formulation

8.1.1 Volterra Model Versus Peierls–Nabarro Model of Dislocations

Let us take another look at the familiar straight edge dislocation originally considered by Volterra. In Fig. 8.1, a cut is introduced from the left along plane $y = 0$ such that the boundary of the cut is along the z axis (out of plane). Assigning the displacement fields on the upper and lower surfaces of the cut to be $u^+(x)$ and $u^-(x)$ respectively, the Volterra dislocation is created by displacing the lower surface with respect to the upper surface by $b\,\mathbf{e}_x$ and then reconnecting the two surfaces together. Choosing the line direction $\boldsymbol{\xi}$ to be along the negative z axis (into the plane), the Burgers vector of this dislocation is $\mathbf{b} = b\,\mathbf{e}_x$.

Let us define $u(x) \equiv u^+(x) - u^-(x)$ as the *disregistry* or *misfit* across the cut plane. Then, in the Volterra dislocation, $u(x)$ is a step function,

$$u(x) = \begin{cases} -b, & x < 0 \\ 0, & x > 0. \end{cases} \tag{8.1}$$

Equivalently, the distribution of disregistry across the plane of the cut is

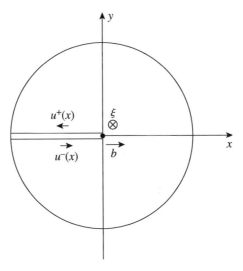

FIG. 8.1. The Volterra edge dislocation inserted along the negative z axis at the origin $x = y = 0$.

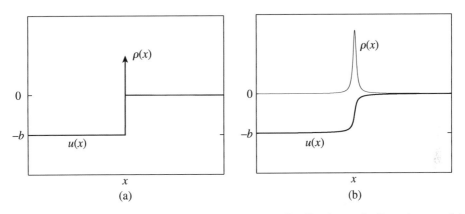

FIG. 8.2. (a) In the Volterra dislocation, the distribution of disregistry $u(x)$ across the cut plane is a step function, while the density of this distribution $\rho(x) = du(x)/dx$ is a delta function. (b) In the PN dislocation, both $u(x)$ and $\rho(x)$ are smooth functions.

described by the disregistry density $\rho(x) \equiv du(x)/dx$, which, for the Volterra dislocation, is a delta function, $\rho(x) = b\,\delta(x)$, Fig. 8.2(a). This simple discontinuous distribution is a useful idealization leading to simple analytic solutions for the elastic fields produced by the Volterra dislocation. However, the same solutions for the stress and strain fields are singular along the dislocation line $(x = y = 0)$ itself. This behavior is unphysical and the PN model removes the singularity by allowing the Burgers vector distribution to spread out, as shown in Fig. 8.2(b). Furthermore, the actual shape of the distribution function $\rho(x)$ is selected in the PN model to take into account the non-linear interactions in the dislocation core.

The basic premise of the PN model is very simple: the range in which the non-linearity of the interatomic interaction is prominent is limited to a small volume around the geometrical center of the dislocation. The idea, reportedly due to E. Orowan, is to treat the non-linear core in a special way, while keeping with the linear elasticity everywhere else. Accordingly, the total energy of a dislocation is written as the sum of two contributions, $E_{tot} = E_{el} + E_{msft}$, in which E_{el} is the elastic energy and E_{msft} is an extra energy term associated with the non-linear misfit in the core. The actual shape of the distribution function $\rho(x)$ is such that it minimizes the total energy E_{tot}. As we shall see below, the elastic energy decreases, while the misfit energy increases with the increasing width of the misfit distribution $\rho(x)$. The shape of this distribution is determined by the best compromise between these two opposing energy terms.

8.1.2   Elastic Energy of the PN Dislocation

Let us consider the elastic energy term first. It can be obtained by considering a reversible path of dislocation insertion into a continuum linear elastic solid and by measuring the work done along the way (see Problem 8.1.1):

$$E_{el} = \frac{1}{2} \int_{-\infty}^{\infty} \sigma_{xy}(x) u(x) \, dx. \tag{8.2}$$

The solution for the stress produced by the singular Volterra dislocation shown in Fig. 8.1 is very simple,

$$\sigma_{xy}^{0}(x) = \frac{\mu b}{2\pi(1-v)} \frac{1}{x}. \tag{8.3}$$

For the PN dislocation whose disregistry density is $\rho(x)$, the stress $\sigma_{xy}(x)$ is simply the convolution of $\sigma_{xy}^{0}(x)$ with $\rho(x)$, i.e.

$$\sigma_{xy}(x) = \frac{\mu}{2\pi(1-v)} \int_{-\infty}^{\infty} \frac{\rho(x')}{x-x'} \, dx'. \tag{8.4}$$

Substituting this expression into eq. (8.2) and integrating by parts, the elastic energy becomes

$$E_{el} = \frac{\mu}{4\pi(1-v)} \int_{-\infty}^{\infty} \int_{-\infty}^{\infty} \frac{\rho(x')u(x)}{x-x'} \, dx \, dx'$$

$$= -\frac{\mu}{4\pi(1-v)} \int_{-\infty}^{\infty} \int_{-\infty}^{\infty} \rho(x)\rho(x') \ln|x-x'| \, dx \, dx' + C, \tag{8.5}$$

where C is a constant whose magnitude is independent of the shape of the disregistry distribution as long as the values of $u(x)$ at two integration limits $x = \pm\infty$ are fixed. More generally

$$E_{el}[u(x)] = -K \int_{-\infty}^{\infty} \int_{-\infty}^{\infty} \rho(x)\rho(x') \ln |x - x'| \, dx \, dx' + C, \qquad (8.6)$$

where $K = \mu/(4\pi)$ for a screw dislocation and $K = \mu/(4\pi(1 - \nu))$ for an edge dislocation.[1] Obviously, the elastic energy is minimized ($E_{el} \to -\infty$) when the dislocation spreads out such that $\rho(x) \to 0$ everywhere, even though the normalization condition $\int_{-\infty}^{\infty} \rho(x) \, dx = b$ is still satisfied. Of course, in this limit the dislocation can no longer be identified. Countering the spreading of the misfit in the PN dislocation is a special misfit energy term.

8.1.3   The Misfit Energy

The PN model relies on a misfit energy term E_{msft} to account, in a limited way, for the non-linear interatomic interactions in the dislocation core. To appreciate where the misfit energy comes from, consider a crystallographic plane dividing the crystal into two halves. Now imagine that the two half-crystals slide with respect to each other uniformly by vector $\mathbf{u}_0$ along the plane. In a continuum solid, such uniform sliding produces no elastic distortion and thus costs no energy for as long as the cut is perfectly planar and the shift is uniform. However, real crystals are made up of discrete atoms whose relative positions across the cut plane depend on the two-dimensional vector $\mathbf{u}_0$. To describe the energy cost incurred as a result of the shift, it is convenient to use the energy γ per unit area of the cut as a function of the shift vector $\mathbf{u}_0$. Function $\gamma(\mathbf{u}_0)$ is commonly referred to as the *generalized stacking fault* (GSF) energy or simply the γ-surface. In the context of the PN model, $\gamma(\mathbf{u}_0)$ is sometimes called the *misfit potential*. Because of the periodic nature of the crystal lattice, $\gamma(\mathbf{u}_0)$ is a periodic function of two components of the shift vector $\mathbf{u}_0$.

The misfit potential $\gamma(\mathbf{u}_0)$ is relatively easy to compute, given an atomistic model (see Problem 8.1.4). For simplicity, let us limit the discussion to the situation where $\mathbf{u}_0$ is along the x axis. Then the period of $\gamma(\mathbf{u}_0)$ along in the x direction is equal to the Burgers vector b of the dislocation under consideration. The simplest

---

[1] The resulting expression (8.6) is equal to the same integrals evaluated from $-R$ to R and taken in the limit $R \to \infty$. In the same limit, $C \to Kb^2 \ln R$. The latter is a manifestation of the well-known result that the elastic energy of an infinite straight dislocation diverges logarithmically with respect to the size of the host solid R. The actual value of C is inconsequential when comparing the energies of different core configurations of the same dislocation. For practical purposes, we can simply regard C as a large positive constant.

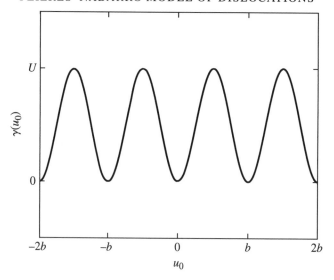

FIG. 8.3. A sinusoidal misfit potential γ-surface in the form of eq. (8.7).

form of $\gamma(u_0)$ is a sinusoid, Fig. 8.3,

$$\gamma(u_0) = \frac{U}{2}[1 - \cos(2\pi u_0/b)]. \qquad (8.7)$$

Even though in the PN model the misfit $u(x)$ varies along the cut plane, the misfit energy is expressed as

$$E_{\text{msft}}[u(x)] = \int_{-\infty}^{\infty} \gamma(u(x))\,\mathrm{d}x. \qquad (8.8)$$

This integral form assumes that the misfit energy density at point x is a function of the local misfit at the same point. This locality assumption is an approximation that is expected to be valid if $u(x)$ varies slowly as a function of x. In dislocations with a narrow core, $u(x)$ changes rapidly from $-b$ to 0 and the validity of this approximation is questionable. Notice that E_{msft}, as defined in eqs. (8.7) and (8.8), is minimized for the Volterra dislocation because its disregistry $u(x)$ is a step function: in this case E_{msft} is zero.

Combining eqs. (8.6) and (8.8), we arrive at an expression for the total energy of the PN dislocation,

$$E_{\text{tot}}[u(x)] = -K \int_{-\infty}^{\infty} \int_{-\infty}^{\infty} \rho(x)\rho(x') \ln|x - x'|\,\mathrm{d}x\,\mathrm{d}x'$$

$$+ \int_{-\infty}^{\infty} \gamma(u(x))\,\mathrm{d}x + C, \qquad (8.9)$$

where $\rho(x) = \mathrm{d}u(x)/\mathrm{d}x$. This completes the formulation of the PN model.

8.1.4   The Analytic Solution

The solution of the PN model is a misfit distribution function $u(x)$ that minimizes the total energy functional eq. (8.9). In general, there are two approaches to finding a solution. The first is to take the variational derivative of the energy functional to obtain an equation that the optimal solution must satisfy and then solve this equation. We will take this approach in this section. The second approach is to obtain a solution by direct minimization of eq. (8.9). This second approach will be discussed in the next section.

Assume that $u(x)$ is the yet unknown solution, subject to the boundary conditions of $u(-\infty)=-b$, $u(\infty)=0$. Let us add to it a small variation $\delta u(x)$ such that $\delta u(-\infty) = \delta u(\infty) = 0$. The corresponding variation in E_{tot} must be zero, leading to the following integro-differential equation,[2] (Problem 8.1.2):

$$0 = \frac{\delta E_{\text{tot}}}{\delta u(x)} = 2K \int_{-\infty}^{\infty} \frac{\rho(x')\,dx'}{x - x'} + \frac{d\gamma(u(x))}{du(x)}. \tag{8.10}$$

This equation has the following physical interpretation. At each position x, the first term on the right-hand side of eq. (8.10) represents the Peach–Koehler force due to the local elastic stress which, in equilibrium, must balance the force produced by the gradient of the misfit potential (the second term). When the misfit potential γ takes the form of eq. (8.7), this equation becomes

$$-2K \int_{-\infty}^{\infty} \frac{\rho(x')\,dx'}{x - x'} = \frac{U\pi}{b} \sin\left(\frac{2\pi u(x)}{b}\right). \tag{8.11}$$

Peierls found the following simple analytical solution of this equation (see Problem 8.1.5):

$$u(x) = \frac{b}{\pi} \arctan\left(\frac{x}{\xi}\right) - \frac{b}{2}, \tag{8.12}$$

$$\rho(x) = \frac{b}{\pi} \frac{\xi}{x^2 + \xi^2}, \tag{8.13}$$

where $\xi = Kb^2/(U\pi)$ can be interpreted as the width of the dislocation core. The stress field of this dislocation along the x axis is

$$\sigma_{xy}(x) = \frac{\mu b}{2\pi(1 - \nu)} \frac{x}{x^2 + \xi^2}, \tag{8.14}$$

which is no longer singular. Notice that ξ is proportional to K but inversely

---

[2] For more discussions on the variational derivatives, see Section 11.1.

proportional to U. Thus, a stronger elastic interaction (large K) produces a wider dislocation core, whereas a stronger misfit interaction (large U) leads to a narrower core.

Summary

- The Peierls–Nabarro model removes the singularity in the Volterra dislocation by allowing the Burgers vector to spread out around the dislocation center.

- The total energy of a PN dislocation is the sum of a linear elastic energy term and a non-linear misfit energy term. The optimal misfit distribution minimizes the PN energy functional.

Problems

8.1.1. Derive the energy functional in eq. (8.2). The elastic energy equals the reversible work spent on creating the dislocation. To compute this work, imagine that traction forces are applied to both sides of the cut plane and introduce displacement fields $u^+(x)$ and $u^-(x)$. The traction forces should exactly balance the internal stress field for the process to be reversible.

8.1.2. Obtain the PN integro-differential equation, eq. (8.10), by taking the first variational derivative of the energy functional in eq. (8.9).

8.1.3. Assume that the density of a screw dislocation is uniformly distributed in an interval of width ξ, i.e. $\rho(x) = b/\xi$ when $x \in [-\xi/2, \xi/2]$ and $\rho(x) = 0$ otherwise. Compute the elastic energy E_{el} of this dislocation using eq. (8.6).

8.1.4. Compute the γ-surface from an atomistic model of FCC copper. In FCC crystals, the γ-surface of the {111} planes is most interesting because dislocations dissociate on {111} planes. Build a small fragment of the perfect FCC crystal in a periodic box defined by

$$\mathbf{c}_1 = 4a[112] \tag{8.15}$$

$$\mathbf{c}_2 = 4a[\bar{1}10] \tag{8.16}$$

$$\mathbf{c}_3 = 4a[111], \tag{8.17}$$

where a is the lattice constant. Perform shift along the {111} plane by redefining vector $\mathbf{c}_3$ so that

$$\mathbf{c}_3 = \mathbf{c}_3 + \delta\mathbf{c}_1. \tag{8.18}$$

Use the EAM glue potential for copper [132] (available in MD + +) to compute the energy as a function of δ, allowing for relaxation of atomic positions as well as c_3 in the direction normal to the (111) planes.

8.1.5. Verify that eq. (8.12) is indeed the solution of the Peierls–Nabarro equation (8.11). Hint: first prove the identity

$$\int_{-\infty}^{\infty} \frac{dt}{(x-t)(t^2 + \xi^2)} = \frac{\pi x}{(x^2 + \xi^2)\xi}.$$

8.1.6. Because the same interatomic interaction is responsible both for the γ-surface and for the elastic constants, the two properties are related to each other in the limit of small strain. Specifically, shear modulus μ is proportional to the second derivative $d^2\gamma/du^2$. To see this, consider a uniform shear strain such that all atomic planes are shifted with respect to their adjacent planes by a small amount u. Let d be the distance between the adjacent planes. The energy increase per unit volume caused by this deformation is $E = \frac{1}{2}(d^2\gamma/du^2) \cdot u^2/d$. In linear elasticity theory, the same energy is expressed as, $E = \frac{1}{2}\mu \cdot (u/d)^2$. Therefore, $d^2\gamma/du^2 = \mu/d$. Assuming that function γ is a sinusoid, as in eq. (8.7), express U and the core width ξ in terms of μ, b and d.

8.1.7. The singularity in the center of the Volterra dislocation continuum elasticity theory is usually removed using a core cut-off radius r_c, so that the dislocation's energy is written as $E = Kb^2 \ln(R/r_c) + E_{\text{core}}(r_c)$. From eq. (8.9) compute the total energy of the Peierls dislocation. Find a value of the core energy E_{core} that matches the energy of the Peierls dislocation.

8.2   Numerical Solutions

Another way to obtain a solution of the PN model is by direct numerical minimization of its energy functional $E_{\text{tot}}[u(x)]$. The first step is to "guess" a functional form for the solution $u(x)$ that should include a few numerical parameters, say $\{\alpha_i\}$. Upon substitution into eq. (8.9), the total energy functional becomes a function of parameters $\{\alpha_i\}$ that can be minimized using numerical methods, such as the CGR algorithm (Section 2.3). The minimum energy obtained in this way provides an upper bound of the true energy of the PN dislocation. The closer the trial function is to the true solution, the tighter the upper bound.

Mathematical formulations possessing such properties are called *variational* and hold considerable advantages over non-variational formulations. In a variational formulation, it is possible to systematically improve the upper bound on the minimal energy by allowing more flexibility in the trial functions. Furthermore, the solutions of variational problems tend to be numerically stable. What is also important for us here is that the direct minimization approach can be applied to energy functionals with more complex misfit potentials for which analytic solutions do not exist. But first, to illustrate how the method works, let us consider the same sinusoidal misfit potential as used before.

8.2.1   The Sinusoid Misfit Potential

It is convenient to limit our choices to trial functions $u(x)$ that automatically satisfy boundary conditions at both ends of the x axis, $u(-\infty) = -b$ and $u(\infty) = 0$, regardless of the values of their numerical parameters. Suppose that the guess solution takes the following form (a very lucky guess!):

$$u(x) = \frac{b}{\pi} \arctan\left(\frac{x}{c}\right) - \frac{b}{2}. \tag{8.19}$$

This function satisfies the boundary conditions by construction and we would like to find the optimal value of parameter c. After this trial function has been plugged into eq. (8.9), the energy functional E_{tot} becomes a function of c, i.e.

$$E_{tot}(c) = -Kb^2 \ln 2c + U\pi c + C, \tag{8.20}$$

which is plotted as a solid line in Fig. 8.4(a). Obviously, the energy minimum is achieved when $c = \xi = Kb^2/(U\pi)$, in agreement with the exact solution.

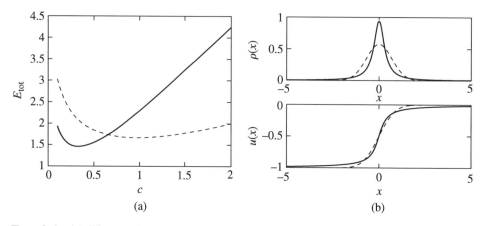

(a)   (b)

FIG. 8.4. (a) The total PN energy as a function of the dislocation width parameter c, with $U = 1$, $K = 1$, $b = 1$ (here constant C is zero). (b) The misfit function $u(x)$ and its derivative $\rho(x)$ obtained for the optimal value of c. The solid line is for $u(x) = (b/\pi)\arctan(x/c) - b/2$ and the dashed line is for $u(x) = (b/2)\,\text{erf}(x/c) - b/2$.

Therefore, when the trial function happens to match the form of the exact solution, subsequent parameter optimization should reproduce the exact solution.

The direct optimization approach is most valuable when the functional form of the exact solution is unknown, which is nearly always the case. Then, the "best" solution within the chosen parametric family of trial functions provides an upper bound on the true E_{tot}. The value of the optimized energy itself provides a measure of the quality of the selected trial function. As an example, let us try another functional form,

$$u(x) = \frac{b}{2} \operatorname{erf}\left(\frac{x}{c}\right) - \frac{b}{2},$$

$$\rho(x) = \frac{b}{\sqrt{\pi} c} \exp\left(-\frac{x^2}{2c^2}\right).$$

In this case, E_{tot} as a function of parameter c is computed by numerical integration. The result is plotted as the dashed line in Fig. 8.4(a). The minimum value of E_{tot} obtained for this trial function is slightly higher than the exact solution. This indicates that this function is probably acceptable as an approximation but is not the optimal functional form. The "best" approximate solution for $u(x)$ and $\rho(x)$ within this functional form is compared to the exact solution in Fig. 8.4(b).

8.2.2 Misfit Potential with Intermediate Minima

The variational property of the energy formulation is invaluable when it comes to more complex problems with realistic misfit potentials $\gamma(u_0)$. If appropriate trial functions are chosen, the quality of the solution can be systematically improved, as quantified by the reduction in the total energy. As an example, consider the more complex misfit potential shown in Fig. 8.5(a) and given by the following expression:

$$\gamma(u_0) = \frac{U}{0.14}\left[a - \frac{\exp(1.09a) - 1}{2}\right] \tag{8.21}$$

$$a \equiv \frac{1 - \cos(2\pi u_0)}{2}.$$

This function not only has global minima at $u_0 = nb$ (n integer), it also has local minima at $u_0 = -(n + 1/2)b$. This mimics the features of γ-surfaces in FCC, HCP and some other crystals (see the next section for a more realistic model). For this misfit potential, it is probably impossible to find an analytic solution of the PN model and we resort to numerical methods.

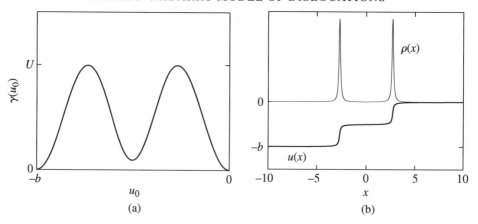

FIG. 8.5. (a) A single period of the model misfit potential $\gamma(u_0)$ given in eq. (8.21). This potential has a local minimum at $u_0 = -b/2$. (b) Optimal numerical solutions for $\rho(x)$ and $u(x)$.

Again, first we should "guess" a suitable functional form for the trial solution $u(x)$. Motivated by the canonical analytical solution, eq. (8.12), let us express $u(x)$ as a superposition of several arctan functions, i.e.

$$u(x) = \frac{b}{\pi} \sum_{k=1}^{m} A_k \arctan \frac{x - x_k}{c_k} - \frac{b}{2}, \tag{8.22}$$

$$\rho(x) = \frac{b}{\pi} \sum_{k=1}^{m} \frac{A_k c_k}{(x - x_k)^2 + c_k^2}, \tag{8.23}$$

subject to the constraint

$$\sum_{k=1}^{m} A_k = 1, \tag{8.24}$$

so that the boundary conditions $u(-\infty) = -b$ and $u(\infty) = 0$ are automatically satisfied.

The next step is to express the total energy as a function of parameters A_k, c_k and x_k. The elastic-energy term can be obtained analytically,

$$E_{\text{el}} = -\frac{Kb^2}{2} \sum_{i=1}^{m} \sum_{k=1}^{m} A_i A_k \ln[(x_i - x_k)^2 + (c_i + c_k)^2] + C, \tag{8.25}$$

whereas the misfit energy is more complicated and should be evaluated numerically.

For faster numerical convergence, it is useful to change the integration variable as follows:

$$x \to L \tan \eta \tag{8.26}$$

$$\int_{-\infty}^{+\infty} dx \to L \int_{-\pi/2}^{\pi/2} \frac{d\eta}{\cos^2 \eta}. \tag{8.27}$$

Here, L is a parameter that defines the length of interval over which numerical integration of the misfit energy requires small spacing along the x axis. This interval should comfortably cover the expected spread of the dislocation core, over which $u(x)$ changes appreciably. In most situations $L = 100b$ should suffice. After this transformation, even a simple Simpson integration [28] over approximately $N = 1000$ equally spaced grid points η_j in the domain $\eta \in [-\pi/2, \pi/2]$ is sufficient for accurate evaluation of the misfit energy integral. The discretized version of the misfit energy integral is

$$\tilde{E}_{\text{msft}} = \sum_{j=1}^{N} \gamma(u(L \tan \eta_j)) \frac{L \Delta \eta}{\cos^2 \eta_j}, \tag{8.28}$$

where $\Delta \eta$ is the spacing between the grid points on the η axis. From eqs. (8.25) and (8.28), E_{tot} can be calculated for any parameter set $\{A_k, x_k, c_k\}$. The derivatives of E_{tot} with respect to $\{A_k, x_k, c_k\}$ can also be obtained, making it possible to use the efficient minimization algorithms such as CGR (Section 2.3).

As an example, consider the case of $K = 1$, $U = 1$, $b = 1$ and a trial form of $u(x)$ containing only two arctan functions. The misfit potential is integrated numerically over $N = 10^4$ equally spaced grid points along the η axis. Starting from the initial values of variables at $A_1 = A_2 = 0.5, c_1 = c_2 = 1, x_1 = -1, x_2 = 1$, a CGR minimization converges to $A_1 = A_2 = 0.5, c_1 = c_2 = 0.08560, -x_1 = x_2 = 2.6958$ and yields the minimum energy of $E_{\text{tot}} = 1.0398$ (constant C is ignored). The optimized $\rho(x)$ and $u(x)$ functions are plotted in Fig. 8.5(b). The most distinct feature of $u(x)$ is that the two arctan functions move apart while leaving the misfit at a plateau value of $u_0 = b/2$ in the middle. A displacement profile like this is typical for dissociated dislocations in FCC, HCP and some other crystals. Each peak of $\rho(x)$ corresponds to a *partial* dislocation while the plateau in the middle corresponds to a stable stacking fault whose energy per unit area is $\gamma(b/2)$. The origin of this dissociation is obviously the existence of the intermediate minimum of $\gamma(u_0)$ at $u_0 = b/2$.

Multiple *local* minima of the total energy may exist. The solution the CGR algorithm converges to may depend on the initial values of parameters A_k, x_k, c_k. For example, starting from a different set of initial values for the positions of two arctan components, $x_1 = x_2 = 0$, the CGR minimization converges to a different solution: $A_1 = A_2 = 0.5$, $c_1 = c_2 = 0.17505$, $x_1 = x_2 = 0$. The energy of this solution is higher, $E_{tot} = 2.0502$, and it corresponds to a non-dissociated dislocation core. This second solution is thermodynamically unstable with respect to spontaneous core dissociation. Given this possibility of multiple solutions, the solution with the lowest energy E_{tot} is usually of interest. The *globally optimal* solution can be obtained using the methods of global minimization, as was discussed in Section 2.3.

To assess the quality of our approximate solution, let us add one more arctan term to the trial function $u(x)$ and search for an improved solution, now in the space of nine parameters. Using the previously optimized values of the first six parameters as initial conditions, let us initialize three new parameters at $A_3 = 0$, $c_3 = 10$, $x_3 = 0$. The new converged solution is $A_1 = A_2 = 0.4976$, $c_1 = c_2 = 0.08125$, $-x_1 = x_2 = 2.7447$, $A_3 = 4.747 \times 10^{-3}$, $c_3 = 6.8082$, $x_3 = 0$, with $E_{tot} = 1.0377$. The reduction of total energy from the previous case ($E_{tot} = 1.0388$) is very small, suggesting that the addition of the third arctan term does not improve the solution appreciably. In fact, the converged value for A_3 is close to zero, indicating that the two-arctan representation of $u(x)$ was already quite accurate.

Summary

- The PN model can be solved numerically using a trial function $u(x)$ with free parameters. The optimal misfit distribution is determined by minimization of the total energy with respect to the parameters.

- The energy obtained by numerical optimization provides an upper bound to the exact solution. It also serves as a measure of the quality of the chosen trial functions.

8.3   Extension to Two Displacement Components

In the two preceding sections, we limited the misfit vector **u** to have only one non-zero component along the Burgers vector. In the geometry where the dislocation line is aligned along the z axis, the Burgers vector is perpendicular to the z axis for an edge dislocation and parallel to it for a screw dislocation. In the PN model, the only difference between these two cases is the elastic prefactor, which is $K = \mu/(4\pi(1-\nu))$ for an edge dislocation and $K = \mu/(4\pi)$ for a screw dislocation. However, the local disregistry **u** does not have to be parallel to the Burgers vector everywhere in the dislocation core. More generally, **u** is a two-dimensional vector in the slip plane (x, z) and the misfit potential is a function of its two components, i.e. $\gamma(u_x, u_z)$.

Taking into account both components of the misfit vector is critical in many important crystal structures, such as face-centered-cubic (FCC) or

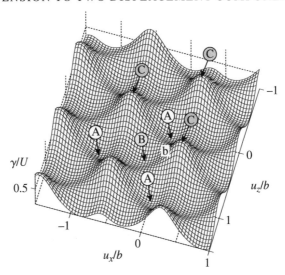

FIG. 8.6. γ-surface of the form of eq. (8.30) for the {111} planes in FCC crystals. The deep minima A, the shallow minima B and the local maxima C form three triangular lattices. Every triangular lattice is at the center of two other lattices—see Fig. 8.7 for a more detailed view.

hexagonal-close-packed (HCP) crystals, where dislocations tend to split into partials. In the preceding section, we observed that a full dislocation can disso- ciate into two partial dislocations if the misfit potential has a local minimum at $u_x = b/2$. In that case, the Burgers vectors of both partials were exactly half of the total Burgers vector, $b_{p1} = b_{p2} = b/2$. However, in real crystals, the Burgers vectors of partial dislocations rarely point in the same direction as the total Burgers vector. This is because the dissociation is ultimately caused by the existence of an intermediate minimum of the misfit potential, but this minimum corresponds to a disregistry vector $\mathbf{u}_{SF}$ that is not parallel to the total Burgers vector $\mathbf{b}$.

   In this section, we consider an extension of the PN model that includes both components of the disregistry vector u_x and u_z. We then use the extended model to describe the dissociation of perfect dislocations into two partials often observed on the {111} planes in FCC crystals. Fig. 8.6 shows the misfit potential to be used in this study. It is constructed to account for the existence of stable stacking faults in FCC crystals (see below).

Stacking Fault and Partial Dislocations in FCC Crystals The two-dimensional misfit potential function must share the symmetry of the {111} planes in FCC crystals. The atomic arrangement on the {111} planes in a perfect FCC crystal is illustrated in Fig. 8.7(a). Three distinct types of atomic planes, marked A, B and C,

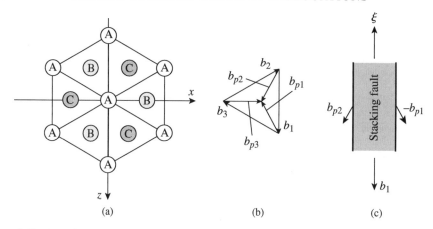

FIG. 8.7. (a) The stacking pattern of the {111} planes in FCC crystals. (b) Perfect
Burgers vectors $\mathbf{b}_1$, $\mathbf{b}_2$, $\mathbf{b}_3$ and partial Burgers vectors $\mathbf{b}_{p1}$, $\mathbf{b}_{p2}$, $\mathbf{b}_{p3}$ on the
{111} plane. (c) A perfect dislocation with total Burgers vector $\mathbf{b}_1$ dissociates
into two partial dislocations with Burgers vectors $-\mathbf{b}_{p1}$ and $\mathbf{b}_{p2}$ bounding an
area of stacking fault.

are stacked upon one another in a repeating sequence ($ABCABC \cdots$) to form the
FCC structure. The atoms in each plane, e.g. plane A, align with the centers of
triangles formed by the atoms on two underlying planes B and C.

Consider a planar cut between plane A (above) and plane C (below) and displace
the atoms above the cut by $\mathbf{u}$ with respect to the atoms below the cut. The misfit
energy must be zero when $\mathbf{u}$ is equal to any one of three perfect Burgers vec-
tors marked $\mathbf{b}_1$, $\mathbf{b}_2$ and $\mathbf{b}_3$ in Fig. 8.7(b). In this case, each A atom moves to
a neighboring A site—the corresponding energy minima of the misfit potential
function are marked A in Fig. 8.6.

Now assume that $\mathbf{u}$ equals either $\mathbf{b}_{p1}$, $\mathbf{b}_{p2}$ or $\mathbf{b}_{p3}$, in which case the atoms above
the cut plane move to different sites, $A \to B$, $B \to C$, $C \to A$. Such a shift pro-
duces a new stacking sequence $\cdots ABC|BCABC \cdots$, where | indicates position
of the cut plane. The energy of interfaces created by shifts like this is often low
because the neighborhood of every atom remains "nearly the same" as before the
shift. Even the atoms immediately above and below the cut have the same number
of nearest neighbors located at the same distance as in the perfect crystal. A small
energy increase results from the second-nearest-neighbor interactions. The interface resulting at the cut plane is an (intrinsic) *stacking fault*. The misfit potential
should have local minima at $\mathbf{u} = \mathbf{b}_{p1}$, $\mathbf{b}_{p2}$ and $\mathbf{b}_{p3}$, and these minima are marked
B in Fig. 8.6. Their energy is equal to the energy of (intrinsic) stacking fault γ_{SF}.
Perfect dislocations in FCC crystals with low stacking-fault energy can dissociate
into two partial dislocations separated by a stacking fault. An example of this is

shown in Fig. 8.7(c), where the decomposition of the perfect dislocation may be written as $\mathbf{b}_1 = -\mathbf{b}_{p1} + \mathrm{SF} + \mathbf{b}_{p2}$. The disregistry vector over the SF area remains close to $\mathbf{u} = \mathbf{b}_{p1}$.

When $\mathbf{u}$ equals $-\mathbf{b}_{p1}$, $-\mathbf{b}_{p2}$ or $-\mathbf{b}_{p3}$, the atoms in plane A immediately above the cut plane move exactly on top of the atoms in plane C immediately below the cut. The resulting repulsion between the atoms in planes A and C causes the energy of these states to become high, resulting in maxima marked C in Fig. 8.6.

To conform to the symmetry of the (111) planes in FCC crystals, the misfit potential should satisfy the following periodicity conditions:

$$\gamma(u_x, u_z) = \gamma(u_x, u_z + b) = \gamma(u_x + b\sqrt{3}/2, u_z + b/2). \tag{8.29}$$

As shown in Fig. 8.6, the misfit energy function $\gamma(u_x, u_z)$ must have minima of zero energy at $(u_x, u_z) = (0, 0)$, $(0, b)$, $(b\sqrt{3}/2, b/2)$, ..., minima of energy $\gamma_{SF} > 0$ at $(u_x, u_z) = (-b\sqrt{3}/6, -b/2)$, $(b\sqrt{3}/3, 0)$, ... and maxima of energy at $(u_x, u_z) = (b\sqrt{3}/6, b/2)$, $(-b\sqrt{3}/3, 0)$, ... For example, the following function satisfies the above conditions:

$$\gamma(u_x, u_z) = U[(1 - \alpha)E_1 + \alpha E_2] \tag{8.30}$$

$$E_1 = [f(2u'_x) + f(u'_x + u_z) + f(u'_x - u_z)]/3$$

$$E_2 = g(2u''_x) \cdot g(u''_x + u_z) \cdot g(u''_x - u_z) \cdot 4/(3\sqrt{3}) + 1/2$$

$$f(x) \equiv \frac{e^{\cos(2\pi x)} - e^{-1/2}}{e - e^{-1/2}}$$

$$g(x) \equiv \sin(\pi x)$$

$$u'_x \equiv u_x/\sqrt{3} - 2/3$$

$$u''_x \equiv u_x/\sqrt{3} - 1/3.$$

The parameter α conveniently defines the magnitude of the stacking fault energy, $\gamma_{SF} = U\alpha/2$. The topography of this function is shown in Fig. 8.6(b) for the case of $\alpha = 0.1$.

Consider again a straight PN dislocation parallel to the z axis. This time, let us allow the disregistry vector to have two components: $u_x(x)$ for the edge component and $u_z(x)$ for the screw component. Assuming that the total Burgers vector of this dislocation is $\mathbf{b}_1 = b\mathbf{e}_z$ (screw), the components of $\mathbf{u}$ must satisfy the following boundary conditions:

$$u_x(-\infty) = 0$$

$$u_x(+\infty) = 0$$

$$u_z(-\infty) = -b$$

$$u_z(+\infty) = 0.$$

In this geometry the elastic fields produced by the edge and screw components of the disregistry vector do not interact, so that the PN energy of this dislocation is a simple extension of the classical one-dimensional model:

$$E_{tot}[u_x(x), u_z(x)] = -K_e \int_{-\infty}^{\infty} \int_{-\infty}^{\infty} \rho_x(x)\rho_x(x') \ln|x - x'| \, dx \, dx'$$

$$- K_s \int_{-\infty}^{\infty} \int_{-\infty}^{\infty} \rho_z(x)\rho_z(x') \ln|x - x'| \, dx \, dx'$$

$$+ \int_{-\infty}^{\infty} \gamma(u_x(x), u_z(x)) \, dx + C, \tag{8.31}$$

where $K_s = \mu/(4\pi)$, $K_e = \mu/(4\pi(1 - \nu))$, and C is the same constant as before.

8.3.1 Numerical Solutions

Motivated by the analytic solution of the classical PN model, let us use the following trial functions:

$$u_x(x) = \frac{b}{\pi}\left(A_1 \arctan \frac{x - x_1}{c_1} + A_2 \arctan \frac{x - x_2}{c_2}\right) \tag{8.32}$$

$$u_z(x) = \frac{b}{\pi}\left(A_3 \arctan \frac{x - x_3}{c_3} + A_4 \arctan \frac{x - x_4}{c_4}\right) - \frac{b}{2},$$

subject to the constraints $A_1 + A_2 = 0$, $A_3 + A_4 = 1$, to satisfy the boundary conditions.

As an example, let us set $K_s = 1$, $\nu = 0.3$, $U = 1$, $b = 1$ and minimize the total energy functional, eq. (8.31), using the familiar CGR algorithm. The initial values of the parameters and their final converged values after the relaxation are given in the table below. The resulting solutions for the components of disregistry are

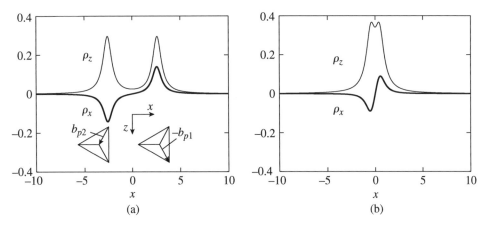

FIG. 8.8. (a) Solution of the two-dimensional PN model at $\alpha = 0.1$, i.e. $\gamma_{SF} = 0.05U$. (b) Solution of the two-dimensional PN model at $\alpha = 0.6$, i.e. $\gamma_{SF} = 0.3U$.

shown in Fig. 8.8(a) in terms of disregistry density functions $\rho_x(x) = du_x(x)/dx$ and $\rho_z(x) = du_z/dx$.

	$-A_1 = A_2$	$c_1 = c_2$	$-x_1 = x_2$	$A_3 = A_4$	$c_3 = c_4$	$-x_3 = x_4$
Initial	0	1	0.1	0.5	0.1	1
Final	0.2539	0.5616	2.5537	0.5	0.5260	2.5761

The resulting solution for the Burgers vector distribution has two peaks, indicating dissociation into two partial dislocations. It is reasonable to attribute the integral disregistry in the interval $-\infty < x < 0$ to the left partial ($\mathbf{b}^{\text{left}}$) and in the interval $0 < x < \infty$ to the right partial ($\mathbf{b}^{\text{right}}$), i.e.

$$b_x^{\text{left}} = \int_{-\infty}^{0} \rho_x(x)\,dx \quad b_x^{\text{right}} = \int_{0}^{\infty} \rho_x(x)\,dx$$

$$b_z^{\text{left}} = \int_{-\infty}^{0} \rho_z(x)\,dx \quad b_z^{\text{right}} = \int_{0}^{\infty} \rho_z(x)\,dx. \tag{8.33}$$

Because of the boundary conditions, $b_x^{\text{left}} + b_x^{\text{right}} = 0$ and $b_z^{\text{left}} + b_z^{\text{right}} = b$. At the same time, the integration yields $b_x^{\text{left}} = -0.2189b$ and $b_z^{\text{left}} = 0.5b$ for the optimized disregistry distribution shown in Fig. 8.8(a), which is very close to $\mathbf{b}_{p2} = (-b\sqrt{3}/6, b/2)$ in Fig. 8.7(b). Likewise, $\mathbf{b}^{\text{right}}$ is found to be very close to $-\mathbf{b}_{p1} = (b\sqrt{3}/6, b/2)$. Therefore, our two-component PN model predicts dissociation of a perfect screw dislocation with Burgers vector $\mathbf{b}_1$ into two partial dislocations with Burgers vectors $-\mathbf{b}_{p1}$ and $\mathbf{b}_{p2}$, exactly as shown in Fig. 8.7(c). The separation between the two partials is approximately $x_4 - x_3 = 5.15$, which is close to the value $X_0 = 5.24$ predicted for the same partials but treated as two singular Volterra dislocations (Problem 8.3.1).

To appreciate the effect of the stacking-fault energy on dislocation core, let us now repeat the same calculation with $\alpha = 0.6$, which corresponds to $\gamma_{SF} = 0.3U$. The optimized disregistry distribution obtained for this higher value of γ_{SF} is shown in Fig. 8.8(b). The separation between two partials is now much smaller, $x_4 - x_3 = 0.97$. Still, this is remarkably close to the prediction for Volterra partial dislocations at $X_0 = 0.87$. This agreement is somewhat surprising because it is difficult to consider two partials as separate dislocations when their separation is comparable to their half widths, $c_3 = c_4 = 0.53$.

8.3.2 Stress Effects

It is of interest to see if our two-component PN model can describe the effects of applied stress on the dislocation core structure. Consider, for example, the effect of stress component σ_{xy} on the dissociated dislocation shown in Fig. 8.7(c). According to the Peach–Koehler formula (Section 1.3), σ_{xy} interacts only with the b_x components of the Burgers vectors of two partials. Because they are equal but of opposite sign, this stress component exerts no net force on the dislocation couple. However, depending on the sign of this stress, it can pull the two partials together or push them apart – this stress component is sometimes called *Escaig stress*.[3]

To include the effects of stress in the PN model, let us first notice that under applied stress, the system should evolve towards the minimum of its total enthalpy. This is the system's energy minus the work done by the applied stress. If $\tau = -\sigma_{xy}$, the enthalpy is

$$H_{\text{tot}}[u_x(x), u_z(x)] = E_{\text{tot}}[u_x(x), u_z(x)] + \tau \int_{-\infty}^{\infty} u_x(x') \, dx', \qquad (8.34)$$

where E_{tot} is the PN energy given in eq. (8.31). In equilibrium, the disregistry vector should be distributed in such a way that minimizes H_{tot}.

As an example, consider the effect of Escaig stress in an FCC crystal with high stacking-fault energy, $\gamma_{SF} = 0.3U$ ($\alpha = 0.6$). The optimal misfit distribution under zero stress has already been shown for a screw dislocation in this material in Fig. 8.8(b). The partials are barely separated from each other. Under an applied stress of $\tau = 0.2$, the partials move farther apart to $x_4 - x_3 = 1.28$, which is comparable to the Volterra model prediction $X_0 = 1.08$. However, agreement between the more accurate PN model and the predictions obtained for Volterra

---

[3] The component of stress that exerts a net force on the total dislocation is often referred to as *Schmid* stress.

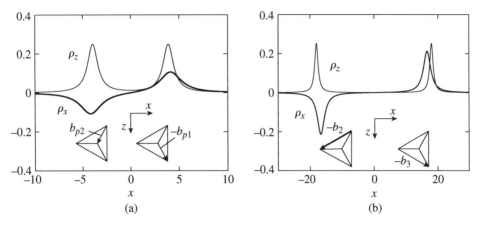

FIG. 8.9. (a) The solution of the two-component PN model at $\alpha = 0.6$ and stress $\tau = 0.732$. (b) A snapshot obtained during CGR relaxation under stress $\tau = 0.733$: the dislocation spontaneously splits into two perfect dislocations moving apart in the opposite directions.

partials becomes progressively worse under increasing Escaig stress. Shown in Fig. 8.9(a) is the distribution obtained under stress $\tau = 0.732$, where two dislocations are separated by $x_4 - x_3 = 7.86$. For the same stress, the Volterra model predicts a much narrower separation of $X_0 = 2.95$.

It turns out that $\tau = 0.732$ is just below the critical stress at which the core of the extended dislocation undergoes a dramatic transformation. While the PN model captures this transition, the Volterra dislocation model does not. Starting from the parameter values obtained by relaxation under $\tau = 0.732$, under a slightly higher stress $\tau = 0.733$ the relaxation takes a very long time, over which the enthalpy drops precipitously. At the same time, the two dislocations are seen to gradually move far apart. To see more clearly what is happening, it is useful to interrupt the relaxation at some iteration, fix the dislocation positions (x_3 and x_4) and relax the remaining parameters under zero stress. The resulting configuration is shown in Fig. 8.9(b). The two dislocations are indeed far away from each other, but what is more interesting is that their edge components have grown significantly. For example, for the dislocation on the left $b_x^{\text{left}} = -0.8495b$ and $b_z^{\text{left}} = 0.5b$, which is no longer consistent with the partial Burgers vector $\mathbf{b}_{p2}$. Instead, this disregistry is very close to the perfect Burgers vector $-\mathbf{b}_2$. Similarly, $\mathbf{b}^{\text{right}}$ has grown to be close to $-\mathbf{b}_3$. This means that the perfect screw dislocation $\mathbf{b}_1$ is predicted to split under stress into two perfect 60° dislocations! This interesting behavior is not an artifact of our PN model, as a very similar behavior has been observed in full atomistic simulations (see [41] and Problem 8.3.4). Whether or not such splitting can occur in a given FCC crystal depends on the shape of its misfit potential γ (see Problem 8.3.5).

Summary

- To describe extended dislocations in FCC, HCP and other similar materials, the PN model is extended to treat both components of the misfit vector simultaneously.

- The core structure of extended dislocations in FCC materials depends on the stacking-fault energy and the applied stress. Under zero stress, the width of the stacking fault predicted by the PN model is in agreement with the estimate given by the continuum theory of Volterra dislocations.

- The theory of Volterra dislocations becomes increasingly inaccurate under high stress. In contrast, the PN model remains qualitatively accurate over a wide range of stress, where it agrees closely with the atomistic calculations.

Problems

8.3.1. Consider a screw dislocation b dissociated into two 30° partials. The Burgers vector of the partials are: $b_x = \pm b\sqrt{3}/6$ and $b_z = b/2$. When the partials are at the equilibrium separation X_0, their elastic repulsion exactly balances the attractive force due to the stacking-fault energy. Observing that the force (per unit length) acting between two edge dislocations is $[\mu/2\pi(1-\nu)]b_x^2/X$ while the force between two screw dislocations is $(\mu/2\pi)b_z^2/X$, show that the equilibrium separation of two 30° partials in a screw dislocation is $X_0 = pKb^2/(\gamma_{\rm SF} - \tau q b)$, where $p = 1/2 - 1/[6(1-\nu)]$, $q = \sqrt{3}/6$, and τ is the xy component of the applied stress. Compute the stress τ_∞ under which the partials should separate completely.

8.3.2. Repeat the calculation leading to Fig. 8.8 for different γ-potentials with $\alpha = 0.2, 0.3, 0.4, 0.5$. Compute and plot the predicted separation between two partials $x_4 - x_3$ as a function of $\gamma_{\rm SF}$. On the same figure, plot the corresponding prediction from the singular elasticity theory of Volterra dislocations. Compute the difference between the two predictions.

8.3.3. Repeat the calculation that leads to Fig. 8.9 for $\alpha = 0.6$ and $\tau = 0.1, 0.2, \ldots, 0.7$. For each α, plot the predicted separation between two partials, $x_4 - x_3$, as a function of τ. Compare the result with the predictions from the singular theory of Volterra dislocations.

8.3.4. In this problem, the task is to observe spontaneous splitting of a perfect dislocation into two perfect dislocations in FCC aluminum, using an atomistic simulation [41].

Build a small fragment of the perfect FCC crystal in a periodic supercell with repeat vectors $c_1 = 6[11\bar{2}]$, $c_2 = 12[111]$, $c_3 = 2[1\bar{1}0]$. Insert a screw dislocation dipole with Burgers vector $\mathbf{b} = [1\bar{1}0]/2$ separated

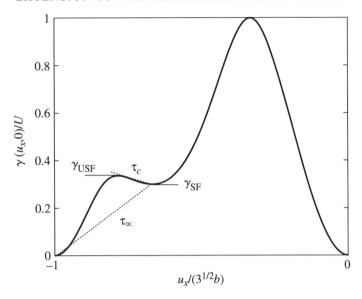

FIG. 8.10. The cross-cut along the u_x axis of the two-dimensional γ-surface computed for $\alpha = 0.6$ (see Problem 8.3.5).

by $\mathbf{a} = \mathbf{c}_2/2$ (see Chapter 3). Relax the structure using the CGR algorithm and the Ercolessi–Adams EAM potential [132]. Apply a stress, σ_{xy}, in increments of 100 MPa, each time followed by a CGR minimization using the Parrinello–Raman boundary condition. Compute the stress τ_c under which the dislocation splits into two perfect dislocations. Use the two-component PN model to predict the critical stress τ_c at which the splitting takes place. Compare the two predicted values.

8.3.5. In FCC crystals, the topography of the γ-surface has much to do with the behavior of the dislocation core under stress. Fig. 8.10 plots γ as a function of u_x for $u_z = 0$. The local minimum in the middle corresponds to a meta-stable stacking fault with the energy γ_{SF}. The local maximum to the left is the so-called *unstable* stacking fault with the energy γ_{USF}. Continuum theory of Volterra dislocations predicts that the separation between two partials should become infinite at stress $\tau_\infty = 2\gamma_{SF}/b_p$. On the other hand, the PN model predicts that the partials should evolve into two perfect dislocations under stress τ_c, which is approximately equal to the maximum slope of the γ-function between γ_{SF} and γ_{USF}. When $\tau_c < \tau < \tau_\infty$, formation of two perfect dislocations preempts the separation of partial dislocations. On the other hand, when $\tau_\infty < \tau < \tau_c$, two dislocations should remain partials but separate indefinitely leaving a wide stacking fault between them. Compute and plot $\gamma(u_x, 0)$ for $\alpha = 0.3$ and $\alpha = 0.7$. Compute τ_c and τ_∞ for both misfit potentials based on the above arguments. Run PN calculations under increasing stress τ and observe what happens in each case.

8.4   Further Extensions

The purpose of this section is to point out directions for possible further development of the PN model that will make it more realistic and extend its range of applicability.

8.4.1   Lattice Resistance to Dislocation Motion

One notable, but sometimes overlooked, deficiency of the original model is that the PN energy does not depend on the dislocation position. Indeed, the energy given in eq. (8.9) is invariant with respect to shifting the misfit field $u(x)$ by an arbitrary distance x_0, i.e. $E_{tot}[u(x - x_0)] = E_{tot}[u(x)]$. In its original formulation, the PN model disregards the discrete nature of the host crystal lattice. Realizing that it is precisely the lattice discreteness that gives rise to the resistance to dislocation motion, Nabarro proposed replacing the continuous integral in the misfit energy term by a discrete sum in which the misfit potential $\gamma(u_0)$ is sampled only at the actual positions of atoms immediately adjacent to the slip plane [87]. The total energy of the PN dislocation thus becomes

$$E_{tot}[u(x)] = -K \int_{-\infty}^{\infty} \int_{-\infty}^{\infty} \rho(x)\rho(x') \ln|x - x'| \, dx \, dx'$$

$$+ \sum_{n=-\infty}^{\infty} \gamma(u(na_p))a_p + C, \tag{8.35}$$

where a_p is the lattice period along the x axis. Assuming that, irrespective of the dislocation position x_0, the disregistry distribution remains the same as the classical solution, i.e. $u(x) = (b/\pi) \arctan[(x - x_0)/\xi)]$, Nabarro evaluated the misfit energy analytically as a function of x_0. The so-computed misfit energy varies periodically as a function of x_0 with period a_p, as it should.

   Rather than assuming that the classical solution remains valid for the discrete form of the misfit energy, it is more consistent to find an optimal solution for the "discrete" functional (8.35). More importantly, it should be possible to allow the distribution $u(x)$ to vary as a function of dislocation position x_0. With these issues taken into account, analytic solutions are unlikely to be applicable, but the numerical methods introduced in this chapter work quite well. As an example, a more consistent calculation of the energy variations with dislocation position is given in Problem 8.4.2, where the (discretized) PN energy is minimized with respect to all other parameters except for the dislocation position x_0. The amplitude of the energy variation, from minimum to maximum, is the *Peierls barrier* (see Section 1.3). Likewise, a more consistent calculation of the Peierls stress is given in Problem 8.4.3. The Peierls stress is identified with the critical stress at which the Peierls barrier is completely removed.

8.4.2 Non-locality of Dislocation Energy

Even if direct atomistic simulations have gradually replaced the PN model for studying dislocation core structure and lattice resistance to dislocation motion, the PN model has recently experienced a revival. This is largely due to the development of accurate first principles methods (Section 2.1) for computing γ-surfaces in real crystals. It was even argued that the PN model based on an accurate γ-surface input may compete with the direct atomistic simulations, whose own validity is limited by the accuracy of empirical interatomic potentials. However, the PN model has its own sources of inaccuracy in the explicit and implicit assumptions made in its formulation.

A major assumption in the original PN model is that the density of misfit energy at point x depends only on the local misfit $u(x)$ at the same point. This assumption of locality is expected to be appropriate when $u(x)$ varies slowly as a function of x. However, in a dislocation whose core is narrow, $u(x)$ changes rapidly within the core. In such a case, the locality assumption may well be inaccurate and a gradient correction to the PN energy may be required [88].

8.4.3 More Complex Dislocation Geometries

While the original PN model was constructed for straight dislocations in which disregistry is confined to a single glide plane, it has been subsequently generalized to treat more complex situations [89]. Extension of the PN model to curved dislocation lines (e.g. planar dislocation loops) leads to a formulation similar to the phase field model to be discussed in Chapter 11 [90, 91, 92]. The PN model has been also extended to deal with situations when the dislocation core spreads into several planes [93]. Finally, the dynamic behavior of dislocations moving at a constant speed has been examined within the PN model [94].

Summary

- To describe the lattice resistance to dislocation motion, lattice discreteness needs to be accounted for in the PN model.

- The locality assumption in the PN model may be inaccurate for dislocations with narrow cores.

Problems

8.4.1. Assuming the sinusoidal form of the misfit potential (8.7) and using $u(x) = b \arctan((x - x_0)/c)/\pi - b/2$ as the trial function, minimize the discretized PN energy (8.35) with respect to parameters x_0 and c by conjugate gradient relaxation. Use $K = 1$, $U = 1$, $b = 1$, and $a_p = 1$. Make sure that the sum is taken over discrete values $x = \ldots, -2a_p, -a_p, 0, a_p, 2a_p, \ldots$. Observe that, with the discretized form of the misfit energy term, the optimal dislocation position x_0 is mid-way between the atomic rows (e.g. $x_0 = a_p/2$). Compare the optimized dislocation width c with that from the analytic solution of the continuum PN model ($\xi = 1/\pi$).

8.4.2. Using the same set-up as in Problem 8.4.1, displace x_0 from its optimal solution in steps of 0.01, each time followed by a complete relaxation of parameter c. Plot c and the total energy as functions of x_0. The difference between the maximum and minimum of the curve $E_{tot}(x_0)$ is the *Peierls barrier* E_{PN} (see Section 1.3). The maximum slope of this curve $(dE_{tot}/dx_0)_{max}$ is related to the *Peierls stress* (see Problem 8.4.3). Compute $E_{tot}(x_0)$ again, but now using the classical solution of the continuum PN model $u(x) = b \arctan((x - x_0)/c)/\pi - b/2$ with c fixed at $1/\pi$. Compare the solution obtained for the semi-discrete PN model, in which c is allowed to relax, with the solution of the continuum PN model, in which $c = 1/\pi$. Repeat calculations for $U = 10$ keeping all other parameters the same. Is the agreement between the continuum and semi-discrete solutions better in this case?

8.4.3. Under applied stress τ the system evolves towards a minimum of its enthalpy $H_{tot} = E_{tot} + \tau \int u(x) \, dx$ (here, $\tau = -\sigma_{xy}$ when $\mathbf{u}$ is along x and $\tau = -\sigma_{yz}$ when $\mathbf{u}$ is along z). Parameterize the disregistry in the same way as in Problem 8.4.2, minimize H_{tot} with respect to c, and plot the result as a function of x_0 for several values of stress τ. When τ reaches a critical value, the barrier to dislocation motion ceases to exist since no single point on the $H_{tot}(x_0)$ curve has a positive slope. The stress at which the barrier disappears is the *Peierls stress* τ_P (see Section 1.3). Determine τ_P by minimizing the semi-discrete version of H_{tot} with respect to both x_0 and c. Ramp up the stress in increments of 0.02 and observe that, above a certain value of stress, minimization no longer converges, while x_0 drifts towards very large values. This is a signal that the system has lost stability under stress exceeding τ_P. Compare $\tau_P b$ with the maximum slope $(dE_{tot}/dx_0)_{max}$ as well as with $E_{PN}\pi/a$, where dE_{tot}/dx_0 and E_{PN} are obtained in Problem 8.4.2.

8.4.4. Examine the effect of lattice discreteness on the Peierls barrier. Intuitively, the barrier should vanish in the limit of $a_p/\xi \to 0$, where $\xi = Kb/(U\pi)$ is the dislocation width obtained in the continuum PN model. Set $K = 1$ and $U = 1$ and compute the Peierls barrier E_{PN} for $a_p = 0.2, 0.4, \ldots, 1.2$, following the same procedure as in Problem 8.4.2. Plot E_{PN} as a function of a_p. Compute E_{PN} for $U = 0.2, 0.4, \ldots, 1.2$ with $K = 1$ and $a_p = 1$. Plot E_{PN} as a function of U. Repeat the same calculations using the classic solution of the continuum PN model and compare the results.

9

KINETIC MONTE CARLO METHOD

The PN model discussed in the preceding chapter is a continuum approach that requires some atomistic input to account for non-linear interactions in the dislocation core. In this chapter, we introduce yet another continuum model that uses atomistic input for a different purpose. The kinetic Monte Carlo (kMC) model[1] does not consider any details of the core structure but instead focuses on dislocation motion on length and time scales far greater than those of the atomistic simulations. The model is especially effective for diamond-cubic semiconductors and other materials in which dislocation motion is too slow to be observed on the time scale of molecular dynamics simulations. The key idea of the kMC approach is to treat dislocation motion as a stochastic sequence of discrete rare events whose mechanisms and rates are computed within the framework of the transition state theory (Chapter 7). Built around its unit mechanisms, the kMC model simulates dislocation motion and predicts dislocation velocity as a function of stress and temperature. This data then can be used to construct accurate mobility functions for dislocation dynamics simulations on still larger scales (Chapter 10). In this sense, kMC serves as a link between atomistic models and coarse-grained continuum models of dislocations.

The kMC approach is most useful in situations where the system evolves through a stochastic sequence of events with only a few possible event types. The method has been used in a wide variety of applications other than dislocations. For example, the growth of solid thin films from vapor or in solution is known to proceed through attachment and diffusion of adatoms deposited on the surface. Based on a finite set of unit mechanisms of the motion of adatoms, kMC models accurately describe the kinetics of growth and the resulting morphology evolution of the epitaxial films [95, 96, 97].

Similar kMC models have been applied to dislocation motion in crystals with high lattice resistance, such as silicon. In these materials, dislocations consist of long straight segments interspersed with atomic-sized kinks, depicted

---

[1] Just as in the Metropolis Monte Carlo approach discussed in Section 2.4, random numbers are used to generate a stochastic sequence of moves in the kinetic Monte Carlo method. However, the two Monte Carlo methods are quite different. Whereas the whole purpose of the Metropolis Monte Carlo is to generate a random sequence of states that sample the equilibrium distribution, kMC simulations are concerned with the system's evolution in real time.

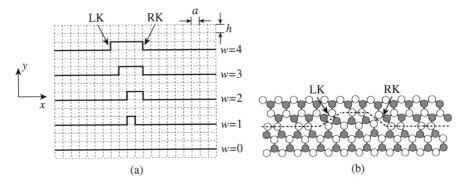

FIG. 9.1. (a) A dislocation with kinks is represented as a collection of horizontal and vertical line segments on a regular grid. The short vertical segments on the lines are kinks. Notations LK and RK are used to differentiate between left and right kinks of the kink pair separated by distance w. The line moves upward through a sequence involving kink pair nucleation and growth. (b) Atomic core structure of a $30°$ partial dislocation in silicon containing a kink pair. The dashed line marks the dislocation core position.

schematically in Fig. 9.1(a) as short vertical segments. As was explained in Section 1.3, dislocation motion proceeds through nucleation and migration of kink pairs and can be described well by a kMC model.

One aspect of kMC simulations that is particular to dislocations is the need to account for the elastic interaction among the segments of the dislocation line. This aspect makes the kMC method numerically demanding. To simplify the discussion, the first section of this chapter describes a kMC model of dislocation motion that ignores the long-range interactions altogether. The effects of elastic interactions between the dislocation segments are dealt with in two subsequent sections.

9.1 Non-interacting Model

In crystals with strong lattice resistance to dislocation motion, dislocation lines extend along certain crystallographic directions (Peierls valleys, Section 1.3), except for locations of atomistic-sized bends or *kinks*. Such a dislocation can be represented by a continuous piecewise straight line on a rectangular grid, as illustrated in Fig. 9.1(a). Here, a and h are the grid spacings in the horizontal (x) and vertical (y) directions, respectively. In the figure, the kinks are represented by vertical segments with zero width. As demonstrated in Fig. 9.1(b), kinks on dislocations in real crystals have finite width. However, the approximation of sharp kinks is appropriate when the kink width is much smaller than other relevant length scales, such as the distance between two kinks in a kink pair. We will continue to use the sharp kink approximation throughout this chapter.

When the stress on a dislocation is higher than its Peierls stress (Section 1.3), the dislocation can move continuously. However, the behavior is different when the stress is below the Peierls stress. Now the dislocation stays at rest in a given lattice position (Peierls valley) most of the time, interrupted by short bursts of thermally assisted nucleation of closely spaced kink pairs, such as shown in Fig. 9.1(b). Under the combined influence of stress and thermal activation, the kinks may diffuse apart along the line making the entire dislocation line translate to the next lattice position. Similar to many other thermally activated processes, kink pair nucleation and, in some cases, kink migration events are stochastic in nature. The line will experience net translation in a certain direction when one of the two directions of kink pair nucleation and propagation is favored by the applied stress. Let us regard the favored direction as forward, and the reverse as backward. In Fig. 9.1, "forward" means upward. Thus the motion of a left kink (LK) to the left or a right kink (RK) to the right both result in the dislocation moving forward.

Transition-state theory expresses the rates of kink pair nucleation and kink migration events in terms of their respective energy barriers, which can be computed using the atomistic simulation techniques discussed in Chapter 7. Let J_+^n and J_-^n be the rates of kink pair nucleation events in the forward and backward directions, and J_+^m and J_-^m for the kink migration events. From transition state theory,

$$J_\pm^n = \nu_0 \exp \left[\frac{-E_{nuc} + TS \pm \sigma abh/2}{k_B T} \right] \tag{9.1}$$

$$J_\pm^m = \nu_0 \exp \left[\frac{-W_m + TS \pm \sigma abh/2}{k_B T} \right], \tag{9.2}$$

where ν_0 is a frequency pre-factor,[2] E_{nuc} and W_m are the appropriate energy barriers (computed under zero stress), T is the temperature, S is the entropy, σ is the local stress on the dislocation line, and b is the Burgers vector. The term TS accounts for the difference between the Helmholtz free energy and the energy (Chapter 6). Here, entropy S is taken to be $3k_B$ [134]. The last term in the numerator represent the work done by the applied stress. The factor $1/2$ reflects the assumption that only half the work is done when the system reaches the transition state, which is midway between the initial and final states.

There are several analytic models, including the kink diffusion model [7], that relate dislocation mobility to the rates of kink nucleation and migration.

---

[2] ν_0 can be approximated by the Debye frequency, which is related to the Debye temperature θ_D by $h\nu_0 = k_B \theta_D$, where h is Planck's constant. A table of Debye temperatures for various elements can be found in [9]. For example, for silicon $\theta_D = 645$ K and $\nu_0 = 1.344 \times 10^{13}$ Hz.

Unfortunately, the existing models rely on various approximations to make analytical solutions possible. The purpose of this chapter is to construct a *numerical* kMC model around the same rate parameters to avoid the additional approximations invoked in the development of the analytical models. As we will show, such a model overcomes certain limitations of the analytical models and can account for more realistic and complicated behaviors, such as the motion of dissociated dislocations [134] and three-dimensional mobility of screw dislocation in BCC metals [98].

In this section, we introduce a simplified kinetic Monte Carlo model in which the long-range elastic interactions among dislocation segments are ignored. This model gives us a chance to define the state of a moving dislocation, consider transitions from one state to another, and introduce the notions of stochastic processes and *Markov chains*. In the next section we will discuss a more realistic kMC model complete with the long-range elastic interactions and the numerical complications these interactions cause.

9.1.1   Model Formulation

The first step in building our model will be to specify the configuration of a dislocation line. In our kMC model, only the y coordinates of the N horizontal segments need to be defined. Neighboring horizontal segments with different y coordinates will be connected by vertical segments. The dislocation line is assumed to be periodic along the x direction. Thus, a N-dimensional vector $\{y_1, y_2, \ldots, y_N\}$ completely defines the instantaneous state of the line. A change in the vertical position of any line segment can be regarded as a transition from one state to another.[3]

As illustrated in Fig. 9.2, each event of kink pair nucleation or kink migration amounts to a transition between two states such that one of N horizontal segments

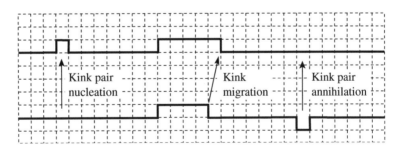

FIG. 9.2. Depending on the positions of its neighboring segments, an upward motion of a horizontal segment corresponds to the events of kink pair nucleation (left), kink migration (middle) or kink pair annihilation (right).

---

[3] This model is a one-dimensional version of the solid-on-solid (SOS) model for thin film growth [97].

(say segment i) moves up or down by h. This can be written as

$$\{y_1, y_2, \ldots, y_i, \ldots, y_N\} \to \{y_1, y_2, \ldots, y_i \pm h, \ldots, y_N\}. \qquad (9.3)$$

Each of the $2N$ possible transitions from a given state is characterized by its own rate,

$$\{r_1^+, r_1^-, r_2^+, r_2^-, \ldots, r_N^+, r_N^-\}, \qquad (9.4)$$

where $r_i^+ (r_i^-)$ is the rate (probability per unit time) for segment i to move up (down) by h. Next, we will relate rates $r_i^\pm$ to the rates of kink pair nucleation and kink migration given in eqs. (9.1) and (9.2).

Even though we decided to ignore for now the long-range elastic interactions between the kinks, it is still necessary to make the rates $r_i^\pm$ for each segment i depend on the positions of its neighboring segments $i-1$ and $i+1$. This is because, depending on the relative positions of segments $i-1$, i and $i+1$, the motion of segment i may correspond to kink pair nucleation, kink migration or kink recombination events. For example, consider a configuration in which $y_i = y_{i-1} = y_{i+1}$. In this case, the motion of segment y_i up or down corresponds to a kink pair nucleation event (on the left in Fig. 9.2) and the rates for this case are $r_i^\pm = J_\pm^n$. However, when $y_{i-1} - 1 = y_i = y_{i+1}$ the motion of segment i upwards by h represents a right kink translation to the right by a (in the middle of Fig. 9.2) and $r_i^+ = J_+^m$. When $y_{i-1} = y_i + 1 = y_{i+1}$, the upward motion of segment i represents kink pair annihilation (on the right in Fig. 9.2). In this case, it makes sense to double the rate to $r_i^+ = 2J_+^m$ since either of the two kinks can migrate, moving segment i forward.

To prevent the lines from becoming excessively jagged, let us only consider transitions that will not result in the formation of kinks with height more than h.[4] This exclusion rule is enforced by setting $r_i^- = 0$ if $y_{i-1} > y_i$ or $y_{i+1} > y_i$ and $r_i^+ = 0$ if $y_{i-1} < y_i$ or $y_{i+1} < y_i$. The following algorithm summarizes our procedure for calculating jump rates $r_i^\pm$ for segment i depending on the relative positions of segments $i-1$, i and $i+1$.

Algorithm 9.1

    1. Compute $d_1 := (y_{i-1} - y_i)/h$, $d_2 := (y_{i+1} - y_i)/h$, $d := d_1 + d_2$.

    2. If $d_1 = 0$ and $d_2 = 0$, assign $r_i^+ := J_+^n$ and $r_i^- := J_-^n$.

    3. If $d > 0$, assign $r_i^+ := J_+^m \cdot |d|$ and $r_i^- := 0$.

---

[4] This rule will become unnecessary once we turn the elastic interactions on in the next section because the elastic interaction makes kinks of like signs strongly repel each other at close distances.

4. If $d < 0$, assign $r_i^+ := 0$ and $r_i^- := J_-^m \cdot |d|$.

5. If $d_1 = -d_2 \neq 0$, then $r_i^+ := 0$ and $r_i^- := 0$.

Together with equations (9.1) and (9.2), this algorithm can be used to compute rates of all allowed transitions from any given state $\{y_1, y_2, \ldots, y_N\}$. The set of allowed transitions together with their rates is sometimes called the *event catalog*.

9.1.2 Markov Chains

Let us take a closer look at the nature of stochastic evolution in our model. To simplify the notation, we will use bold symbols, such as $\mathbf{i}$ or $\mathbf{j}$, to represent the state of the entire system, i.e. $\{y_1, y_2, \ldots, y_N\}$. Furthermore, we shall assume that the rates of all possible transitions from a given state are completely determined by the state itself. A stochastic sequence of states produced by transitions that have this property is called a *Markov* stochastic process or a *Markov chain*. The statistical properties of a Markov chain are completely specified by its transition rate matrix $J(\mathbf{i}, \mathbf{j})$, each element of which is the transition rate from state $\mathbf{i}$ to state $\mathbf{j}$.[5] The total escape rate from state $\mathbf{i}$ is

$$R(\mathbf{i}) = \sum_{\mathbf{j}} J(\mathbf{i}, \mathbf{j}), \qquad (9.5)$$

where the sum is over all states accessible by transitions from state $\mathbf{i}$. For a Markov chain, the escape rate from state $\mathbf{i}$ does not change as long as the system still remains in $\mathbf{i}$. Therefore, the system's residence time $\tau_{\mathbf{i}}$ in state $\mathbf{i}$ is a random variable that satisfies the exponential distribution

$$f(\tau_{\mathbf{i}}) = R(\mathbf{i})e^{-R(\mathbf{i})\tau_{\mathbf{i}}}. \qquad (9.6)$$

Once the system escapes from state $\mathbf{i}$, the probability that it will arrive at state $\mathbf{j}$ is

$$\pi(\mathbf{i}, \mathbf{j}) = \frac{J(\mathbf{i}, \mathbf{j})}{\sum_{\mathbf{k}} J(\mathbf{i}, \mathbf{k})}. \qquad (9.7)$$

Equations (9.6) and (9.7) form the basis for the following kinetic Monte Carlo algorithm.

9.1.3 Kinetic Monte Carlo Algorithm

In the Metropolis Monte Carlo algorithm (Section 2.4), a trial move is often rejected. However, in the widely used kinetic Monte Carlo algorithm [99] every attempted move is successful, leading to a change in the system's state. This enhanced

---

[5] This is analogous to transition probability matrix $\pi(\mathbf{R}, \mathbf{R}')$ in the Metropolis Monte Carlo algorithm (Section 2.4), except that now J is a probability per unit time.

efficiency is achieved by always selecting an increment of time over which the next transition takes place. Given a current state $\{y_1, \ldots, y_N\}$, the total rate of escape from this state is

$$R = \sum_{i=1}^{N} (r_i^+ + r_i^-). \tag{9.8}$$

The time τ that the system resides in the current state before the next transition is randomly selected from the exponential distribution, $f(\tau) = Re^{-R\tau}$. At time τ, the system moves to one of its $2N$ accessible states. Which state the system arrives at is selected according to the contributions of each transition to the total transition rate R. By repetition of this process, a stochastic trajectory in the space of states $\mathbf{i}$ is generated. Statistical properties of the model system are then obtained by averaging over one or several such trajectories. The following is a possible implementation of this efficient algorithm.

Algorithm 9.2

1. Initialize time $t := 0$. Initialize the state of the dislocation as a straight line, e.g. $y_i := 0$ for $i = 1, \ldots, N$.

2. Use Algorithm 9.1 and eqs. (9.1) and (9.2) to compute $2N$ rates of transitions out of the current state.[6] Reorder the computed transition rates into a one-dimensional array $\{r_j\}$, $j = 1, \ldots, 2N$, where $r_{2i-1} \equiv r_i^+$, $r_{2i} \equiv r_i^-$, $i = 1, \ldots, N$.

3. Compute the total transition rate $R := \sum_{j=1}^{2N} r_j$.

4. Draw a random number ξ_1 uniformly distributed in $[0, 1]$ and obtain $\Delta t := -R^{-1} \ln \xi_1$. This maps the uniform distribution for random variable ξ_1 onto the exponential distribution $f(\Delta t) = Re^{-R\Delta t}$. Increment time $t := t + \Delta t$.

5. For each transition $k = 1, 2, \ldots, 2N$, compute the partial sum $s_k := \sum_{j=1}^{k} r_j/R$. (Obviously, $s_{2N} = 1$.)

6. Draw another random number ξ_2 from a uniform distribution in $[0, 1]$ and find k such that $s_{k-1} < \xi_2 \leq s_k$ ($s_0 = 0$). Label event k as selected.

7. Execute the selected event k. Let $p := \mod(k, 2)$, $n := (k + p)/2$. If k is odd, move segment n forward, $y_n := y_n + 1$. If k is even, move segment n backward, $y_n := y_n - 1$.

8. If the maximum number of iterations is reached, stop. If not, go to step 2.

[6] To simplify bookkeeping, here we keep the number of transitions constant ($2N$) and enforce the exclusion rules by setting the rates of excluded transitions to zero.

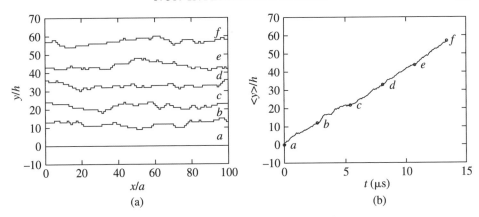

FIG. 9.3. (a) Snapshots of dislocation configurations taken at different times during
kMC simulation at $T = 1000$ K, $\sigma = 30$ MPa. (b) The instantaneous mean dislocation position as a function of time in the same simulation. Positions marked
a–f correspond to the snapshots in (a).

Let us now use this algorithm to simulate a dislocation of total length $L = 100a$
moving under stress $\sigma = 30$ MPa at temperature $T = 1000$ K. The following activation parameters are selected close to the values cited in the literature for the partial dislocations in silicon: $E_{nuc} = 1.4$ eV, $W_m = 1.2$ eV,
$\nu_0 = 1.344 \times 10^{13}$ Hz [134]. The geometric parameters correspond to a 30° partial
dislocation oriented along the [110] direction in the $(1\bar{1}1)$ plane with $a = 3.84$ Å,
$h = a\sqrt{3}/2$ and $b = a/6[121]$.[7] Fig. 9.3(a) shows several snapshots of dislocation
configurations attained during one such simulation. After a brief transient period,
steady forward motion sets in where the total rate of kink pair nucleation events
is balanced by the rate of kink annihilations. In this regime, the line contains on
average ~30 kinks. Fig. 9.3(b) is a plot of the instantaneous mean dislocation
position $\langle y \rangle$ as a function of time. The slope of this curve is the average dislocation velocity, $v \approx 1.4 \times 10^{-3}$ m s^{-1}. This value is markedly higher than the
dislocation velocities observed in silicon under similar conditions. The reason for
this discrepancy will be explained in the next section. To reduce variations in the
predicted velocity and to estimate its statistical errors, several kMC runs using
different sequences of random numbers can be performed.

To streamline our discussion, the algorithm described above is made
intentionally very simple. It turns out that in most situations of interest, dislocations remain straight over rather long distances. In such cases, our representation
of dislocation configurations by the positions of its unit segments is unnecessarily

---

[7] In simulations of partial dislocations, it is possible to account for the force associated with the
stacking fault by adding an appropriate contribution to the stress σ seen by the dislocation segments.
The latter stress is applied here along the [110] direction on the $(1\bar{1}1)$ plane.

detailed. Obviously, a more economical representation would be only to specify locations along the line where y_i's change (by either $+1$ or -1). This means that the dislocation's state is now specified by the location (i) and type ($+1$ or -1) of the kinks. A kMC algorithm based on this "sparse" representation requires more careful bookkeeping because the total number of degrees of freedom (the kinks) is no longer conserved. Specifically, it would be necessary to add kinks to the list when they are nucleated and delete them from the list when a kink pair annihilates; this is something we did not have to worry about in Algorithm 9.2. This is very similar to the difference between the phase field methods based on regular grids (Chapter 11) and the line-tracking method for dislocation dynamics simulations (Chapter 10), to be discussed later. Interface-tracking methods are often more computationally efficient but require considerably more work to keep track of the constantly changing set of degrees of freedom. The computational efficiency of Algorithm 9.2 can be further improved by organizing its event list into a hierarchical structure to expedite the searches for the selected events in step 6 [96, 97].

Summary

- Kinetic Monte Carlo (kMC) is an efficient numerical approach to simulations of systems whose evolution proceeds through a sequence of stochastic events with only a few possible event types. The method is suitable for simulations of the thermally activated motion of dislocations by kink mechanisms.

- The kMC method derives from the mathematical theory of stochastic processes with its notions of states, transition events and transition rates. In kMC, statistical measures of a system's evolution are obtained by averaging over a set of stochastic (sample) trajectories.

- At each step of the kMC algorithm, the waiting time until the next transition and the destination state are randomly sampled from appropriate distributions.

Problems

9.1.1. Algorithms 9.1 and 9.2 are implemented in the Matlab code available at the book website. Use the code to run two simulations similar to the one depicted in Fig. 9.3 but with two different values of the kink migration barrier, $W_m = 0.4\,\text{eV}$ and $0.8\,\text{eV}$. For each case, compute the average dislocation velocity and the number of kinks.

9.1.2. Run a series of kMC simulations with the same parameter values as in Section 9.1 but reset the temperature to $T = 300\,\text{K}$ and stress to $\sigma = 10\,\text{MPa}$. Run simulations for different values of the dislocation length: $L/a = 20$, 50, 100, 200, 500. Plot dislocation velocity as a function of length L and find a cross-over length L^* such that the velocity becomes independent of the dislocation length for $L > L^*$.

9.1.3. In the kink diffusion model [7], dislocation velocity is computed from the nucleation rate of kink pairs J and kink drift velocity v_k. First, the average distance X to which a kink pair expands before it annihilates with other kinks is given by

$$X = \sqrt{2v_k/J}. \tag{9.9}$$

The velocity for an infinitely long dislocation is then obtained as

$$v = hXJ = h\sqrt{2v_k J}. \tag{9.10}$$

The velocity for a dislocation with finite length L is

$$v = hXJ \frac{L}{L+X} = h\sqrt{2v_k J} \frac{L}{L+X}. \tag{9.11}$$

Expressed in terms of the activation parameters of our kMC model, the values of parameters entering the kink diffusion model are $J = J_+^n - J_-^n$ and $v_k = (J_+^m - J_-^m)a$. Use the kink diffusion model to compute dislocation velocity as a function of dislocation length and compare with the simulation results obtained in Problem 9.1.2. Compare the average number of kinks measured in the kMC simulations in Problem 9.1.1 with the predictions of the kink diffusion model.

9.1.4. Starting with a perfectly straight dislocation of length $L = 10^5 a$, run a kMC simulation for 10^6 steps at $T = 600$ K, $\sigma = 30$ MPa and all other parameters the same as in the example discussed in Section 9.1. Define a function $\Delta y(x_0, l)$ as the maximum difference between the vertical (y) positions of the dislocation segments in a portion of dislocation line bounded by $x = x_0$ and $x = x_0 + l$. Obtain the average span $\Delta y(l)$ by averaging over x_0. Function $\Delta y(l)$ characterizes the roughness of the dislocation line and can be approximated by a power law, $\Delta y(l) \propto l^\nu$. Measure the exponent ν by plotting $\Delta y(l)$ on a log–log scale. Continue the simulation for another 10^6 steps and observe any changes in the roughness.

9.2   Dealing with Elastic Interactions

9.2.1   Model Formulation

Let us now make our kMC model more realistic and include the elastic interactions between dislocation segments. The good news is that, in doing so, we will get rid of the *ad hoc* exclusion rules used in Algorithm 9.1. The bad news is that accounting for the elastic interactions makes kMC simulations much more computationally demanding. As before, we use bold symbols, such as **i** and **j**, for compact notation of dislocation configurations $\{y_1, y_2, \ldots, y_N\}$. We will also continue to use italic symbols, such as m and n, for individual dislocation segments. For every state **i**, the interactions among all (horizontal and vertical) segments define the state's energy $E(\mathbf{i})$. Analytic expressions for $E(\mathbf{i})$ will be presented shortly.

In the kMC method, the kinetics of dislocation motion is completely specified by the matrix of transition rates, $J(\mathbf{i}, \mathbf{j})$. To account for possible effects of elastic interactions on dislocation motion, we will now express the transition matrix in terms of the energy function $E(\mathbf{i})$. Suppose state **i** is the same as state **j** except that the y coordinate of segment m is different by $\pm h$. Define an incremental energy difference

$$\Delta E_\pm(\mathbf{i}, m) \equiv E(\mathbf{j}) - E(\mathbf{i}). \tag{9.12}$$

Based on the transition state theory, the following form of the transition-rate matrix provides a physically consistent description of the energetics of nucleation and migration of dislocation kinks:

$$J(\mathbf{i}, \mathbf{j}) = \nu_0 \exp\left(-\frac{\Delta E_\pm(\mathbf{i}, m)/2 + W_m - TS}{k_B T}\right). \tag{9.13}$$

Equation (9.13) is an attempt to treat kink nucleation and kink migration events on an equal footing. The term $W_m - TS$ contributes an activation barrier that is the same for all unit events. The kink nucleation and kink migration events are only differentiated by their corresponding change in the total energy $\Delta E_\pm(\mathbf{i}, m)$. This turns out to be a very good approximation for dislocation kink mechanisms in silicon, as verified by detailed atomistic calculations [100]. Despite this unifying assumption, the rates for kink pair nucleation and kink migration events are still very different because of the elastic interaction between dislocation segments. The new kMC simulation proceeds in almost exactly the same way as in the previous non-interacting model, Algorithm 9.2, except for step 2, where the $2N$ transition rates should be now computed using eq. (9.13).

9.2.2   Expressions for the Elastic Energy of a Kinked Dislocation

In eq. (9.13), the kink pair nucleation and kink migration rates are expressed in terms of the change of total energy. Here we discuss a method to compute this energy using the non-singular continuum theory of dislocations [101]. The major advantage of this theory is that it contains no singularities; this greatly simplifies its numerical implementation. The singularities present in the classical theory are removed by smearing out the Burgers vector distribution around the dislocation center, somewhat similar to the Peierls–Nabarro model discussed in Chapter 8. The result is a set of analytically simple, self-consistent and non-singular expressions for the energy, force and stress fields of an arbitrary dislocation network. More details of this approach will be discussed in Chapter 10. Here we only list the results used in the present kMC model.

Since our model of dislocation consists of horizontal and vertical segments on the two-dimensional glide plane, the expressions for their interactions are considerably simpler than in the most general case. The only two distinct geometries we need to consider are shown in Fig. 9.4.[8] The interaction energy of two horizontal segments described in Fig. 9.4(a) is

$$
\begin{aligned}
W_{hh} = \frac{\mu}{4\pi} \left(b_{1x}b_{2x} + \frac{b_{1y}b_{2y}}{1-\nu} \right) & [f(x_{14}, y_{13}, r_c) - f(x_{24}, y_{13}, r_c) \\
& - f(x_{13}, y_{13}, r_c) + f(x_{23}, y_{13}, r_c)] \\
+ \frac{\mu}{4\pi} b_{1x}b_{2x} [g(x_{14}, y_{13}, r_c) & - g(x_{24}, y_{13}, r_c) \\
- g(x_{13}, y_{13}, r_c) & + g(x_{23}, y_{13}, r_c)],
\end{aligned} \qquad (9.14)
$$

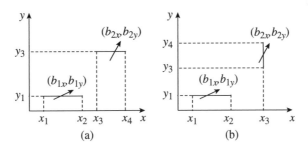

FIG. 9.4. (a) Two horizontal segments on the glide plane. The line sense directions of two segments are defined as $x_1 \to x_2$ and $x_3 \to x_4$ and their Burgers vectors are $\mathbf{b}_1 = (b_{x1}, b_{y1})$ and $\mathbf{b}_2 = (b_{x2}, b_{y2})$, respectively. (b) A horizontal and a vertical segment on the glide plane. Their line sense directions are $x_1 \to x_2$ and $y_3 \to y_4$ and their Burgers vectors are $\mathbf{b}_1 = (b_{1x}, b_{1y})$ and $\mathbf{b}_2 = (b_{2x}, b_{2y})$, respectively.

---

[8] The interaction term for two horizontal segments also works for two vertical segments by 90° rotation of the coordinates.

where

$$f(x, y, r_c) = x \ln(R_a + x) - R_a, \tag{9.15}$$

$$g(x, y, r_c) = \frac{r_c^2 R_a}{2(y^2 + r_c^2)}, \tag{9.16}$$

$$R_a \equiv \sqrt{x^2 + y^2 + r_c^2}, \tag{9.17}$$

$$x_{ij} \equiv x_j - x_i, \tag{9.18}$$

$$y_{ij} \equiv y_j - y_i \tag{9.19}$$

and r_c is the dislocation radius parameter in the non-singular theory. Here, μ is the shear modulus and ν is the Poisson's ratio of the crystal.

The interaction energy between a horizontal and a vertical segment shown in Fig. 9.4(b) is

$$
\begin{aligned}
W_{hv} = \ & \frac{\mu}{4\pi} \left(-b_{1x}b_{2y} + \frac{1 - 2\nu}{1 - \nu} b_{1y}b_{2x} \right) [h(x_{13}, y_{14}, r_c) - h(x_{23}, y_{14}, r_c) \\
& \qquad\qquad\qquad\qquad\qquad\qquad - h(x_{13}, y_{13}, r_c) + h(x_{23}, y_{13}, r_c)] \\
& + \frac{\mu}{4\pi} \left(b_{1x}b_{2y} + \frac{2\nu}{1 - \nu} b_{1y}b_{2x} \right) [k(x_{13}, y_{14}, r_c) - k(x_{23}, y_{14}, r_c) \\
& \qquad\qquad\qquad\qquad\qquad\qquad - k(x_{13}, y_{13}, r_c) + k(x_{23}, y_{13}, r_c)],
\end{aligned}
\tag{9.20}
$$

where

$$h(x, y, r_c) = x \ln(R_a + y) + y \ln(R_a + x), \tag{9.21}$$

$$k(x, y, r_c) = \frac{r_c}{2} \arctan\left(\frac{xy}{r_c R_a} \right). \tag{9.22}$$

The above expressions remain non-singular even when the segments overlap. In particular, when two segments completely overlap each other, i.e. $x_3 = x_1$, $x_4 = x_2$, $y_2 = y_1$, the resulting interaction term is twice the self-energy of this segment. Therefore, the self-energy and the interaction energies are described by the same equations, which is another useful feature of the non-singular theory.

The total elastic energy of our dislocation is the sum of the interaction energies of all segment pairs (including self interactions):

$$E(\mathbf{i}) = \frac{1}{2} \sum_{m=1}^{N_s} \sum_{n=1}^{N_s} W(m, n) - \sigma_{xz} b_x A(\mathbf{i}), \tag{9.23}$$

where N_s is the total number of segments in state **i**. The last term on the right-hand side is minus the work done by the applied stress σ_{xz} and $A(\mathbf{i})$ is the area of the glide plane enclosed between state **i** and some reference state, e.g. $y = 0$. When a periodic boundary condition is applied along the x direction, eq. (9.23) should also include the interactions between segment m and (an infinite number of) images of segment n. Here for simplicity we truncate the range of elastic interactions and apply the *minimum image convention* (Section 3.2), meaning that segment m interacts only with the nearest image of segment n. The error introduced by this truncation decreases with the increasing length of the dislocation line.[9]

9.2.3 Energy of a Kink Pair

The non-singular expressions for the dislocation energy depend on the choice of the cut-off parameter r_c, which still needs to be determined.[10] For our purpose here, it makes sense to determine r_c by comparing the energy of a kink pair predicted by the non-singular continuum theory with experimental data.

Consider a dislocation of length $L = 100a$ containing a single kink pair of width w. The non-singular theory prediction of the kink pair energy, E_{kp}, which is the excess energy of the kinked dislocation with respect to the perfectly straight dislocation, is plotted in Fig. 9.5(a). In this case, we have chosen $r_c = 0.125a$, for which the kink pair energy in the limit of $w \to \infty$ is 1.4 eV, which matches with the experimental data [102]. Therefore, we will use this value of r_c in the following simulations.

Figure 9.5(a) also plots the prediction of E_{kp} from the classical (singular) elasticity theory of dislocations [7] as a dashed line,

$$E_{\text{kp}}(w) = E_{\text{kp}}(\infty) - \frac{B}{w} \tag{9.24}$$

$$B = \frac{\mu h^2}{8\pi a} \left(b_x^2 \frac{1+\nu}{1-\nu} + b_y^2 \frac{1-2\nu}{1-\nu} \right),$$

where b_x and b_y are screw and edge components of the Burgers vector. It is satisfying to see that the classical (singular) and non-singular theories agree with each other very well, except at $w = 0$. At $w = 0$, the classical (singular) solution diverges, whereas the non-singular theory converges to the correct value, $E_{\text{kp}} = 0$. In the kMC model w only takes integer values, corresponding to different states of the system. The periodic energy oscillation for w between two integers, shown

---

[9] Strictly speaking, only the energy difference $E(\mathbf{j}) - E(\mathbf{i})$ converges, whereas the total energy of state $E(\mathbf{i})$ diverges in the limit $L \to \infty$. Fortunately, our kMC model relies on the energy differences, not the energy itself.

[10] More precisely, we need to determine the value of r_c for which the core energy E_c (Section 5.2) vanishes. Otherwise, E_c will need to be included in the total energy expressions.

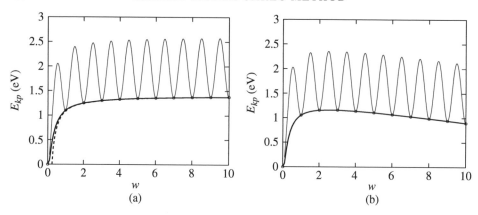

FIG. 9.5. (a) The energy of a kink pair on a partial dislocation in silicon as a function of kink pair width w. The thick solid line is the energy function computed using the non-singular elastic solutions with cut-off parameter $r_c = 0.125a$. This value of r_c leads to $E_{kp}(\infty) = 1.4\,\text{eV}$, which matches the experimental value. The dashed line is the energy function eq. (9.24). (b) The same kink pair energy under stress $\sigma_{xz} = 1.2\,\text{GPa}$. The circles mark the integer values of kink pair width. The thin solid line shows the periodic energy modulations accounting for the kink migration barrier.

in Fig. 9.5(a), illustrates the energy barrier between adjacent states. Application of external stress adds a constant slope to the energy landscape of the kink pair, as shown in Fig. 9.5(b).

9.2.4   Computing Energy Changes due to Segment Motion

An important step of the kMC algorithm is the computation of rates for $2N$ possible transitions out of a given state $\mathbf{i}$. The key component is to obtain the energy differences between the present state $\mathbf{i}$ and $2N$ "candidate" states $\mathbf{j}$, $E(\mathbf{j}) - E(\mathbf{i})$, as follows from eq. (9.13). Because dislocation interactions are long-range, the interactions between all pairs of dislocation segments must be included to compute the elastic energy of a given dislocation state $\mathbf{i}$. As a result, the kMC model accounting for the elastic interaction is much more computationally expensive than the non-interacting model described in the previous section. Thus, a successful application of the present interacting model depends critically on the efficient evaluation of the elastic energy in eq. (9.23). In a naive approach, it takes $\mathcal{O}(N^2)$ operations to evaluate the total energy of state $\mathbf{j}$. Therefore, it would take $\mathcal{O}(N^3)$ operations to evaluate all $2N$ energy differences. The purpose for the reminder of this section is to show how this can be accomplished in $\mathcal{O}(N)$ operations.

First, the difference between states $\mathbf{j}$ and $\mathbf{i}$ is small. Within our kMC model, only incremental moves are allowed, one segment at a time. One such move is illustrated in Fig. 9.6(a), where segment m moves upward by h. It is easy to show that $\Delta E_+(\mathbf{i}, m) \equiv E(\mathbf{j}) - E(\mathbf{i})$ is precisely the interaction energy $W(\mathbf{p}, \mathbf{i})$ between unit dislocation loop $\mathbf{p}$ located at $x = ma$ and N dislocation segments in state $\mathbf{i}$, as shown in Fig. 9.6(b). Loop $\mathbf{p}$ must have the same Burgers vector as dislocation $\mathbf{i}$ and its line sense is as shown in Fig. 9.6(b). Given that incremental loop $\mathbf{p}$ consists of only four segments, this method allows the computation of $E(\mathbf{j}) - E(\mathbf{i})$ in $\mathcal{O}(N)$ operations. The total number of operations per kMC step now becomes $\mathcal{O}(N^2)$, since it is still necessary to place the incremental loop at $2N$ different locations to compute all $2N$ rates. We are now one step away from the desired $\mathcal{O}(N)$ algorithm.

Consider the progress of a kMC simulation starting from state $\mathbf{i}$. Before the first step is completed, the energy differences $\Delta E_\pm(\mathbf{i}, m)$ are computed as described above. At the end of this step, one of $2N$ events is selected for execution. Assume that unit segment n moves upward by $+h$, as illustrated in Fig. 9.7(a), and label the resulting new state $\mathbf{k}$. For the second kMC step, it is again necessary to compute

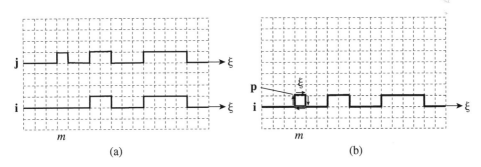

FIG. 9.6. The energy difference between two states $\mathbf{j}$ and $\mathbf{i}$ is equal to the interaction energy of incremental loop $\mathbf{p}$ with all dislocation segments in the initial
configuration $\mathbf{i}$.

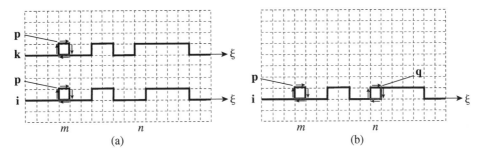

FIG. 9.7. (a) The difference between the interaction energy of dislocation $\mathbf{k}$ with
loop $\mathbf{p}$ and the interaction energy of dislocation $\mathbf{i}$ with loop $\mathbf{p}$ is equal to the
interaction energy between loop $\mathbf{p}$ and loop $\mathbf{q}$ in (b).

$2N$ energy changes, $\Delta E_\pm(\mathbf{k}, m)$, due to the long-range character of dislocation interaction. Every entry in the new list of energy changes will be different from that in the old list.

It turns out that when $m \neq n$, the difference between $\Delta E_\pm(\mathbf{k}, m)$ and $\Delta E_\pm(\mathbf{i}, m)$ is precisely the interaction energy between two dislocation loops, loop $\mathbf{p}$ at $x = ma$ and loop $\mathbf{q}$ at $x = na$, as illustrated in Fig. 9.7(b). This results in an efficient algorithm in which the energy differences are *updated* by adding the incremental interactions between loop $\mathbf{q}$ and loops $\mathbf{p}$ to the energy differences computed in the previous step. Only two entries in the list, $\Delta E_\pm(\mathbf{k}, n)$, need to be computed anew using the method described previously. Because each of the loop–loop interactions takes only $\mathcal{O}(1)$ operations to compute, the entire list $\Delta E_\pm(\mathbf{k}, m)$ is updated in $\mathcal{O}(N)$ operations. To enhance computational efficiency even further, the loop–loop interaction terms can be precomputed and stored in a look-up table for subsequent use in the kMC simulations.

9.2.5   Kinetic Monte Carlo Simulation Results

To put the new model to test, let us run a kMC simulation in which the dislocation radius is set to $r_c = 0.125b$. All other parameters are the same as in Section 9.1 except for a higher temperature $T = 1200\,\mathrm{K}$ and stress $\sigma_{xz} = 1.2\,\mathrm{GPa}$. Figure 9.8(a) shows several snapshots of dislocation configurations attained during this simulation. Note that the line remains much smoother than in the non-interacting case shown in Fig. 9.3(a), even though the temperature and stress in the present example are higher. Fig. 9.8(b) shows the instantaneous mean

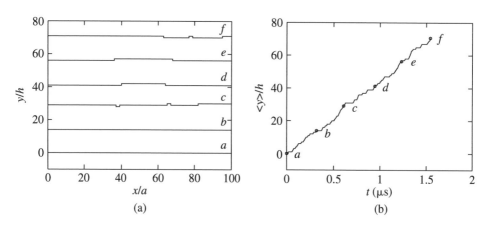

(a)   (b)

FIG. 9.8. (a) Snapshots of dislocation configurations taken at different times during the kMC simulation at $T = 1200\,\mathrm{K}$, $\sigma = 1.2\,\mathrm{GPa}$. (b) The instantaneous mean dislocation position $\langle y \rangle$ as a function of time in the same simulation. Positions marked (a–f) correspond to the snapshots in (a).

dislocation position as a function of time. The average dislocation velocity obtained from the slope is $v \approx 1.5 \times 10^{-2} \, \mathrm{m \, s^{-1}}$. The reason for using a very high temperature and applied stress[11] in this simulation is that, once the elastic interactions are turned on, "successful" kink pair nucleation becomes a rare event in the kMC simulation. This is because the kinks in a newly nucleated kink pair attract each other, resulting in nearly immediate mutual annihilation. Because of this, the kMC simulation becomes very inefficient at producing net dislocation motion, unless the temperature and/or external stress are very high. A method to address this problem is discussed in the next section.

Summary

- The long-range elastic interaction among dislocation segments affects the transition rates in the event catalog.

- Taking advantage of the non-singular continuum theory, it becomes possible to update all energy differences and their associated transition rates in just $\mathcal{O}(N)$ operations.

9.3 Kink Pair Nucleation as a Rare Event

If we were to run the same kMC simulation as in Section 9.2 but at a lower temperature and/or stress, the dislocation might not move at all, even after a large number of simulation steps (Problem 9.3.1). Just by simply watching a stochastic sequence generated in such a simulation, it should be immediately clear what the problem is. Even though kink pairs do nucleate on the dislocation line in every other simulation step, the same kinks recombine (disappear) in the very next step. This sequence repeats for many cycles, producing no net dislocation motion. The cause of this behavior is the elastic interaction between dislocation kinks, the same interaction that we turned on in the preceding section to make kMC simulations of dislocation motion more realistic. As can be seen from the kink pair energy landscape shown in Fig. 9.5(a), two kinks in a newly formed embryonic kink pair (width $w=1$) strongly attract each other. Thus, in the next kMC step, the kinks are most likely to recombine, returning the dislocation to its original "un-kinked" state $w=0$. There is a small chance for the embryonic kink pair to expand to state $w=2$. However, even if such an unlikely event does take place, the kink pair is still more likely to return to state $w=1$ than to expand further to state $w=3$. Consequently, the probability for a kink pair to escape the fate of self-recombination is very small. Such repetitive nucleation–recombination cycles do not contribute to dislocation

---

[11] As a reference point, the melting temperature of silicon is $T_m = 1693$ K and the Peierls stress of dislocations in silicon is in the range of several GPa.

motion and the resulting simulation may produce nothing of interest.[12] After a large number of kMC steps, a kink pair may, by chance, grow to a sufficiently large width so that its further expansion becomes more likely than recombination. Because only such infrequent events can result in net dislocation motion, standard kMC algorithms, such as Algorithm 9.2, become extremely inefficient.

It is worth noting that the behavior described above is just one manifestation of the "rare-event" problem that pervades many different types of numerical simulations. We have already encountered this problem in Chapter 7, where embryonic kink pair nucleation and kink migration processes were themselves rare events on the picosecond time scale of MD simulations. It is precisely this inefficiency of direct MD simulations that made us resort to the kMC method in the first place. In kMC both kink pair nucleation and kink migration events constitute just one single step in the simulation, which is a significant advantage over MD. It is interesting that in trying to overcome one rare-event problem we arrived at yet another type of rare event, which is the nucleation of a kink pair of sustainable width through a sequence of kink pair nucleation and expansion steps. Hierarchical rare events and the associated multiple time scales play an important role in many other interesting situations in nature, such as in the kinetics of protein folding. Nucleation of dislocation kink pairs presents a relatively simple case for which we can discuss methods for boosting the efficiency of stochastic simulations.

9.3.1 Sustainable Kink Pair Nucleation

For our simulation to proceed with reasonable efficiency, let us try to modify the kMC algorithm to skip the unsuccessful nucleation events altogether. Instead, we would like the modified simulation to generate only kink pairs with a prescribed width n (say $n = 10$) giving the resulting kinks a better chance to run away from each other. To be able to do that, it is necessary to compute the rate at which such *sustainable* kink pairs should be nucleated. Computing this rate is the focus of this section.

First, consider a simplified situation of a single kink pair nucleating and expanding on an otherwise straight dislocation under constant external stress. This situation is equivalent to a one-dimensional random walk on the energy landscape shown in Fig. 9.5. Next, we will obtain the nucleation rate of kink pairs of width n using three different approaches. The first two develop recursive formulae for *survival probabilities* and *average first-passage times* and then relate the results to the sought rate based on intuitive arguments. These first two approaches predict very similar, but not identical, values for the kink pair nucleation rate. The third approach leads

---

[12] Strictly speaking, these cyclic events may have some effect on dislocation motion by producing stress fluctuations everywhere on the dislocation line. However, provided the segments stay in state $w = 0$ most of the time, the overall effect of states with $w > 0$ is negligible.

to a rigorous solution for the probability distribution of the first passage times and clarifies the issues raised in the first two approaches.

9.3.2 Survival Probability

Consider a discrete one-dimensional random walk along the w axis, where $w = 0, 1, \ldots$ is the width of the kink pair, as shown in Fig. 9.9(a). The energy landscape of this walk is specified by the following function:

$$E(w) = \begin{cases} 2E_k - B/w - (\sigma bah) \cdot w, & w > 0 \\ 0, & w = 0, \end{cases} \tag{9.25}$$

where the term B/w describes the attraction between two kinks of the kink pair and term $(\sigma bah)w$ is the work done by the applied stress.

To be consistent with the model described in Section 9.2, let us define the rates of forward and backward jumps from state w as follows:

$$r_+(w) = \nu_0 \exp\left[-\frac{[E(w+1) - E(w)]/2 + W_m - TS}{k_B T}\right] \cdot [1 + H(w)]$$

$$r_-(w) = \nu_0 \exp\left[-\frac{[E(w-1) - E(w)]/2 + W_m - TS}{k_B T}\right] \cdot [1 + H(w-1)],$$

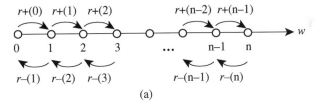

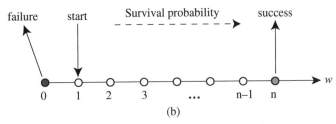

FIG. 9.9. (a) One-dimensional random walk along the w axis. Parameters $r_\pm(w)$ are the rates of forward (+) and backward (−) moves from state w. (b) The walk starts from state $w = 1$ and is terminated if it reaches either state $w = 0$ (failure) or state $w = n$ (success). The survival probability is defined as the probability that the walker will visit state $w = n$ before it visits state $w = 0$.

where $H(w) = 1$ if $w \geq 1$ and $H(w) = 0$ if $w \leq 0$. $H(w)$ accounts for the fact that, once an embryonic kink pair ($w = 1$) is nucleated, its width can subsequently change by moving either one of its two kinks. Therefore,

$$r_0(w) \equiv r_+(w) + r_-(w) \tag{9.26}$$

is the total rate of escape from state w and

$$F(w) \equiv \frac{r_+(w)}{r_0(w)} \tag{9.27}$$

is the probability that, upon escape from state w, the walker will move "forward" to state $w + 1$. With these definitions, the rate of embryonic kink pair nucleation is simply $J(1) = r_+(0)$ and our task is to find $J(n)$, the rate at which kink pairs of width n appear.

For $n > 1$, $J(n)$ is smaller than $J(1)$ because not all embryonic kinks will expand to width n before recombination. Thus we can write

$$J(n) = J(1) \cdot p_s(1 \to n), \tag{9.28}$$

where $p_s(i \to j)$ is the *survival probability*, which is defined as the probability that a kink pair of width $w = i$ expands to $w = j$ before it shrinks to $w = 0$.[13] It turns out that to solve for $p_s(1 \to n)$, it is necessary and sufficient to consider the problem of a one-dimensional discrete random walk between two absorbing boundaries [103], as shown in Fig. 9.9(b). The walk starts at state 1 and terminates either at 0 or at n. If the walker visits 0 before it visits n, such a walk results in *failure*. Conversely, if the walker visits n before visiting 0, the walk is counted as a *success*. Both boundaries are absorbing because the walk is terminated as soon as one of two ends of the interval is reached.

Consider a walk starting from state 1. Before the walker reaches state n, it has to visit state $n - 1$. Likewise, to arrive at $n - 1$, it must visit $n - 2$ before that. Hence,

$$p_s(1 \to n) = \prod_{k=1}^{n} p_s(k \to k + 1). \tag{9.29}$$

To obtain $p_s(k \to k + 1)$, note that there are only two mutually exclusive possibilities to reach $k + 1$ from k. First, with probability $F(k)$, the walker can jump from k directly to $k + 1$. Second, with probability $1 - F(k)$ the walker may first jump backward to $k - 1$ and, because it made a "wrong" move, it has to subsequently correct itself by coming back to k, with probability $p_s(k - 1 \to k)$ and start all over

---

[13] Note that $p_s(0 \to i) = 0$ and $p_s(i \to i) = 1$ for $i > 0$.

again. This reasoning immediately leads to the following recursive equation:

$$p_s(k \rightarrow k+1) = F(k) + [1 - F(k)] \cdot p_s(k-1 \rightarrow k) \cdot p_s(k \rightarrow k+1). \quad (9.30)$$

Therefore,

$$p_s(k \rightarrow k+1) = \frac{F(k)}{1 - [1 - F(k)] \cdot p_s(k-1 \rightarrow k)}, \quad \text{for } w \geq 1. \quad (9.31)$$

Using the initial condition $p_s(0 \rightarrow 1) = 0$, probabilities $p_s(k \rightarrow k+1)$ can be computed sequentially for $k = 1, 2, \ldots$. The desired survival probability $p_s(1 \rightarrow n)$ is then obtained from eq. (9.29) for arbitrary n [104].

As an example, let us compute the probability for an embryonic kink pair (random walk) to expand (survive) from state $w = 1$ to state $n = 10$. Let us set the temperature at $T = 1000$ K and use the potential energy landscape shown in Fig. 9.10(a) and defined by eq. (9.25) with $E_k = 0.7$ eV, $\sigma = 100$ MPa and $b = a/2$ and all other parameters the same as before. The computed survival probabilities, $p_s(w \rightarrow 10)$ and $p_s(w \rightarrow w+1)$, are plotted in Fig. 9.10(b) as a function of initial width w. The resulting probability to expand to $n = 10$ is $p_s(1 \rightarrow 10) = 6.00125 \times 10^{-5}$ meaning that, in the straightforward algorithm described in Section 9.2, only about 6 out of 10^5 embryonic kink pairs can be expected to expand to width 10. Since the rate for embryonic kink nucleation from eq. (9.26) is $J(1) = 1.5839 \times 10^7$ s^{-1}, the rate for nucleation of sustainable kink pairs of width 10 is $J(10) = J(1) \cdot p_s(1 \rightarrow 10) = 29.82016$ s^{-1}.

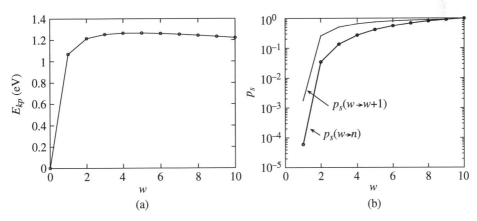

FIG. 9.10. (a) Kink pair energy landscape with $\sigma = 100$ MPa. (b) Survival probabilities $p_s(w \rightarrow 10)$ (thick line) and $p_s(w \rightarrow w+1)$ (thin line) for $n = 10$ at $T = 1000$ K.

9.3.3   Average Time of First Arrival

The same problem of nucleation of sustainable kink pairs can be analyzed in terms of the average time of first arrival to state n [105]. Consider a one-dimensional random walk on the same potential-energy landscape. However, this time the walk starts at $w = 0$, continues even after returning back to state $w = 0$ and is terminated only when it reaches state $w = n$. This corresponds to a random walk with one reflecting $(w = 0)$ and one absorbing end $(w = n)$, as shown in Fig. 9.11(a). Consider first the time t the walker spends in state w before jumping to one of its neighboring states. Assuming that this time satisfies an exponential distribution, $f(t) = r_0(w)e^{-r_0(w)t}$, the average residence time in state w is

$$\tau(w) = \int_0^\infty t \cdot r_0(w)\, e^{-r_0(w)t}\, dt = \frac{1}{r_0(w)}. \tag{9.32}$$

Defining $t(i \to j)$ as the average time of first arrival from i to j, let us now derive a recursive equation for $t(k \to k + 1)$ using the same arguments as in the earlier derivation for p_s. Because the walker has two ways to reach $k + 1$ from k, the following recursive relationship holds:

$$t(k \to k + 1) = F(k)\tau(k) + [1 - F(k)]$$
$$\times [\tau(k) + t(k - 1 \to k) + t(k \to k + 1)]. \tag{9.33}$$

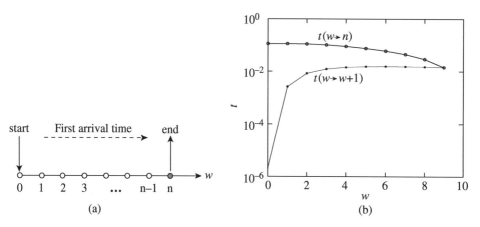

(a)   (b)

FIG. 9.11. (a) A random walk starts at state $w = 0$ and is terminated only when it reaches state $w = n$. The task is to compute the average time of first arrival $t(0 \to n)$ from state $w = 0$ to state $w = n$. (b) Average first arrival time $t(w \to 10)$ (thick line) and $t(w \to w + 1)$ (thin line) as a function of the initial state w computed under the same conditions as in Fig. 9.10.

Thus,

$$t(k \to k+1) = \frac{\tau(k) + [1 - F(k)]t(k-1 \to k)}{F(k)} \tag{9.34}$$

with an obvious initial condition $t(0 \to 1) = \tau(0) = 1/r_+(0)$. Then, the average time of first arrival to state n is simply the sum

$$t(0 \to n) = \sum_{k=1}^{n} t(k-1 \to k). \tag{9.35}$$

Applying these expressions to the walks on the energy landscape shown in Fig. 9.10(a), the results are plotted in Fig. 9.11(b). The desired average time of first arrival is $t_s(0 \to 10) = 3.35346 \times 10^{-2}$. Assuming that a relationship similar to eq. (9.32) holds for $t_s(0 \to 10)$, we obtain for the rate of sustainable kink nucleation $J(10) = 1/t_s(0 \to 10) = 29.81998\ \text{s}^{-1}$. This rate is close but not identical to the value for $J(10)$ obtained earlier from the analysis of survival probability.[14] The reason for this discrepancy is explained below.

9.3.4 Probability Distribution for the First Arrival Time

The discrepancy in the predictions for $J(10)$ from the above two approaches, however small, points to the fact that the nucleation of a kink pair of width greater than one is not a Poisson process. Its first arrival time does not satisfy the exponential distribution. To clarify this point, let us obtain the distribution function of the first arrival time to state n and compare it to the previous results.

Consider again a one-dimensional random walk with one reflecting and one absorbing end, as shown in Fig. 9.11(a). Define $P(k, t)$ as the probability for the walker to be in state k at time t. At $t = 0$, the initial condition for the walker is

$$P(k, 0) = \delta_{k,0}. \tag{9.36}$$

At $t > 0$ and $0 < k < n$, the probability $P(k, t)$ evolves according to the following ordinary differential equation (ODE):

$$\frac{dP(k, t)}{dt} = r_+(k-1)P(k-1, t)$$
$$+ r_-(k+1)P(k+1, t) - r_0(k)P(k, t). \tag{9.37}$$

---

[14] While the difference observed here is very small, by changing the parameters, such as T, σ or n, the noted discrepancy can become much larger.

With the reflecting and absorbing boundary conditions at 0 and n taken into account, the set of ODEs can be rewritten in the following matrix form [105, 106],

$$\frac{d}{dt} \begin{bmatrix} P(0,t) \\ P(1,t) \\ \vdots \\ P(n-1,t) \\ P(n,t) \end{bmatrix} = -M \cdot \begin{bmatrix} P(0,t) \\ P(1,t) \\ \vdots \\ P(n-1,t) \\ P(n,t) \end{bmatrix}, \tag{9.38}$$

where

$$M = \begin{bmatrix} r_0(0) & -r_-(1) & 0 & \cdots & & & 0 \\ -r_+(0) & r_0(1) & -r_-(2) & \cdots & & & 0 \\ & & \ddots & & & & \vdots \\ & & & -r_+(n-2) & r_0(n-1) & 0 \\ & & & 0 & -r_+(n-1) & 0 \end{bmatrix}. \tag{9.39}$$

The solution for this set of ODEs can be written as

$$P(k,t) = \sum_{i=0}^{n} c_{k,i} a_i e^{-\lambda_i t}, \tag{9.40}$$

where coefficients a_i are determined by the initial condition and $c_{k,i}$ satisfy the normalization condition $\sum_{k=0}^{n} |c_{k,i}|^2 = 1$. Substituting this trial solution in the ODEs, it is easy to see that the λ_i are the eigenvalues of matrix M and $c_{k,i}$ are the corresponding eigenvectors. After λ_i and $c_{k,i}$ are solved, the initial condition is enforced by solving the following equations for coefficients a_i,

$$\sum_{i=0}^{n} c_{k,i} a_i = \delta_{k,0}. \tag{9.41}$$

Once the probability $P(k,t)$ is solved for, the density $j(n,t)$ of the probability distribution for the first arrival time to state n is immediately obtained by noticing that state n can be visited only through state $n-1$,

$$j(n,t) = P(n-1,t)r_+(n-1) = r_+(n-1) \sum_{i=0}^{n} c_{n-1,i} a_i e^{-\lambda_i t}. \tag{9.42}$$

The distribution of the first arrival times is a sum of exponential functions and not just a single exponential. In general, for composite events that involve a sequence of stochastic transitions, the distribution of waiting times is not exponential even when each transition in the sequence obeys an exponential distribution. For such events, the average time of first occurrence is no longer equal to the inverse rate of such an event occurring. This is why the rate obtained by solving for the survival probabilities is not equal to the inverse of the average first arrival time. Qualitatively, the reason for this discrepancy is simple; a composite event does not occur instantly but takes some time to fully develop through a sequence of elementary steps, interspersed with their own individual waiting periods.

Algorithm 9.2 relies on the assumption that every one of $2N$ stochastic events in the event catalog is a Poisson process and its waiting time satisfies an exponential distribution. If we were to include in the event list any composite event, for example nucleation of kink-pairs of width $n > 1$, the resulting modified algorithm might be more efficient but would no longer be exact. As an illustration, Fig. 9.12 shows the numerical results for $j(10, t)$ computed with the same parameters as before. The scale of the horizontal axis is expanded in the narrow interval $[0, 2 \times 10^{-7}]$ to show that the probability of arriving in state $n = 10$ at $t = 0$ is zero. For comparison, the same figure shows the plot of the exponential function

$$\tilde{j}(n, t) = J(n)e^{-J(n)t}, \tag{9.43}$$

with the nucleation rate $J(10)$ computed from the solution for the survival probability. $\tilde{j}(n, t)$ provides a very good approximation for the exact solution $j(n, t)$,

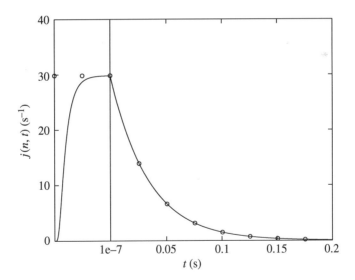

FIG. 9.12. The rate of first arrival to state $n = 10$ as a function of time t obtained from the exact solution for $j(n, t)$ (solid line). The narrow time interval $[0, 2 \times 10^{-7}]$ is expanded to show the difference between the exact function and the approximate exponential distribution (the circles).

except for the narrow region near zero, explaining why two predictions for $J(10)$ were so close to each other. The discrepancy in the nucleation rates predicted by the previous two approaches can be viewed as a measure of the error introduced in this approximation.

Fortunately, under conditions where the brute-force algorithm is least efficient, the distribution of the first arrival time can be made very close to an exponential function by choosing an appropriate n. A large n will certainly enhance the probability for kink pairs to survive, thus making the algorithm more efficient. But, for excessively large n, the deviation from the exponential distribution can become unacceptably large as the kink pairs now take a longer time to expand to the specified width n. The good news is that, when n is chosen such that the probability of the kink pairs of width $w < n$ expanding remains smaller than 1, the actual expansion time from 1 to n is negligible compared with the total waiting time (see Problem 9.3.2).

9.3.5   An Enhanced kMC Simulation

If a kMC simulation is to be performed at a fixed temperature, it usually pays to precompute the kink pair nucleation rates $J(n)$ at this temperature and for a range of stress values σ. The results can be then stored in a table for subsequent look-up and interpolation. In the following discussion, we write $J(n; \sigma)$ to emphasize the functional dependence of $J(n)$ on stress σ. As an example, Fig. 9.13(a) shows $J(n; \sigma)$ computed at two temperatures, 1000 K and 1200 K, for stress values σ ranging from -2 GPa to 2 GPa. On this plot, the results obtained from the solution

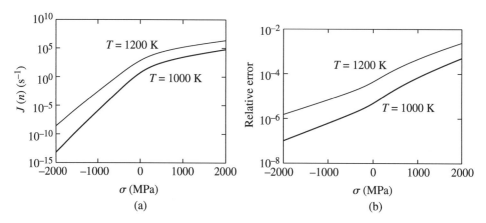

FIG. 9.13. (a) Nucleation rate $J(n)$ of kink pairs of width $n = 10$ as a function of local stress σ_{xz} computed at $T = 1000$ K (thick line) and $T = 1200$ K (thin line). (b) The relative discrepancy between two predictions for $J(n)$.

for the survival probability are indistinguishable from those obtained from the average first arrival time; their relative differences are shown in Fig. 9.13(b).

We can now modify our kMC algorithm to take advantage of the tabulated rates $J(n; \sigma)$. All events that correspond to the nucleation of embryonic kink pairs will be replaced by "super-events" of the nucleation of sustainable kink pairs of width n. Now let us go through the details of this method.[15] Let $\mathbf{i}$ represent the current dislocation configuration and consider its horizontal segment m. If $y_{m-1} = y_m = y_{m+1}$, then motion of segment m up or down corresponds to an embryonic kink pair nucleation.

Earlier in Section 9.2, we described efficient methods for computing energy changes $\Delta E_{\pm}(\mathbf{i}, m)$ caused by segment m moving by $\pm h$. To retain this efficiency and using the tabulated values of $J(n, \sigma)$, we can rewrite the results for $2N$ values of $\Delta E_{\pm}(\mathbf{i}, m)$ in terms of an *effective local stress* as follows:

$$\sigma_{xz}(m) \equiv \frac{\Delta E_-(\mathbf{i}, m) - \Delta E_+(\mathbf{i}, m)}{2ab_x h}. \tag{9.44}$$

The above effective stress subsumes both the local stress caused by all dislocation segments in the current state $\mathbf{i}$ and the external stress. The nucleation rate $r_m^{\pm}$ for a sustainable kink pair at site m is simply

$$r_m^{\pm} = J(n; \pm\sigma_{xz}(m)). \tag{9.45}$$

This should replace the nucleation rate of an embryonic kink pair in the event catalog.

Another change required in the kMC algorithm is that it is necessary to move all n horizontal segments in the neighborhood of the chosen segment m, once a kink pair nucleation event is selected. The list of energy changes $\Delta E_{\pm}(\mathbf{i}, m)$ will then need to be updated accordingly.[16]

As a test for the self-consistency of the enhanced approach just described, let us run again the same kMC simulation described in Section 9.2 (Fig. 9.8). However, this time we only allow nucleation of kink pairs of width $n = 5$. Several snapshots of the dislocation's configuration and a plot of its instantaneous mean position as a

---

[15] Here we introduce a rule that sustainable kink pairs can develop only at sites m where the dislocation contains at least n consecutive horizontal segments, to give room for the new kink pair to expand. If for a given site m this condition is not satisfied, the rate for kink pair nucleation at this site is set to zero.

[16] Another approximation implied here is that local stress remains constant over the time it takes for an embryonic kink pair to expand to width n. This turns out to be a fairly good approximation because, under the conditions considered here, this expansion time is usually very small compared to the time over which the dislocation configuration and the local stress change significantly.

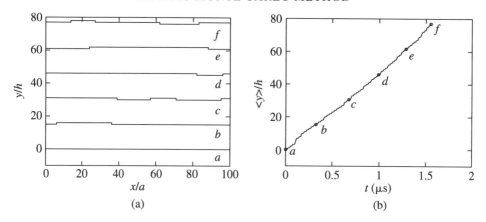

FIG. 9.14. (a) Snapshots of dislocation configurations attained during a kMC simulation at $T = 1200\,\mathrm{K}$, $\sigma = 1.2\,\mathrm{GPa}$. (b) The instantaneous mean dislocation position as a function of time. Symbols (a–f) correspond to the snapshots in (a).

function of time are shown in Fig. 9.14(a) and (b). The average dislocation velocity obtained from this simulation is $v \approx 1.6 \times 10^{-2}\,\mathrm{m\,s^{-1}}$, which is the same, within statistical errors, as the value obtained in the brute-force simulation in Section 9.2. It can be verified that the predicted dislocation velocity is insensitive to the specific choice of n, as long as n is not too large (Problem 9.3.3). However, in the new simulation, it only takes about half the number of steps to move the dislocation by the same distance, compared to the brute-force approach. This speed-up is modest because the temperature and stress are high; these were selected to allow the brute-force algorithm to produce some dislocation motion over a reasonable number of simulation steps. At lower temperature and/or stress, speed-ups by orders of magnitude can be achieved (Problem 9.3.4).

Summary

- Turning on the elastic interaction between dislocation segments makes two kinks in an embryonic kink pair attract each other, resulting in their almost certain and nearly immediate recombination. This makes kMC simulations of dislocation motion very inefficient in many situations of interest.

- It is possible to boost the efficiency of kMC simulations of dislocation motion by predicting the rates of nucleation of sustainable kink pairs. The effective rates for such composite events can be computed by analyzing the kink pair nucleation and expansion process in terms of discrete one-dimensional random walks.

Problems

9.3.1. Run a kMC simulation using the algorithm described in Section 9.2 with the same parameters as in Fig. 9.8 but at $T = 1000$ K and $\sigma_{xz} = 100$ MPa, for 10^5 steps. Plot the instantaneous mean dislocation position as a function of time and estimate the average dislocation velocity.

9.3.2. In this section, we wrote down a set of ODEs for the probability distribution function $P(k, t)$ for one-dimensional random walks in an interval with one reflecting (0) and one absorbing boundary (n). Modify the ODEs for the case where both boundaries are absorbing. Assuming the walker starts at state 1, obtain the probability distribution function for the time of its first arrival at state n. Compute the average first arrival time of such "successful" walkers. Derive the recursive formula for the average first arrival time $t(1 \to n)$ for this case and compare it to $1/J(n)$. $J(n)$ is the rate of nucleation of kink pairs of width n computed from the solution for the survival probabilities, as described in Section 9.3. For numerical calculations, use $n = 10$, $T = 1000$ K, $\sigma_{xz} = 30$ MPa, and all other parameters the same as in Section 9.3.

9.3.3. Rerun the kMC simulation described in Fig. 9.14 using two different values for the width of sustainable kink pairs $n = 3$ and $n = 10$. Plot the instantaneous mean dislocation position as a function of time and estimate the average dislocation velocity. Does the average dislocation velocity depend on the choice of n?

9.3.4. Rerun the kMC simulation described in Problem 9.3.1, but using the enhanced kMC algorithm in which only kink pairs of width $n = 10$ are nucleated. Plot the instantaneous mean dislocation position as a function of time and estimate the average dislocation velocity. Repeat the simulation with $n = 5$ and $n = 15$ and comment on how the predicted dislocation velocity changes.

10

LINE DISLOCATION DYNAMICS

In the preceding chapters we have discussed several computational approaches focused on the structure and motion of single dislocations. Here we turn our attention to collective motion of many dislocations, which is what the method of dislocation dynamics (DD) was designed for. Typical length and time scales of DD simulations are on the order of microns and seconds, similar to *in situ* transmission electron microscopy (TEM) experiments where dislocations are observed to move in real time.[1] In a way, DD simulations can be regarded as a computational counterpart of *in situ* TEM experiments. One very valuable aspect of such a "computational experiment" is that one has full control of the simulation conditions and access to the positions of all dislocation lines at any instant of time. Provided the dislocation model is realistic, DD simulations can offer important insights that help answer the fundamental questions in crystal plasticity, such as the origin of the complex dislocation patterns that emerge during plastic deformation and the relationship between microstructure, loading conditions and the mechanical strength of the crystal.

So far, two approaches to dislocation dynamics simulations have emerged. In the line DD method to be discussed in this chapter, dislocations are represented as mathematical lines in an otherwise featureless host medium. An alternative approach is to rely on a continuous field of eigenstrains, in which regions of high strain gradients reveal the locations of the dislocation lines. This representation leads to the phase field DD approach, which will be discussed in Chapter 11.

Line DD has certain similarities with the models discussed in the previous chapters, but, at the same time, is rather different from all of them. For example, the representation of dislocations by line segments in line DD method is similar to the kinetic Monte Carlo (kMC) model of Chapter 9. However, having to deal with multiple dislocations on large length and times scales necessitates a more economical treatment of dislocations in the line DD method. Thus, line DD usually relies on less detailed discretization of dislocation lines and treats dislocation motion as deterministic. Likewise, both line DD and the molecular dynamics (MD) method (Chapter 4) involve numerical integration of equations of motion satisfying a set of initial and boundary conditions. However, there are several aspects that make

---

[1] Links to a sampling of *in situ* TEM movies of dislocation motion are available at the book web site.

line DD significantly more complicated than MD. First, while the total number of atoms is fixed during an MD simulation, the total number of dislocation segments is usually not a constant during a line DD simulation. Furthermore, as the dislocations move around, it becomes necessary to change the connectivity of line segments in order to enable dislocation reactions. Keeping track of the evolving line topology presents a challenge for line DD that has no counterpart in MD.[2]

In this chapter, we lay out the details of numerical implementation of the line DD method. The material covered here may be a bit more complicated than most other chapters in this book. However, working our way through the numerical aspects, let us remember that the algorithms to be discussed in this chapter, even if complicated, are introduced to improve computational efficiency of line DD simulations, whereas the physical model we are dealing with here is simple. In a nutshell, all we will be doing is computing forces on dislocation lines and moving the lines in response to these forces.

The chapter is organized as follows. Its first three sections cover the aspects of line DD simulations that do not involve any changes in the line topology and are similar to MD. Section 10.1 introduces a nodal representation of the dislocation network and describes how to compute the forces on the nodes. Section 10.2 explains how to compute nodal velocities in response to these forces, thus completing the nodal equations of motion. Section 10.3 then discusses numerical integrators for solving the nodal equations of motion. The last three sections deal with issues more specific to the line DD method. Section 10.4 explains how to handle topological changes taking place during a simulation. Section 10.5 describes the techniques for enabling line DD simulations on massively parallel computers. Finally, Section 10.6 discusses a case study in which line DD simulations are used to predict the stress–strain behavior of a crystal in a virtual deformation test.

10.1   Nodal Representation and Forces

10.1.1   Nodal Representation of Dislocation Networks

Figure 10.1 illustrates a simple nodal representation used here to describe dislocation networks of arbitrary topology. The dislocation lines are represented

---

[2] Two-dimensional versions of line DD exist that do not involve topological changes. In a 2D model, all dislocations are assumed to be infinite, straight and parallel to each other. Thus, the dynamics of the dislocation lines is reduced to the motion of points confined to the glide planes. This is an interesting model that captures some of the physics of crystal plasticity [107]. However, in reality dislocations are almost never aligned along a single direction and their behavior is strongly affected by line tension and dislocation intersections in three dimensions. These and other important aspects cannot be accounted for in the 2D DD models and here we will focus on the more realistic line DD models in three dimensions.

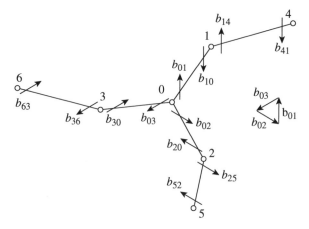

FIG. 10.1. Dislocation network is represented by a set of nodes (circles) connected by straight segments. $\mathbf{b}_{ij}$ is the Burgers vector of the directed segment connecting node i to node j.

by a set of nodes connected to each other by straight segments. Each segment has a non-zero Burgers vector. Given the freedom of choice between two opposite line directions (Section 1.2), our convention here is to define $\mathbf{b}_{ij}$ with respect to the line sense pointing from node i to node j. Likewise, $\mathbf{b}_{ji}$ is the Burgers vector of the same segment but defined with respect to the opposite line sense, i.e. from node j to node i. Obviously, $\mathbf{b}_{ij} + \mathbf{b}_{ji} = 0$. Under this convention, the conservation of Burgers vector means that the Burgers vectors of all segments pointing out from the same node must sum up to zero, i.e. $\sum_k \mathbf{b}_{ik} = 0$, where the sum is over all nodes k connected to node i. These sum rules provide useful checks for topological self-consistency during line DD simulations. While there is no limit on the maximum number of nodes that can be connected to a given node, each node must be connected to at least two other nodes, because dislocation lines cannot terminate inside the crystal. Also, to avoid redundancy, no two nodes can be connected by more than one segment.

Alternative representations of dislocation networks are certainly possible. For example, similar to the kMC model in Chapter 9, lines of arbitrary curvature and orientation can be represented by small segments of two (or more) fixed directions, such as the zigzagging "edge–screw" representation in [108]. Smoother representations by circular arcs [109, 110] and cubic splines [111] have also been used.[3] Selection of a particular representation for the dislocation network reflects one's style and taste. We prefer the nodal representation for its simplicity and generality.

---

[3] Our representation can be considered as the simplest form of splines, i.e. linear splines, in which the nodes themselves are the control points.

The degrees of freedom in this model are the coordinates of all nodes and the Burgers vectors of all segments connecting the nodes: $\{\mathbf{r}_i, \mathbf{b}_{ij}\}$. The nodes can assume arbitrary positions in space, and the connectivity of a given node i is fully specified by its non-zero Burgers vectors $\mathbf{b}_{ij}$. In Sections 10.1, 10.2 and 10.3, we introduce methods for evolving nodal positions $\mathbf{r}_i$ while assuming that the Burgers vectors $\mathbf{b}_{ij}$ do not change. The motion of the nodes may lead to situations that demand changes in the nodal connectivity, i.e. the Burgers vectors of the segments—these will be dealt with in Section 10.4.

10.1.2   Energy and Forces

The next important step in defining the model is to introduce the total energy of the dislocation network as a function of the nodal positions and Burgers vectors, i.e. $E_{\text{tot}}(\{\mathbf{r}_i, \mathbf{b}_{ij}\})$. The force on node i is then defined as minus the derivative of the total energy with respect to its position $\mathbf{r}_i$, i.e.

$$\mathbf{f}_i = -\frac{\partial E_{\text{tot}}(\{\mathbf{r}_i, \mathbf{b}_{ij}\})}{\partial \mathbf{r}_i}. \tag{10.1}$$

In defining the nodal force as above, let us always keep in mind an important, even if obvious, condition that the derivatives are taken while the node connectivity remains unchanged. To simplify the notation, let C represent the entire dislocation network, i.e. $C \equiv \{\mathbf{r}_i, \mathbf{b}_{ij}\}$, so that the total energy can be rewritten as $E_{\text{tot}}(C)$.

As was discussed in Section 5.2, the total energy of a dislocation can be partitioned into elastic and a core-energy contributions, i.e.

$$E_{\text{tot}}(C) = E_{\text{el}}(C, r_c) + E_{\text{core}}(C, r_c), \tag{10.2}$$

where r_c is a cut-off parameter in linear elasticity theory introduced to eliminate the singularity in the dislocation core. Accordingly, the nodal force is made up of two contributions, the elastic and core terms,

$$\mathbf{f}_i = \mathbf{f}_i^{\text{el}} + \mathbf{f}_i^{\text{core}}, \tag{10.3}$$

where $\mathbf{f}_i^{\text{el}}$ and $\mathbf{f}_i^{\text{core}}$ are the negative derivatives of E_{el} and E_{core}, respectively.

10.1.3   Elastic Energy and Force Contributions

Assuming that the solid is elastically isotropic, linear elasticity theory expresses the elastic energy of a dislocation network as the following double line integral [7]:

$$E_{\text{el}}(C) = \frac{\mu}{16\pi} \oint_C \oint_C b_i b'_j \partial_k \partial_k R\, dx_i dx'_j - \frac{\mu}{8\pi} \oint_C \oint_C \epsilon_{ijq} \epsilon_{mnq} b_i b'_j \partial_k \partial_k R\, dx_m dx'_n$$

$$+ \frac{\mu}{8\pi(1-\nu)} \oint_C \oint_C \epsilon_{ikl} \epsilon_{jmn} b_k b'_m \partial_i \partial_j R\, dx_l dx'_n. \tag{10.4}$$

Here, μ and ν are the shear modulus and Poisson's ratio, $R \equiv \|\mathbf{x} - \mathbf{x}'\|$, $\partial_i \partial_j R \equiv \partial^2 R / \partial x_i \partial x_j$. $\mathbf{b}$ and $\mathbf{b}'$ are the Burgers vectors at locations $\mathbf{x}$ and $\mathbf{x}'$, respectively. Each line integral $\oint_C$ is taken over all segments in the dislocation network exactly once. The direction of each line segment is arbitrary but the Burgers vector has to be defined with respect to the chosen line direction.

It is easy to see that both the integrand and the resulting integral in eq. (10.4) are infinite, because the derivatives $\partial_i \partial_j R$ diverge as $\|\mathbf{x} - \mathbf{x}'\| \to 0$. This is unacceptable in a numerical implementation. A common approach to get rid of this unwanted singularity is to truncate the integrals in eq. (10.4) so that the integrand is set to zero whenever R becomes smaller than some cut-off radius r_c. An alternative approach, to be followed here, is to replace every R inside eq. (10.4) with $R_a \equiv \sqrt{R^2 + r_c^2}$ [101], i.e.

$$
\begin{aligned}
E_{\mathrm{el}}(C, r_c) = {} & \frac{\mu}{16\pi} \oint_C \oint_C b_i b'_j \partial_k \partial_k R_a \, \mathrm{d}x_i \, \mathrm{d}x'_j \\
& - \frac{\mu}{8\pi} \oint_C \oint_C \epsilon_{ijq} \epsilon_{mnq} b_i b'_j \partial_k \partial_k R_a \, \mathrm{d}x_m \, \mathrm{d}x'_n \\
& + \frac{\mu}{8\pi(1-\nu)} \oint_C \oint_C \epsilon_{ikl} \epsilon_{jmn} b_k b'_m \partial_i \partial_j R_a \, \mathrm{d}x_l \, \mathrm{d}x'_n .
\end{aligned} \tag{10.5}
$$

Obviously, the modified energy reduces to the original one in the limit $r_c \to 0$. There are two advantages to this approach. First, it completely removes the singularity because the derivatives $\partial_i \partial_j R_a$ become finite and smooth everywhere. Second, the derivatives of R_a can be integrated analytically over straight segments. This leads to closed-form expressions that look very similar to and are as simple as the ones obtained in the original singular theory. To save space, these energy expressions are not given here but the corresponding Matlab functions are available on the book's web site. Physically, replacing R with R_a amounts to replacing a singular distribution of the Burgers vector on the dislocation line with a smooth distribution of Burgers vector *centered around* the dislocation line. In the following, we will refer to this approach as the non-singular continuum theory of dislocations [101].

Based on the analytic expression for the elastic energy, the elastic contribution to the nodal force can be obtained by taking the derivative

$$
f_i^{\mathrm{el}} = -\frac{\partial E_{\mathrm{el}}}{\partial r_i}. \tag{10.6}
$$

However, for reasons that will become clear later, we will take a different approach to computing the nodal forces. In Section 1.3, we observed that the dislocation experiences a force per unit length that is proportional to the local stress through the Peach–Koehler formula,

$$
\mathbf{f}^{\mathrm{PK}}(\mathbf{x}) = (\sigma(\mathbf{x}) \cdot \mathbf{b}) \times \xi(\mathbf{x}), \tag{10.7}
$$

where $\sigma(\mathbf{x})$ is the local stress at point $\mathbf{x}$ on the dislocation line, $\mathbf{b}$ is the Burgers vector, and $\xi(\mathbf{x})$ is the local tangent vector of the dislocation line. Within the non-singular continuum theory of dislocations, the internal stress field generated by the dislocation lines themselves can be written down explicitly,

$$\sigma_{\alpha\beta}(\mathbf{x}) = \frac{\mu}{8\pi} \oint_C \partial_i \partial_p \partial_p R_a \left(b_m \epsilon_{ima} dx'_\beta + b_m \epsilon_{imb} dx'_\alpha \right)$$

$$+ \frac{\mu}{4\pi(1-\nu)} \oint_C b_m \epsilon_{imk} \left(\partial_i \partial_\alpha \partial_\beta R_a - \delta_{\alpha\beta} \partial_i \partial_p \partial_p R_a \right) dx'_k. \quad (10.8)$$

Equations (10.7) and (10.8) provide analytic expressions for the Peach–Koehler force due to the dislocation lines themselves. The physical meaning of the Peach–Koehler force can be explained by considering the following thought process. Imagine that the shape of the dislocation line changes by $\delta\mathbf{r}(\mathbf{x})$, i.e. every point $\mathbf{x}$ on the dislocation moves by $\delta\mathbf{r}(\mathbf{x})$. In the limit of small $\delta\mathbf{r}(\mathbf{x})$, the corresponding change of the elastic energy is

$$\delta E_{\text{el}} = -\oint_C \mathbf{f}^{\text{PK}}(\mathbf{x}) \cdot \delta\mathbf{r}(\mathbf{x}) \, dL(\mathbf{x}). \quad (10.9)$$

To relate the PK force to force on node i, let us remember that in our model the segments are constrained to remain straight between every pair of connected nodes. Let us define a shape function $N_i(\mathbf{x})$ for every node i in such a way that $N_i(\mathbf{x})$ is non-zero only when $\mathbf{x}$ lies on a segment connected to node i. On a given segment i–j, let $N_i(\mathbf{x})$ decrease linearly from one at node i to zero at node j, i.e.

$$N_i(\mathbf{x}) = \frac{\|\mathbf{x} - \mathbf{r}_j\|}{\|\mathbf{r}_i - \mathbf{r}_j\|}, \quad (10.10)$$

as illustrated in Fig. 10.2(b). Now consider a virtual displacement of node i by $\delta\mathbf{r}_i$ causing the line to change its shape by $N_i(\mathbf{x})\delta\mathbf{r}_i$, as shown in Fig. 10.2(a). The corresponding change in the elastic energy is

$$\delta E_{\text{el}} = -\oint_C \mathbf{f}^{\text{PK}}(\mathbf{x}) \cdot N_i(\mathbf{x})\delta\mathbf{r}_i \, dL(\mathbf{x}), \quad (10.11)$$

and the elastic force on node i is

$$\mathbf{f}_i^{\text{el}} = -\frac{\delta E_{\text{el}}}{\delta\mathbf{r}_i} = \oint_C \mathbf{f}^{\text{PK}}(\mathbf{x}) \, N_i(\mathbf{x}) \, dL(\mathbf{x}). \quad (10.12)$$

Hence, the elastic part of the nodal force is equal to the PK force integrated with weights $N_i(\mathbf{x})$ over all segments connected to node i. Taken together, eqs. (10.12), (10.7) and (10.8) furnish expressions for computing the elastic contribution to the force on node i.

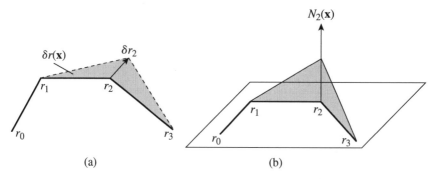

FIG. 10.2. The force on node 2 is computed as a weighted average of the Peach–Koehler force on segments 1–2 and 2–3. (a) Consider a virtual displacement of node 2 by $\delta\mathbf{r}_2$. As a result of this displacement, each point on segments 1–2 and 2–3 is displaced by $\delta\mathbf{r}(\mathbf{x}) = N_2(\mathbf{x})\delta\mathbf{r}_2$, where the shape function $N_2(\mathbf{x})$, shown in (b), is equal to one at node 2 and decreases linearly to zero at two neighboring nodes 1 and 3. The work performed by the stress is obtained by integrating the PK force over the swept area (shaded). The force on node 2 is the derivative of this work with respect to $\delta\mathbf{r}_2$.

Notice that the internal stress field, eq. (10.8), is an integral over the entire dislocation network. This integral can be written as the sum of line integrals over the straight segments making up the network, i.e.

$$\sigma_{\alpha\beta}(\mathbf{x}) = \sum_{(k-l)} \sigma_{\alpha\beta}(\mathbf{x}; k-l). \tag{10.13}$$

Here, $\sigma_{\alpha\beta}(\mathbf{x}; k-l)$ is the stress field at point $\mathbf{x}$ produced by the straight segment connecting node k and node l and the sum is taken over all segments in the dislocation network. Analytic expressions for $\sigma_{\alpha\beta}(\mathbf{x}; k-l)$ have been derived [101] but are omitted here to save space.

In order to arrive at the final expressions for the nodal force, the line integral in eq. (10.12) has to be evaluated. Fortunately, this integral is to be taken only over the segments connected to node i, because the weighting function $N_i(\mathbf{x})$ is zero everywhere else. This leads to the following form for the elastic force on node i:

$$\mathbf{f}_i^{el} = \sum_j \sum_{(k-l)} \mathbf{f}_i^{el}(i-j; k-l), \tag{10.14}$$

where terms $\mathbf{f}_i^{el}(i-j; k-l)$ represent the force on node i due to the elastic interaction between segments i–j and k–l; the first sum is over all nodes j connected to node i and the second sum is over all line segments of the network. Analytic expressions for $\mathbf{f}_i^{el}(i-j; k-l)$ have been derived and implemented into Matlab codes that are available at the book's web site.

We would like to emphasize again that the two ways of computing nodal forces, i.e. by direct differentiation of E_{el} and by the integration of the Peach–Koehler force, are equivalent to each other. They produce identical results, as has been verified both analytically and numerically [101]. The benefit of using the Peach–Koehler formula for computing the nodal forces is that it relies on the local stress $\sigma_{\alpha\beta}(\mathbf{x})$ regardless of the origin of this stress. Hence, it is applicable in a variety of situations, including stress produced by external loads or by other crystal defects, e.g. cracks, phase inclusions, etc. The main purpose of the derivation following eq. (10.6) was to gain this numerical convenience.

Given the long-range nature of the dislocation stress fields, in principle all segments of the network interact with each other. Thus, typically, the calculation of the elastic components of the nodal forces is the most time consuming part of a line DD simulation. At the same time, such calculations require very little input that is material specific. If isotropic elasticity is assumed, the only material parameters used are the shear modulus μ and Poisson's ratio ν. Otherwise, the machinery described in this section remains the same from one material to another, say from silicon to iron. For this reason, it is advantageous to have this *generic* part of the line DD simulation written and optimized separately from the rest of the code.

10.1.4   Core Energy and Force Contributions

According to eqs. (10.2) and (10.3), the dislocation core energy contributes additional terms to the total energy and nodal forces. The need for such terms arises because the elastic energy $E_{\text{el}}(C, r_c)$, in general, does not completely describe the energy of real dislocations. Even though $E_{\text{el}}(C, r_c)$ is no longer singular in our approach, it is still a result from linear elasticity theory. Yet the effect of non-linear interatomic interactions in the dislocation core cannot be fully reproduced by linear elasticity theory alone.

Suppose that the total energy $E_{\text{tot}}(C)$ of an arbitrary dislocation network C could be obtained from an atomistic model that, in the present context, is regarded as exact. Given that the elastic energy $E_{\text{el}}(C, r_c)$ does not account for the the non-linear interactions in the core, let us define the core energy $E_{\text{core}}(C, r_c)$ as the difference between $E_{\text{tot}}(C)$ and $E_{\text{el}}(C, r_c)$. Thus, the core energy is introduced to account for whatever is left unaccounted for in the continuum theory. Fortunately, this contribution is usually small and local (or short-range), and can be written as a single integral along the dislocation network:

$$E_{\text{core}}(C, r_c) = \oint_C E_{\text{c}}(\mathbf{x}; r_c) \, dL(\mathbf{x}), \qquad (10.15)$$

where $E_{\text{c}}(\mathbf{x}; r_c)$ is the energy per unit length of the line at point $\mathbf{x}$ on the network. By comparison, the effect of the elastic interaction is *non-local* (or long-range), since $E_{\text{el}}(C, r_c)$ involves a double integral, as in eq. (10.5). The locality of the core energy means also that the integrand in eq. (10.15) depends only on the local line direction at point $\mathbf{x}$ on the network.[4] Usually, the line direction is specified by two orientation angles θ and ϕ.

If we accept that eq. (10.15) holds for an arbitrary dislocation structure, then the easiest way to compute the core energy function $E_c(\theta, \phi; r_c)$ is to compare the atomistic and elastic energies for a set of parallel, infinitely long and straight dislocations of different orientations. This is precisely what we did in Section 5.2.

Suppose that the core energy function $E_c(\theta, \phi; r_c)$ has been constructed based on such calculations. Then the core energy for a dislocation consisting of straight segments is given by the following:

$$E_{\text{core}}(C, r_c) = \sum_{(i-j)} E_c(\theta_{i-j}, \phi_{i-j}; r_c) \, \|\mathbf{r}_i - \mathbf{r}_j\|, \qquad (10.16)$$

where the sum is over all segments $(i - j)$ and θ_{i-j} and ϕ_{i-j} are the orientation angles of segment $(i - j)$. When r_c is chosen to be comparable to or larger than b, the core energy density function is usually positive for all dislocation orientations. Such a positive contribution introduces a force on the nodes that tends to reduce the length of the segments. In addition, because E_c usually varies with segment orientation, it also introduces a torque that tends to rotate the segments towards low-energy orientations (Problem 10.1.1). For the most part, the core energy contribution to the nodal forces has been ignored in line DD simulations, partly because there is insufficient atomistic data.

10.1.5 Periodic Boundary Conditions

The methods described so far in this section are applicable to an arbitrary set of dislocations in an infinite elastic solid. However, periodic boundary conditions (PBC) are often used in line DD simulations, especially for the simulations of crystal plasticity in the bulk. This is because PBC conveniently eliminate the often unwanted surface effects that are otherwise inevitable due to the finite size of the simulation cell. Even though line DD simulations can handle much larger material volumes (of the order of microns) than the atomistic models, these are still very small compared with the typical scales of laboratory experiments.

In the nodal representation used here, the implementation of PBC in line DD is quite similar to the atomistic case (Section 3.2). Whenever there is a node at position $\mathbf{r}$, there are also nodes at $\mathbf{r} + n_1\mathbf{c}_1 + n_2\mathbf{c}_2 + n_3\mathbf{c}_3$, where $\mathbf{c}_1$, $\mathbf{c}_2$ and $\mathbf{c}_3$ are the repeat vectors of the periodic supercell and n_1, n_2 and n_3 are integers. Now suppose that node i and node j are connected by a segment with Burgers vector $\mathbf{b}_{ij}$. Which of the multiple copies of nodes i and j are actually connected to each

---

[4] If it were not for this locality property, linear elasticity theory of dislocations would not be as useful as it is.

other? An obvious solution is through the *minimum image convention*, which in this context states that, among all copies of node i and node j, the nearest pairs are the ones that are actually connected. Provided that all dislocation segments remain comfortably shorter than half of the simulation cell size, no ambiguity should arise.

Another complication arises because dislocation interactions are long range. Consequently, every segment interacts not only with every other segment, but also with an infinite number of their images. A similar situation appears in the atomistic models of ionic crystals (such as NaCl), where every atom interacts with every other atom through the long range Coulomb interaction. The *Ewald* method is often used in atomistic simulations to combine PBC with long-range interactions. For line DD simulations, we use a method that is equivalent to Ewald's but, in our view, is a bit easier to understand and implement.[5] Our approach is to precompute the stress field produced by the images of a given segment as a function of segment position, orientation and Burgers vector, and store the resulting image stress on a grid for subsequent look-up and interpolation in the line DD simulations. It turns out that computing the image stress for tabulation is not a trivial task, precisely because of the long-range character of the dislocation stress field. Similar to the situation discussed in Chapter 5, a brute-force summation over all image segments is *conditionally convergent*, so its result depends on the order of summation, no matter how many terms are included in the sum [59]. The solution developed in Chapter 5 is equally applicable here. This is discussed in more detail in Problem 10.1.4.

Summary

- An arbitrary dislocation network can be conveniently represented by a set of nodes connected by straight segments.

- The force on every node includes both local core and non-local elastic contributions. The elastic part of the nodal force can be computed by integrating the total stress over the segments connected to the node.

Problems

10.1.1 Assume that the core energy per unit length of a dislocation line is

$$E_{\mathrm{c}}(\theta; \mathbf{r}_c) = (E_{\mathrm{c}}^{\mathrm{e}} \sin^2 \theta + E_{\mathrm{c}}^{\mathrm{s}} \cos^2 \theta) \ln \mathbf{r}_c, \qquad (10.17)$$

where θ is the angle between the Burgers vector and the line direction of the dislocation, $\mathbf{r}_c$ is the cut-off parameter and $E_{\mathrm{c}}^{\mathrm{e}}$ and $E_{\mathrm{c}}^{\mathrm{s}}$ are constants.

---

[5] The Ewald method exploits the $1/r$ scaling of the Coulomb interaction to maximize the method's computational efficiency. Even though in principle a similar approach can be used in the line DD, the corresponding expressions for the elastic interaction terms are far too complex. An effective approach to compute periodic corrections was proposed in Ref [133].

For a segment of arbitrary orientation and length L, derive the core contribution to the forces on the segment's end nodes.

10.1.2. Use the Matlab codes provided at the book's web site to compute the elastic forces on the nodes of a circular dislocation loop. Construct a loop by connecting N nodes positioned at $\mathbf{r}_i = (R \cos \phi_i, R \sin \phi_i, 0)$, $\phi_i = 2\pi i / N, i = 1, \ldots, N$. Assign the same Burgers vector $\mathbf{b} = (0, 0, 1)$ to each segment. Assume that $R = 10$, $\mu = 1$, $\nu = 0.3$ and $r_c = 0.1$. As a result of the symmetry of this problem, the force on every node should have the same magnitude and point to the center of the loop. The sum of the magnitudes of all nodal forces gives a numerical estimate of the derivative of the loop energy with respect to its radius R. Compute and plot the numerical estimate of dE/dR as a function of N for $N = 6, 12, 32, 64, 128$ and compare it to the analytical solution

$$E = 2\pi R \frac{\mu b^2}{4\pi (1 - \nu)} \left(\ln \frac{8R}{r_c} - 1 \right) + \mathcal{O} \left(\frac{r_c^2}{R^2} \right).$$

10.1.3. Consider a situation where the distance between point $\mathbf{x}$ and segment $A-B$ is much larger than the length of this segment. In this case, the stress field produced by segment $A-B$ is well approximated by the stress field of a segment of differential (infinitesimal) length, i.e.

$$\sigma_{\alpha\beta}(\mathbf{x}) = \frac{\mu}{8\pi} \partial_i \partial_p \partial_p R_a \left(b_m \epsilon_{ima} dl'_\beta + b_m \epsilon_{im\beta} dl'_\alpha \right)$$

$$+ \frac{\mu}{4\pi (1 - \nu)} b_m \epsilon_{imk} \left(\partial_i \partial_\alpha \partial_\beta R_a - \delta_{\alpha\beta} \partial_i \partial_p \partial_p R_a \right) dl'_k, \quad (10.18)$$

where $d\mathbf{l} = \mathbf{r}_B - \mathbf{r}_A$. The expression above is the same as the integrand in eq. (10.8). Notice that the stress field of a differential segment is proportional to its tensorial "charge" $\mathcal{N} = \mathbf{b} \otimes d\mathbf{l}$ usually called Nye's tensor. In this problem, let us benchmark the quality of this approximation.

Consider a dislocation segment with the Burgers vector $\mathbf{b} = (1, 1, 1)$ and with its two end nodes at $\mathbf{r}_A = (0, 0, -1)$ and $\mathbf{r}_B = (0, 0, 1)$. Use the Matlab codes at the book's web site to compute all six components of stress produced by this segment at point $\mathbf{x} = (x, 0, 0)$ and plot the stress components as functions of x from $x = 1$ to $x = 100$. Over the same interval, plot the stress produced by a differential segment at the origin $(0, 0, 0)$, with $d\mathbf{l} = (0, 0, 2)$. Find the distance from the origin beyond which the relative difference between two solutions is below 5 per cent.

10.1.4. Direct tabulation of the image stress of a general dislocation segment would result in a very large table: for each grid point, this table would contain six components of stress precomputed as functions of dislocation

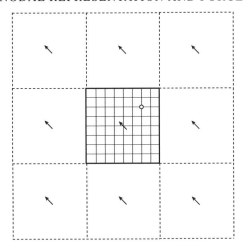

FIG. 10.3. A dislocation segment in the center of a simulation cell under PBC. The stress field from all image segments can be computed on a regular grid in the primary cell and stored for future use.

segment length, orientation and Burgers vector. Fortunately, the distance between the field point and the image segments is, by definition of the images, larger than half of the simulation box size, as shown in Fig. 10.3. Therefore, it is possible to approximate the image segments as differential segments, as long as their length is much smaller than the simulation box size. From eq. (10.18), the stress components of a differential segment are linear combinations of third derivatives of R.[6] It is possible to reduce the size of the table even further by exploiting the fact that there are only 10 distinct third derivatives of R, e.g. $\partial_x \partial_x \partial_y R$, $\partial_x \partial_y \partial_z R$, etc. Now, for each grid point it is sufficient to compute and store 10 numbers for the image sums of $\partial_i \partial_j \partial_k R$. These can be used to reconstruct the image stress produced by an arbitrary differential segment.

To simplify our notation, let $F(\mathbf{R})$ be a particular third derivative of R, i.e. $F(\mathbf{R}) = \partial_i \partial_j \partial_k R$, where $R = \|\mathbf{R}\|$. For each grid point at $\mathbf{r}$, our task is to compute the following sum:

$$F^{\text{img}}(\mathbf{r}) = \sum_{R} {}' F(\mathbf{r} - \mathbf{R}), \qquad (10.19)$$

where the summation $\sum'$ is over all image segments at lattice points $\mathbf{R} = n_1 \mathbf{c}_1 + n_2 \mathbf{c}_2 + n_3 \mathbf{c}_3$ but excluding the origin $(n_1 = n_2 = n_3 = 0)$.

---

[6] When $R \gg r_c$, as is usually the case, the differences between R and $R_a = \sqrt{R^2 + r_c^2}$ can be ignored.

(a) Following the analysis in Chapter 5, show that the sum in eq. (10.19) is not absolutely convergent.

(b) Show that $F^{img}(\mathbf{r})$ can be written as

$$F^{img}(\mathbf{r}) = F_0^{img}(\mathbf{r}) + \mathbf{g} \cdot \mathbf{r} + \phi_0, \qquad (10.20)$$

where $\mathbf{g}$ and ϕ_0 depend on the order of summation but $F_0^{img}(\mathbf{r})$ does not.

(c) Extract $F_0^{img}(\mathbf{r})$ from $F^{img}(\mathbf{r})$. $F_0^{img}(\mathbf{r})$ is the value to be stored in the table.

10.2 Nodal Mobility Functions

In the preceding section, we developed expressions for the forces on the dislocation nodes. Here we discuss how the nodes respond to these forces, leading to nodal equations of motion. It is useful to compare the response of the nodes in a line DD model with that of atoms in a molecular dynamics (MD) model (Section 2.5). In MD the atoms move in response to force $\mathbf{f}$ according to Newton's second law, i.e. $d^2\mathbf{r}_i/dt^2 = \mathbf{f}_i/m$. The nodal response in a line DD simulation is very different and much more complicated.

For example, in the line DD method most nodes are introduced with the sole purpose of representing continuous dislocation lines: let us refer to these as the *discretization* nodes.[7] Obviously, motion of a discretization node along the line direction has no physical significance. By comparison, in MD atomic motions in all three directions are equally meaningful. Furthermore, unlike atoms, a dislocation's response to forces generally depends on the orientation of the force with respect to the glide plane (Section 1.3); motion out of the glide plane (climb) is usually much more difficult than motion within the glide plane (glide). A dislocation's response to forces may also depend on the line orientation and varies significantly from one material to another. Similar to the core energy, how dislocations respond to forces is a function of the non-linear atomistic interactions in the dislocation core and, as such, is beyond the scope of linear elasticity theory. The response functions have to be imported into the line DD model in the form of external material inputs, either from atomistic simulations (Chapter 4) or from experiments.

All material-specific aspects of a line DD simulation can be collected in a *material module*, a significant part of which specifies how the dislocations should respond to forces.[8] Organizing the code in such a modular fashion—keeping generic parts

---

[7] Other nodes represent locations on the network where three or more dislocation meet: these are the *physical* nodes.

[8] Other parameters in the material module include the elastic constants and the core energy function used for force calculations.

separate from the material module—reduces the effort of setting up DD simulations for a new material. All one has to do is develop and supply a new material module, while the remainder of the program does not have to change. In this section we discuss what goes into constructing mobility functions for the line DD simulations.

The effects of inertia on dislocation motion can become important under conditions of very high strain rates, such as during shock propagation. In most other situations, inertia can be safely ignored. This means that, to a good approximation, there is no need to worry about the acceleration and masses of the dislocations. Motion in this regime is often called *over-damped* motion, where the force determines the instantaneous velocity, leading to a first-order differential equation of motion.[9] Let us call the function relating the nodal force to the nodal velocity the *mobility function*:

$$\mathbf{v}_i = \mathbf{M}(\{\mathbf{f}_j\}). \tag{10.21}$$

For reasons that will become clear later, velocity $\mathbf{v}_i$ of node i must be assumed to depend on the forces on other nodes. In addition, mobility function $\mathbf{M}$ depends on the orientations of the dislocation segments, as well as on the type of material being simulated, even though such dependencies are not written out explicitly in eq. (10.21). The purpose of this section is to develop explicit forms for the mobility function and the algorithms for computing the nodal velocities.

10.2.1 A Linear Mobility Model

Consider an arbitrary dislocation network C and assume that the driving force $\mathbf{f}^{\text{drive}}(\mathbf{x})$ on all points $\mathbf{x}$ of the network is known. The driving force is minus the variational derivative of the total energy with respect to an infinitesimal change in the dislocation shape $\delta \mathbf{r}(\mathbf{x})$,[10] i.e.

$$\delta E_{\text{tot}} = -\oint_C \mathbf{f}^{\text{drive}}(\mathbf{x}) \cdot \delta \mathbf{r}(\mathbf{x}) \, dL(\mathbf{x}). \tag{10.22}$$

As a simple model for the viscous drag motion, let us assume that the dislocation experiences a drag force per unit length of the line that is proportional to the local velocity:

$$\mathbf{f}^{\text{drag}}(\mathbf{x}) = -\mathcal{B}(\boldsymbol{\xi}(\mathbf{x})) \cdot \mathbf{v}(\mathbf{x}), \tag{10.23}$$

where the drag coefficient tensor $\mathcal{B}$ is shown to depend on the local tangent vector $\boldsymbol{\xi}$ at point $\mathbf{x}$ of the dislocation line. In the over-damped regime considered here,

---

[9] This behavior is also observed in direct MD simulations (Section 4.5) of dislocation motion under a constant driving force. The dislocation velocity quickly settles down to a constant value after a short transition period. Therefore, on a time scale larger than this transition period, the dislocation velocity responds to the driving force instantaneously.

[10] In the absence of core energy, $\mathbf{f}^{\text{drive}}(\mathbf{x})$ would be the same as the $\mathbf{f}^{\text{PK}}(\mathbf{x})$.

the drag force should exactly balance the total driving force at every point $\mathbf{x}$ on the network C, i.e.

$$\mathbf{f}^{\text{drag}}(\mathbf{x}) + \mathbf{f}^{\text{drive}}(\mathbf{x}) = 0. \tag{10.24}$$

For this condition to hold, local velocity at every point $\mathbf{x}$ on the dislocation must obey the following equation:

$$\mathcal{B}(\boldsymbol{\xi}(\mathbf{x})) \cdot \mathbf{v}(\mathbf{x}) = \mathbf{f}^{\text{drive}}(\mathbf{x}). \tag{10.25}$$

However, dislocations in our line DD model are constrained to be piecewise straight so that every point $\mathbf{x}$ on a given straight segment is "slaved" to its two end nodes. This means, in particular, that the velocity at an arbitrary point $\mathbf{x}$ on the dislocation can be written in terms of the nodal velocities as

$$\mathbf{v}(\mathbf{x}) = \sum_i \mathbf{v}_i N_i(\mathbf{x}), \tag{10.26}$$

where $N_i(\mathbf{x})$ is the shape function defined in the previous section. Practically, the sum involves only two nodes connected by the segment containing point $\mathbf{x}$, since shape functions of other nodes are zero on this segment.

This constraint makes it impossible for condition (10.24) to be satisfied everywhere on the dislocation network. Instead, let us require that this condition is satisfied in a weaker sense, namely

$$\oint_C N_i(\mathbf{x}) \left[\mathbf{f}^{\text{drag}}(\mathbf{x}) + \mathbf{f}^{\text{local}}(\mathbf{x}) \right] \mathrm{d}L(\mathbf{x}) = 0 \quad \text{for all } i, \tag{10.27}$$

leading to the following equation for each node i:

$$\oint_C N_i(\mathbf{x}) \mathcal{B}(\boldsymbol{\xi}(\mathbf{x})) \cdot \mathbf{v}(\mathbf{x}) \mathrm{d}L(\mathbf{x}) = \mathbf{f}_i, \tag{10.28}$$

where $\mathbf{f}_i$ is the nodal force derived in Section 10.1. Substituting eq. (10.26) in the above expression produces

$$\sum_j \mathbf{B}_{ij} \cdot \mathbf{v}_j = \mathbf{f}_i \quad \text{for all } i, \tag{10.29}$$

where

$$\mathbf{B}_{ij} \equiv \oint_C N_i(\mathbf{x}) \mathcal{B}(\boldsymbol{\xi}(\mathbf{x})) N_j(\mathbf{x}) \mathrm{d}L(\mathbf{x}). \tag{10.30}$$

Thus, we have obtained a set of linear equations relating the velocities of all nodes to the nodal forces. The solution of these equations has a nice property: the rate of energy dissipation integrated over the dislocation network exactly matches the discrete sum over the nodes [112]:

$$\dot{W} \equiv - \oint_C \mathbf{f}^{\text{drag}}(\mathbf{x}) \cdot \mathbf{v}(\mathbf{x}) \, dL(\mathbf{x}) = \sum_i \mathbf{f}_i \cdot \mathbf{v}_i. \tag{10.31}$$

This justifies our choice for the weak condition in eq. (10.27).

The nodal velocities can now be obtained from the nodal forces by solving the linear set of equations (10.29) with $\mathbf{B}_{ij}$ defined by eq. (10.30) in terms of the orientation-dependent drag coefficient tensor $\mathcal{B}(\boldsymbol{\xi})$. In the following, we present two possible functional forms for $\mathcal{B}(\boldsymbol{\xi})$, taking into account some of the realistic aspects of dislocation motion in FCC and BCC metals, respectively.

10.2.2   A Mobility Model for FCC Metals

As has already been noted, motion of a discretization node along the local line direction produces no change in the dislocation shape: such motion should not incur any drag force. This can be accounted for by making the drag coefficient tensor $\mathcal{B}(\boldsymbol{\xi})$ proportional to $(\mathbf{I} - \boldsymbol{\xi} \otimes \boldsymbol{\xi})$, where $\mathbf{I}$ is the identity tensor. Also, as was discussed earlier in Section 1.3, dislocation motion out of the glide plane (climb) is usually much more difficult than motion within the glide plane (glide). This anisotropy can be captured by assigning a large drag coefficient B_c for the climb direction. At low temperatures, B_c may become so large that dislocation motion is practically constrained to the glide planes. In such cases, a more efficient and numerically stable approach is first to ignore the glide/climb anisotropy in the construction of tensor $\mathcal{B}$. Then, after the nodal velocities are obtained by solving eq. (10.29), their climb components should be projected out to ensure that the resulting velocity of node i satisfies the following glide constraints

$$\mathbf{v}_i \cdot \mathbf{n}_{ij} = 0 \tag{10.32}$$

for all segments i–j that are connected to node i, where $\mathbf{n}_{ij}$ is the normal vector of the glide plane for segment i–j.

For a non-screw segment, the glide plane is parallel to the segment direction $\boldsymbol{\xi}_{ij}$ and its Burgers vector $\mathbf{b}_{ij}$ so that

$$\mathbf{n}_{ij} = \frac{\mathbf{b}_{ij} \times \boldsymbol{\xi}_{ij}}{\|\mathbf{b}_{ij} \times \boldsymbol{\xi}_{ij}\|}. \tag{10.33}$$

However, for a screw segment $\boldsymbol{\xi}_{ij}$ and $\mathbf{b}_{ij}$ are parallel and the glide plane is not defined. This is a reflection of the fact that, in principle, conservative motion of a screw segment is not confined to any particular plane (Section 1.3).

In some crystals even screw dislocations tend to glide in certain well-defined crystallographic planes, e.g. {111} planes in FCC metals. The reason for this confinement is that dislocations in FCC metals can spontaneously dissociate into pairs of partial dislocations. Such dissociation lowers the dislocation's energy and is only possible on {111} planes (Section 8.3). Two partial dislocations produced by the dissociation of a screw dislocation are no longer screws but have a well-defined glide plane that coincides with the dissociation plane. Therefore, it may be a good idea to assign a glide plane normal $\mathbf{n}_{ij}$ for all segments in an FCC crystal, even for screw dislocations. Occasionally, even in FCC crystals, screw dislocations find a way to overcome this confinement, and move out of their glide planes and disso-ciate into the second glide plane. These relatively infrequent transitions are called *cross-slip*. To allow the screws to cross-slip, $\mathbf{n}_{ij}$ can be treated as an extra degree of freedom in the model. When a segment's orientation becomes screw, its normal vector $\mathbf{n}_{ij}$ can be allowed to change (flip) occasionally from one $\langle 111 \rangle$ direction to the other, inducing a cross-slip event.

Lastly, because the atomistic core structure depends on the line orientation, so does dislocation mobility. For dislocation segments confined to their glide planes, the line orientation is fully specified by the character angle θ between $\boldsymbol{\xi}$ and the Burgers vector $\mathbf{b}$. Assuming that the drag coefficient varies smoothly from screw (B_s) to edge (B_e) orientation, a possible form for the drag tensor that accounts for the above-mentioned physical aspects is

$$\mathcal{B}(\boldsymbol{\xi}) = (B_s \cos^2 \theta + B_e \sin^2 \theta) \, (\mathbf{I} - \boldsymbol{\xi} \otimes \boldsymbol{\xi}). \tag{10.34}$$

The nodal velocities can now be computed using the following algorithm.

Algorithm 10.1

1. Compute the matrix of drag coefficients $\mathcal{B}_{ij}$ using eq. (10.30) and eq. (10.34).

2. Solve for nodal velocities $\mathbf{v}_i$ from eq. (10.29).

3. Orthogonalize $\mathbf{v}_i$ with respect to the normals $\mathbf{n}_{ij}$ of all segments $i-j$ connected to i.

The mobility model introduced here applies to all nodes, regardless of their con-nectivity (number of neighboring nodes). The linear relationship between velocities and forces is a reasonable model for dislocations in FCC metals, because their intrinsic lattice resistance (i.e. Peierls stress) to dislocation motion is usually small. In other materials, the intrinsic resistance can be large, resulting in a non-linear velocity–force relation sometimes described by a power law. In this case, eq. (10.29) shall be replaced by a set of non-linear equations that have to be solved iteratively (see Problem 10.2.1).

The mobility model described here can be further simplified. For example, if the drag coefficient does not depend on the character angle, eq. (10.34) becomes

$$\mathcal{B}(\boldsymbol{\xi}) = B \, (\mathbf{I} - \boldsymbol{\xi} \otimes \boldsymbol{\xi}). \tag{10.35}$$

Furthermore, assuming that all nodes connected to node i have velocities that are close to the velocity of node i, the latter can be approximated as

$$\mathbf{v}_i \approx \left(\sum_j \mathbf{B}_{ij} \right)^{-1} \cdot \mathbf{f}_i. \tag{10.36}$$

With further simplifying assumptions (see Problem 10.2.2), this can be reduced to

$$\mathbf{v}_i \approx \frac{\mathbf{f}_i}{B \sum_j L_{ij}/2}, \tag{10.37}$$

where L_{ij} is the length of segment $i-j$ and the sum is over all nodes j connected to node i. Mobility functions similar to eq. (10.37) are widely used in line DD simulations. Here this function is shown to be an approximation of the more rigorous form, eq. (10.29).

10.2.3 A Mobility Model for BCC Metals

In BCC metals, screw dislocations do not dissociate into partial dislocations. Therefore, for simulations of BCC crystals, it may be better not to assign glide-plane normals $\mathbf{n}_{ij}$ to screw segments. Instead, let the screw dislocations have the same mobility in all directions perpendicular to the line, i.e.

$$\mathcal{B}(\boldsymbol{\xi}) = B_s (\mathbf{I} - \boldsymbol{\xi} \otimes \boldsymbol{\xi}), \quad \text{when } \boldsymbol{\xi} \parallel \mathbf{b}, \tag{10.38}$$

where B_s is a drag coefficient for the motion of screw dislocations. This isotropic mobility for screws mimics the so-called "pencil-glide" behavior of dislocations observed in BCC metals at elevated temperatures.

At the same time, the drag coefficient tensor for the non-screw segments should remain anisotropic with respect to glide and climb. Let us define the drag coefficient tensor for a pure edge dislocation ($\theta = 90°$) as

$$\mathcal{B}(\boldsymbol{\xi}) = B_{eg} (\mathbf{m} \otimes \mathbf{m}) + B_{ec}(\mathbf{n} \otimes \mathbf{n}), \quad \text{when } \boldsymbol{\xi} \perp \mathbf{b}, \tag{10.39}$$

where B_{eg} and B_{ec} are the drag coefficients for motion in glide and climb directions, respectively, with $B_{eg} \ll B_{ec}$. The unit vectors are defined as $\mathbf{n} = \mathbf{b} \times \boldsymbol{\xi}/\|\mathbf{b} \times \boldsymbol{\xi}\|$ and $\mathbf{m} \equiv \mathbf{n} \times \boldsymbol{\xi}$. To completely specify the drag coefficient tensor for all segment orientations, we need a function that smoothly interpolates between two limits, pure screw ($\theta = 0°$) and pure edge ($\theta = 90°$) orientations. A possible form for such an interpolation function is [113]

$$B(\xi) = B_g(\mathbf{m} \otimes \mathbf{m}) + B_c(\mathbf{n} \otimes \mathbf{n}), \quad \text{when } \xi \times \mathbf{b} \neq 0 \tag{10.40}$$

$$B_g = \left[B_{eg}^{-2} \|\mathbf{b} \times \xi\|^2 + B_s^{-2} (\mathbf{b} \cdot \xi)^2 \right]^{-1/2}$$

$$B_c = \left[B_{ec}^{2} \|\mathbf{b} \times \xi\|^2 + B_s^{2} (\mathbf{b} \cdot \xi)^2 \right]^{1/2}.$$

In the limit $\xi \times \mathbf{b} \to 0$, the above function becomes the same as eq. (10.38).

Summary

- The equation of nodal motion requires a mobility function to relate nodal velocity to nodal force. The first-order equation of over-damped motion is valid in most situations of practical interest.

- The mobility functions are material-specific.

- In general, the velocity of a node can depend on forces on the other nodes. A local mobility function is an approximation in which the velocity of a node depends only on the force on the same node.

Problems

10.2.1. Assume that the dislocation velocity–force relation is given by the following power law:

$$\mathbf{f}^{\text{drag}} = -(v/v_0)^{m-1}(B \cdot \mathbf{v}), \tag{10.41}$$

where $v = \|\mathbf{v}\|$, and v_0 and m are constants. Derive equations for the nodal velocities analogous to eq. (10.29).

10.2.2. Assume that the dislocation lines are very smooth so that orientations of the neighboring segments are very close to each other. Show that in this limit the nodal force on a discretization node is nearly perpendicular to the orientations of two segments connected to the node. Show that, in the same limit, the mobility function given in eq. (10.36) is approximated by eq. (10.37).

10.3 Time Integrators

Taken together, the nodal forces and the mobility model comprise the equations of motion for the nodes:

$$\mathbf{v}_i \equiv \frac{d\mathbf{r}_i}{dt} = \mathbf{g}_i(\{\mathbf{r}_j\}), \tag{10.42}$$

where function $\mathbf{g}_i$ subsumes both the nodal force and the mobility model. This is a first-order ordinary differential equation (ODE), for which the simplest numerical integrator is the so-called *Euler forward* method:

$$\mathbf{r}_i(t + \Delta t) = \mathbf{r}_i(t) + \mathbf{g}_i(\{\mathbf{r}_j(t)\})\Delta t. \tag{10.43}$$

Unfortunately, the Euler forward method can encounter problems when applied to line DD simulations. Most notably, it is unclear *a priori* how large the time step Δt can be.[11] A time step too small is obviously inefficient but a time step too large can make the numerical integration inaccurate and even unstable. As the dislocation lines evolve, the optimal time step is probably changing too. Below, we describe an algorithm that automatically adjusts the time step to meet a prespecified level of accuracy.

A numerical integrator with a higher order of accuracy is the *trapezoid* method,[12]

$$\mathbf{r}_i(t + \Delta t) = \mathbf{r}_i(t) + \frac{\mathbf{g}_i(\{\mathbf{r}_j(t)\}) + \mathbf{g}_i(\{\mathbf{r}_j(t + \Delta t)\})}{2}\Delta t. \tag{10.44}$$

This method is more difficult to use because the unknowns $\mathbf{r}_i(t + \Delta t)$ appear on both sides of the equation. The unknowns can only be obtained by solving a set of, generally non-linear, equations. The trapezoid method is an *implicit* method, whereas the Euler forward method is an *explicit* method. Obviously, an implicit method requires more work than an explicit method at every time step. Yet implicit methods usually offer better numerical stability.

Because $\mathbf{g}_i$ is a complicated function in the line DD model, solving the implicit equation (10.44) can be difficult. A compromise can be reached by combining the two methods described above. Specifically, let us substitute the estimate for the new nodal position from the Euler forward method into the right-hand side of eq. (10.44), i.e.

$$\mathbf{r}_i^P(t + \Delta t) = \mathbf{r}_i(t) + \mathbf{g}_i(\{\mathbf{r}_j(t)\})\,\Delta t \tag{10.45}$$

$$\mathbf{r}_i(t + \Delta t) = \mathbf{r}_i(t) + \frac{\mathbf{g}_i(\{\mathbf{r}_j(t)\}) + \mathbf{g}_i(\{\mathbf{r}_i^P(t + \Delta t)\})}{2}\Delta t. \tag{10.46}$$

This is the so-called *Euler–trapezoid* method, which is an explicit method. $\mathbf{r}_i^P(t + \Delta t)$ and $\mathbf{r}_i(t + \Delta t)$ are called *predictor* and *corrector*, respectively. The Euler–trapezoid method is more accurate and stable than the Euler forward method, although it requires twice as many evaluations of function $\mathbf{g}_i$ at each time step. Whenever the difference between the predictor and the corrector becomes too large,

[11] This is unlike the situation in the molecular dynamics simulations of solids, where the intrinsic frequencies of atomic vibrations naturally define the magnitude of the time steps.

[12] A higher accuracy results because the trapezoid method approximates the derivatives by the centered difference, whereas the Euler forward method is based on a non-centered difference.

it is an indication that integration is not accurate and may go unstable if continued with the current time step Δt. When this is observed, the time step should be reduced. This integration procedure with self-adjusting time steps is described in more detail in the following algorithm. Two input parameters of this algorithm are the maximum allowed time step Δt_{max} and the desired accuracy ϵ.

Algorithm 10.2

1. Initialize time step $\Delta t := \Delta t_{\text{max}}$.

2. $\Delta t_0 := \Delta t$.

3. Compute predictor $\mathbf{r}_i^{\text{P}}(t + \Delta t)$ and corrector $\mathbf{r}_i(t + \Delta t)$ from eq. (10.45) and eq. (10.46).

4. If $\max_i(\|\mathbf{r}_i^{\text{P}}(t + \Delta t) - \mathbf{r}_i(t + \Delta t)\|) > \epsilon$, reduce the time step $\Delta t := \Delta t/2$ and go to 3.

5. $t := t + \Delta t$.

6. If $\Delta t = \Delta t_0$, increase the step to $\Delta t := \min(1.2\,\Delta t, \Delta t_{\text{max}})$.

7. Return to 2, unless total number of cycles is reached.

In this algorithm, whenever the difference between predicted and corrected positions for any node exceeds threshold ϵ, the time step is decreased by half and new positions for all nodes are recomputed with the reduced time step. When no time-step reduction is necessary, the algorithm attempts to increase Δt by 20 per cent for the next cycle (but not to exceed a preset value Δt_{max}). Obviously, there is much room for improving the efficiency of this algorithm. For example, because the difference between predicted and corrected positions varies from node to node, there is no need to repeat calculations at a smaller time step for those nodes whose new positions have already converged, i.e. where the difference between the predictor and the corrector is already smaller than ϵ. Allowing time steps to vary from node to node based on the nodal convergence behavior can greatly enhance the computational efficiency but requires a more complicated algorithm. In situations when different variables of the system require very different time steps, the ODEs are said to be *stiff*, as is the case here. More sophisticated numerical methods for integrating stiff ODEs can be found in [28].

10.3.1 An Example: Frank–Read Source

The algorithms described so far allow us to perform simple line DD simulations. Let us consider a Frank–Read source as an example. The initial condition for this simulation is chosen to be a rectangular dislocation loop in the y–z plane, with loop height 40 (along z, in arbitrary units) and width 20 (along y), as shown in Fig. 10.4(a). The Burgers vector is $\mathbf{b} = [100]$, parallel to the x axis. To imitate a

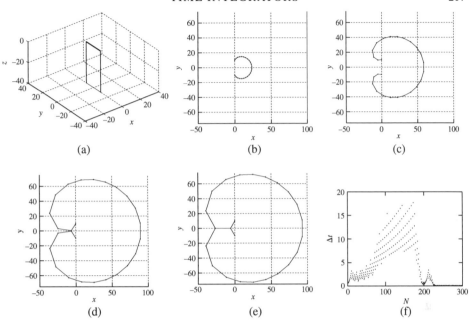

FIG. 10.4. Line DD simulation of a Frank–Read source. (a) Initial condition: a rectangular loop in the y–z plane with the Burgers vector parallel to the x axis. Snapshots at (b) cycle $N = 100$, (c) cycle $N = 150$, (d) cycle $N = 192$, and (e) cycle $N = 300$. (f) Time step Δt at each cycle.

Frank–Read source, only the top part of the loop is allowed to move in the x–y plane, while the other three sides of the rectangle are held fixed. Let us fix four corner nodes of the rectangular loop and discretize the top part into 25 smaller segments of equal length. Only 24 inner nodes are allowed to move within the x–y plane. Let us set the elastic constants to $\mu = 1$, $v = 0.3$, the core radius to $r_c = 0.1$, the core energy to zero, and external stress to $\sigma_{xz} = 0.1$.

For simplicity, let us use the mobility function described in eq. (10.37) with drag coefficient $B = 1$ and employ Algorithm 10.2 for the integration of the nodal equations of motion with $\Delta t_{\max} = 20$ and $\epsilon = 0.01$. Figures 10.4(b)–(e) show four snapshots of the mobile part taken at different times in the simulation. Fig. 10.4(f) is a plot of the time step Δt at each integration cycle.

In Fig. 10.4(b), the dislocation line bows out under applied stress. The line length continues to increase and, in Fig. 10.4(c), the line loops around its two pinning points. Going from (a) to (c), the discretization segments become longer while the time step Δt continues to grow, as shown in Fig. 10.4(f). Then, at about cycle 192, two of the nodes collide with each other causing the time step to drop precipitously. From then on, the time step remains small to the end of this simulation where two dislocation segments with opposite Burgers vectors are seen to overlap with each other, as shown in Fig. 10.4(e).

Several aspects of this simulation are not satisfactory and need to be improved. In a more realistic simulation, collision of two dislocations with opposite Burgers vectors would have led to their recombination. However, in this simulation all dislocation segments continue to exist even though some of them increasingly overlap with each other. A related problem is that, in order to accurately describe the mutual interaction between the redundant (overlapping) segments, the algorithm selects very small time steps. This would be unnecessary if these two segments were allowed to recombine, as they should. To account for dislocation recombination in a line DD simulation, it is necessary to make changes in the line topology, which will be discussed in the next section.

At the beginning of this simulation, the dislocation line is short and straight, so that 24 nodes are more than sufficient (even wasteful) to represent it. However, at the later stages in the same simulation, the same 24 nodes seem to be insufficient to accurately describe the shape of the entire dislocation line, as sharp corners become visible in Fig. 10.4(d). This behavior is a potential source of inefficiency and inaccuracy. A good way to keep the quality of line representation more or less constant during the simulation is to enable the algorithm to add and delete nodes, i.e. to "remesh". This will also be described in the following section.

Summary

- The numerical integrator for the nodal equations of motion should be chosen to balance stability, accuracy, and efficiency. It is often advantageous to adjust the time step automatically during a line DD simulation.

10.4   Topological Changes

The simple simulation example presented in the previous section showed that, to be faithful to the physics of dislocation behavior and for the sake of numerical efficiency, it is important for the line DD simulations to be able to handle topological changes, i.e. changes in the connectivity of the nodes. For example, should a line change its length and/or curvature during the simulation, it may become necessary to adjust the number of nodes used to represent this dislocation line. Another interesting situation demanding a change in nodal topology arises when two dislocation lines meet in space, which may result in recombination or zipping the lines together into a junction. The types of topological changes encountered in a line DD simulation can be many and the task of accounting for all the different cases properly is complicated. This complexity compounds when line DD simulations are performed on a parallel computer with distributed memory: here, in addition to managing the evolving topology of the nodes, the processors must also negotiate and communicate any topological changes taking place across the boundaries of neighboring computational domains. These and other issues of parallel implementation of the line DD method will be discussed in Section 10.5.

The set of algorithms presented in this section was developed with full awareness of the complexity at hand. Whenever possible, we have tried to keep the algorithms simple and logically transparent even, perhaps, at the expense of their elegance. Fortunately, within the nodal representation adopted here, arbitrarily complex topological changes can be produced by combinations of just two elementary topological operators: *merge* (two nodes merge into one node) and *split* (one node splits into two nodes). Implementation of these two operators is relatively straightforward. We just need to ensure is that, for every node and segment involved in a split or merge rearrangement, the sum rules for the Burgers vectors remain satisfied in the end. This section describes the steps necessary to implement these two basic operators in a line DD simulation.

10.4.1 Remeshing

To refine or coarsen line representation, it is necessary to *add* nodes to or *delete* nodes from the existing set of nodes. For example, in the geometry shown in Fig. 10.5, it may be desirable to add a new node E between nodes A and B and/or to delete node D between nodes B and C. Logically, *add* and *delete* are special instances of *split* and *merge* operators applied to nodes with only two arms. For example, adding node E can be regarded as splitting node A into A and E. Likewise, deleting node D can be regarded as merging nodes D and B into B. However, given their simplicity, in the following we will first describe algorithms for *add* and *delete* operators and defer the introduction of the more general *split* and *merge* algorithms until later. The following algorithm manipulates the nodal data as required to *add* node E between A and B. The new node is placed at the mid-point of segment A–B.

Algorithm 10.3 (add node)

1. Allocate node E and assign its position as $\mathbf{r}_E := (\mathbf{r}_A + \mathbf{r}_B)/2$.

2. Allocate new Burgers vectors $\mathbf{b}_{AE} := \mathbf{b}_{AB}$, $\mathbf{b}_{EA} := \mathbf{b}_{BA}$, $\mathbf{b}_{BE} := \mathbf{b}_{BA}$, $\mathbf{b}_{EB} := \mathbf{b}_{AB}$.

3. Deallocate old Burgers vectors $\mathbf{b}_{AB}$ and $\mathbf{b}_{BA}$.

Conversely, the following algorithm deletes node D between nodes B and C.

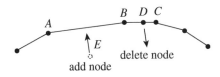

FIG. 10.5. To improve representation of dislocation lines such as this one, it may be necessary to *add* a new node E between A and B or to *delete* node D between B and C.

Algorithm 10.4 (delete node)

1. Allocate new Burgers vectors $\mathbf{b}_{BC} := \mathbf{b}_{BD}$, $\mathbf{b}_{CB} := \mathbf{b}_{CD}$.

2. Deallocate old Burgers vectors $\mathbf{b}_{BD}$, $\mathbf{b}_{DB}$, $\mathbf{b}_{CD}$, $\mathbf{b}_{DC}$. Deallocate node D.

Now that the operators for *add* and *delete* are in place, how shall we use them to improve the representation of the network geometry? The answer depends on one's definition of the "optimal" representation. One simple possibility is to try to maintain the distances between the neighboring nodes within an upper and lower bound. This can be achieved by always adding a new node in the middle of a segment when the segment becomes too long (e.g. $>l_{max}$) and deleting the node should one of its arms (connected segments) become too short (e.g. $<l_{min}$). To avoid repeatedly adding and deleting nodes on the same segment, the bounds should be chosen such that $l_{max} > 2l_{min}$.

A remeshing algorithm based solely on the segment lengths may not be efficient. This is because in the regions of high local curvature, it is necessary to discretize the lines into fine segments, whereas the same is not required in the regions of low curvature. Let us now consider a slightly more sophisticated approach and demand that, for any directization node (a node with only two neighbors), the orientations of two sequential segments connected to the node should not differ too much. The following algorithm enforces this requirement by monitoring the area of the triangle formed by node (N_0) and its two neighbors (N_1 and N_2). The inputs for this algorithm are l_{min}, l_{max}, A_{min}, and A_{max}.

Algorithm 10.5 (remesh)

1. If node N_0 has more than two arms, go to 7. Otherwise, find nodes N_1 and N_2 connected to node N_0. Obtain positions $\mathbf{r}_0$, $\mathbf{r}_1$, $\mathbf{r}_2$ and velocities $\mathbf{v}_0$, $\mathbf{v}_1$, $\mathbf{v}_2$ of nodes N_0, N_1, N_2.

2. Compute the area of the triangle formed by nodes N_0, N_1 and N_2, $A := \|(\mathbf{r}_1 - \mathbf{r}_0) \times (\mathbf{r}_2 - \mathbf{r}_0)\|$. Compute the rate of change of the triangle area, dA/dt.

3. If $\|\mathbf{r}_2 - \mathbf{r}_1\| < l_{max}$, $A < A_{min}$ and $dA/dt < 0$, use Algorithm 10.4 to *delete* node N_0, exit.

4. If $\|\mathbf{r}_1 - \mathbf{r}_0\| < l_{min}$ or $\|\mathbf{r}_2 - \mathbf{r}_0\| < l_{min}$, *delete* node N_0, exit.

5. If $A > A_{max}$ and $\|\mathbf{r}_1 - \mathbf{r}_0\| \geq l_{min}$, use Algorithm 10.3 to *add* a new node between N_0 and N_1.

6. If $A > A_{max}$ and $\|\mathbf{r}_2 - \mathbf{r}_0\| \geq l_{min}$, *add* a new node between N_0 and N_2.

7. If any arm of node N_0 is longer than l_{max}, *add* a new node in the middle of the arm.

To remesh the entire dislocation network, Algorithm 10.5 should be applied sequentially to all nodes. In this algorithm, a node with three or more arms will never be deleted. Such nodes are physical entities where several dislocation lines merge together and should not be altered by the remesh algorithm. Because the new node is always added at the midpoint of the segment, the area of the triangle formed by this node and its two neighbors is exactly zero at the instance of addition. If it were not for the $dA/dt < 0$ condition in step 3, the new node would be deleted the very next time the *remesh* was carried out. Because the new node always moves out of the straight line connecting its two neighbors, the derivative dA/dt is positive, allowing the new node to survive an immediate deletion.

10.4.2 Split and Merge Operators

The need to account for dislocation reactions calls for more general *split* and *merge* operators. We would like to emphasize that neither *split* nor *merge* decides under what conditions and exactly what kind of topological operation is to be performed. As will be discussed later in this section, it takes other algorithms to make such decisions. What *split* and *merge* do is topological bookkeeping. They make sure that, given the initial topology, the connectivity of the nodes in the resulting topology is correct. The implementation and use of *split* and *merge* operators are detailed below.

To fully define a *split* operation, it is not only necessary to specify which node to split, but also how the existing arms of this node should be divided between two new nodes. As an example, consider the configuration in Fig. 10.6(a) in which node 0 has n arms. Suppose the decision is made to split node 0 into two nodes 0 and a. Assume also that, in this split, node a is entitled to keep arms $1 \ldots s$ while the new node 0 retains arms $s + 1, \ldots, n$ (Fig. 10.6(b)). The following algorithm describes how this can be accomplished. For the case $s = 1$, the *split* operation is equivalent to adding a new node on segment 0–1.

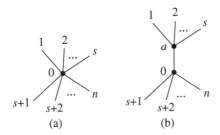

(a)   (b)

FIG. 10.6. *Split* node 0 into two nodes 0 and a. The new node a inherits arms $1 \ldots s$ while node 0 keeps the remaining arms $s + 1 \ldots n$. Depending on the balance of the Burgers vectors, new nodes 0 and a may or may not be connected.

Algorithm 10.6 (split node)

1. Allocate new node a. Allocate new Burgers vectors $\mathbf{b}_{ai} := \mathbf{b}_{0i}$, $\mathbf{b}_{ia} := \mathbf{b}_{i0}$, $i = 1, \ldots, s$.

2. Deallocate Burgers vectors $\mathbf{b}_{0i}$, $\mathbf{b}_{i0}$, $i = 1, \ldots, s$.

3. Evaluate the balance of the Burgers vectors, $\Delta\mathbf{b} := \sum_{i=1}^{s} \mathbf{b}_{ai}$.

4. If $\Delta\mathbf{b} \neq 0$, allocate new Burgers vectors $\mathbf{b}_{0a} := \Delta\mathbf{b}$ and $\mathbf{b}_{a0} := -\Delta\mathbf{b}$.

The *merge* operator can be regarded as the reverse of *split*. This can be visualized as going back from Fig. 10.6(b) to Fig. 10.6(a). If, before *merge*, the two nodes are connected to each other, and at least one of them has only two arms, then *merge* is equivalent to the *delete* operation described earlier. The outcome is more complicated if, prior to *merge*, two nodes about to be merged share one or more common neighbor node, as illustrated in Fig. 10.7(a). In this example, prior to *merge*, node b is connected to both node 0 and node a. After merging, node 0 is connected to node b twice, as shown in Fig. 10.7(b). This violates the condition adopted earlier in Section 10.1 that a pair of nodes cannot be connected by more than one segment. As shown in Fig. 10.7(c), we merge two overlapping segments into one or remove them altogether, depending on the sum of their Burgers vectors. The following algorithm describes the *merge* procedure in detail. In this case, we consider the merging of nodes 0 and a, where node a has s arms before *merge* and disappears after *merge*.

Algorithm 10.7 (merge nodes)

1. Deallocate Burgers vectors $\mathbf{b}_{0a}$ and $\mathbf{b}_{a0}$ if they exist.

2. Allocate new Burgers vectors $\mathbf{b}_{0i} := \mathbf{b}_{ai}$, $\mathbf{b}_{i0} := \mathbf{b}_{ia}$, $i = 1, \ldots, s$.

3. Deallocate Burgers vectors $\mathbf{b}_{ai}$, $\mathbf{b}_{ia}$, $i = 1, \ldots, s$. Deallocate node a.

4. Double-loop through all segments connected to node 0. If two different segments i and j are found to connect node 0 to the same neighbor node

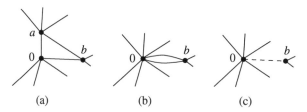

(a)   (b)   (c)

FIG. 10.7. (a) Before *merge*, nodes 0 and a may have a common neighbor b that will result in a double connection between nodes 0 and b after *merge*. (c) This double connection is replaced by a single connection or completely removed, depending on the sum of the Burgers vectors of two overlapping segments.

b, compute $\Delta \mathbf{b} := \mathbf{b}_{0i} + \mathbf{b}_{0j}$. Remove Burgers vectors $\mathbf{b}_{0i}$, $\mathbf{b}_{0j}$, $\mathbf{b}_{i0}$, $\mathbf{b}_{j0}$. If $\Delta \mathbf{b} \neq 0$, allocate links $\mathbf{b}_{0i} := \Delta \mathbf{b}$ and $\mathbf{b}_{i0} := -\Delta \mathbf{b}$. If node b is left with no connected neighbors, deallocate node b. If node 0 is left with no connected neighbors, deallocate node 0.

10.4.3   When and How to Use *merge*

As stated above, the purpose of *split* and *merge* operators is to do bookkeeping of nodal connections. In order to enable consistent physical treatment of dislocation reactions, it is also necessary to specify a set of rules defining when and how to apply these two operators, such that all possible topological switches are described properly.

Our simple rule for invoking the *merge* operator is whenever two nodes or two segments are found close to each other. Consider Fig. 10.8(a) as an example. Assume that d is the minimum distance between segments 1–2 and 3–4 and P and Q are the points of closest approach (an algorithm for finding d is described below). Let us define the two segments to be *in contact* if $d < r_a$, where r_a is a collision distance parameter (subscript a stands for annihilation).

Let us now see what has to happen in the line DD model in order to enable a possible reaction between two segments 1–2 and 3–4. The first step is to add new two nodes at points P and Q on the two colliding segments.[13] Because the new nodes just introduced are within contact radius r_a of each other, the *merge* is called upon to merge them into new node P', as shown in Fig. 10.8(b). To complete the

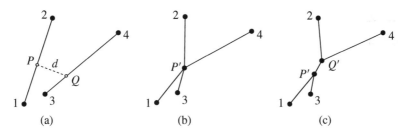

FIG. 10.8. (a) The minimum distance d between two unconnected segments 1–2 and 3–4 is reached at points P and Q. Two segments are considered to be in contact if $d < r_a$. New nodes are introduced at points P and Q. (b) Nodes P and Q are merged into a single node P'. (c) Node P' splits into two new nodes P' and Q', possibly leading to a topology different from that in (a).

[13] This is assuming that points P and Q do not coincide with the end nodes of two colliding segments. If P or Q does coincide with one of the end nodes, the appropriate end node should be used in place of P or Q in the following topological rearrangements.

dislocation reaction, the new node P' should now be allowed to split into two new nodes, such as P' and Q' in Fig. 10.8(c).

Finding minimum distance between two straight segments Given four nodes $\mathbf{r}_1$, $\mathbf{r}_2$, $\mathbf{r}_3$, $\mathbf{r}_4$, the minimum distance d between segments 1–2 and 3–4 can be obtained by solving the following minimization problem:

$$d^2 = \min_{\alpha, \beta \in [0,1]} F(\alpha, \beta), \tag{10.47}$$

where

$$F(\alpha, \beta) \equiv \|[(1 - \alpha)\mathbf{r}_1 + \alpha\mathbf{r}_2] - [(1 - \beta)\mathbf{r}_3 + \beta\mathbf{r}_4]\|^2. \tag{10.48}$$

Because $F(\alpha, \beta)$ is quadratic in α and β and positive definite, it has a single minimum. The optimal values of α and β can be obtained by solving the following set of linear equations:

$$\frac{\partial F}{\partial \alpha} \equiv \mathbf{r}_{12} \cdot \mathbf{r}_{31} + (\mathbf{r}_{12} \cdot \mathbf{r}_{12})\alpha - (\mathbf{r}_{12} \cdot \mathbf{r}_{34})\beta = 0 \tag{10.49}$$

$$\frac{\partial F}{\partial \beta} \equiv \mathbf{r}_{34} \cdot \mathbf{r}_{31} + (\mathbf{r}_{34} \cdot \mathbf{r}_{12})\alpha - (\mathbf{r}_{34} \cdot \mathbf{r}_{34})\beta = 0, \tag{10.50}$$

where $\mathbf{r}_{ij} \equiv \mathbf{r}_j - \mathbf{r}_i$. When the solution of these equations is outside domain $\alpha, \beta \in [0, 1]$, the minimum should be searched for on the boundaries of this domain. This is done by substituting $\alpha = 0, 1$ into the above equations and solving for β and substituting $\beta = 0, 1$ and solving for α. The lowest value of the four values of F is the minimum. In the special case when $\mathbf{r}_{12} \parallel \mathbf{r}_{34}$, the equations become degenerate but the minimum distance can be found on the boundary of domain $\alpha, \beta \in [0, 1]$ using the same procedure as described above.

10.4.4   When and How to *split* a Node

Let us regard any node with four or more neighbors, such as node P' in Fig. 10.8(b), as a *multi-arm node*. To enable dislocation reactions, we would like to allow such nodes to split into two nodes, as in Fig. 10.8(c). Topologically, there are multiple ways for a multi-arm node to split. In particular, the 4-node shown in Fig. 10.8(b) can split in three topologically distinct ways: (12)(34), (13)(24), and (14)(23). Which one of them should we choose?

To justify our choice for the splitting outcome, let us observe that in the over-damped motion regime considered here our dislocation network should evolve towards the minimum of its free energy. In fact, if it were not for the mobility anisotropy and/or glide constraints discussed in Section 10.2, the dislocation nodes would have moved in the direction of nodal forces, which is along the steepest-descent path. However, the ability of the nodes to respond to their forces and, hence, to minimize the energy is affected by the mentioned constraints. The effectiveness of the network's descent towards the underlying minimum of free energy is quantified by the energy reduction rate that, in the over-damped regime, is also the rate of heat production $\dot{Q}$. It can be said that the network evolves in such a way as to maximize $\dot{Q}$ while satisfying all of its constraints. If this is applied to the *split* algorithm, it makes sense to expect that, given a choice of topological states to evolve into, the network should select the outcome that maximizes $\dot{Q}$.

Consider a multi-node i experiencing nodal force $\mathbf{f}_i$. The contribution of this node to the energy dissipation rate is

$$\dot{Q}_i = \mathbf{f}_i \cdot \mathbf{v}_i, \tag{10.51}$$

where $\mathbf{v}_i$ is the velocity of the same node obtained by solving the equations of over-damped motion. Now suppose that node i splits into two nodes P and Q such that node P retains $1, \ldots, s$ of the original neighbors and node Q holds on to all the remaining neighbors. Let us now compute the forces on two new nodes $\mathbf{f}_P$ and $\mathbf{f}_Q$ and solve for their velocities $\mathbf{v}_P$ and $\mathbf{v}_Q$. The contribution of two new nodes to the energy dissipation rate is

$$\dot{Q}_{PQ} = \mathbf{f}_P \cdot \mathbf{v}_P + \mathbf{f}_Q \cdot \mathbf{v}_Q. \tag{10.52}$$

Splitting of node i into P and Q should be preferable to node i staying together if $\dot{Q}_{PQ} > \dot{Q}_i$.[14] Similarly, the energy dissipation rate for all other possible splitting outcomes should be computed and the mode with the highest dissipation rate should be selected.

The above prescription implies force calculations for all candidate nodes created under all possible splitting outcomes. Fortunately, all the terms necessary for computing such forces are available at the time when the total force $\mathbf{f}_i$ is computed. Recall that the force on node i is the sum of contributions from many segment–segment interactions (Section 10.1). Let us partition all such contributions by the segments connected to node i, i.e.

$$\mathbf{f}_i = \sum_j \mathbf{f}_i(i - j), \tag{10.53}$$

where $\mathbf{f}_i(i - j)$ includes the interactions of segment i–j with all segments in the network. If eq. (10.53) is compared with eq. (10.14), the fractional force due to segment $i - j$ is[15]

$$\mathbf{f}_i(i - j) = \sum_{(k-l)} \mathbf{f}_i^{\text{el}}(i - j, k - l). \tag{10.54}$$

---

[14] Assuming that splitting of node i does not affect the velocities of other nodes of the network.

[15] This is assuming that the core energy is zero. Otherwise, the new segment connecting nodes P and Q will contribute an additional force to nodes P and Q.

The forces on the two candidate nodes P and Q are simply

$$\mathbf{f}_P = \sum_{j=1}^{s} \mathbf{f}_i(i-j), \qquad (10.55)$$

$$\mathbf{f}_Q = \sum_{j=s+1}^{n} \mathbf{f}_i(i-j). \qquad (10.56)$$

Here, node P is assumed to have inherited the first s segments of the parent node i while the rest of the segments went to node Q.

10.4.5 A Complete Line DD Algorithm

We have now laid out all the major ingredients of a line DD simulation, including the integration of nodal equations of motion and changes in the nodal topology. A complete algorithm combining all these ingredients is given below.

Algorithm 10.8 (line DD simulation)

1. Initialize time $t := 0$.

2. Use Algorithm 10.2 (Euler–Trapezoid method) to compute nodal forces and velocities, and to advance nodal positions by one time step: the magnitude of time step Δt is automatically adjusted to a prespecified accuracy level ϵ. Increment time $t := t + \Delta t$.

3. For each multi-arm node, compute the rates of energy dissipation for all of its possible dissociation outcomes and *split* the nodes if required.

4. For all pairs of unconnected segments, find the minimum separation d. For pairs with $d < r_a$, introduce new nodes at the locations of closest approach.

5. Use Algorithm 10.7 to *merge* all pairs of nodes within r_a of each other.

6. Split every node with more than three neighbors into two nodes if this increases the local energy dissipation rate.

7. *Remesh* the entire dislocation network using Algorithm 10.5.

8. Return to step 2, unless the total number of cycles is reached.

10.4.6 Frank–Read Source Revisited

With the additional capability for handling topological changes, it is instructive to revisit the Frank–Read source first discussed in Section 10.3. The new simulation is performed using Algorithm 10.8 implemented into code DDLab written in Matlab and available at the book's web site. The parameters for line remeshing and dislocation collisions in this simulation were $r_a = 1, l_{\min} = 1, l_{\max} = 10, A_{\min} = l_{\min}^2 \sqrt{3}/4,$

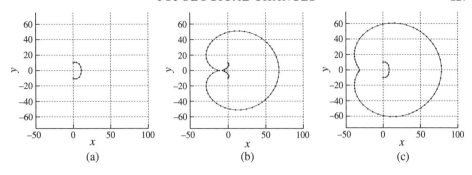

FIG. 10.9. A simulation of the Frank–Read source using Algorithm 10.8 allowing line remeshing and dislocation reactions. The settings are the same as
in Fig. 10.4. (a) A snapshot at cycle $N = 20$, (b) at cycle $N = 100$, (c) at
cycle $N = 150$.

$A_{max} = 16A_{min}$ while all the other parameters were the same as in the earlier
example discussed in Section 10.3. As seen in Fig. 10.9(a), during the early stages
of the simulation, the Frank–Read source is rather short and represented with a
relatively small number of nodes. As the segment bows out under stress, its length
becomes increasingly longer but the remeshing algorithm maintains a high quality
representation of the line by adding more nodes. At cycle $N = 100$, recombination
of two dislocation segments takes place creating a closed dislocation loop detached
from the original Frank–Read source, Fig. 10.9(b). The loop continues to expand
while the Frank–Read source returns to its initial position, Fig. 10.9(c). The source
is now ready to repeat the cycle and to emit another concentric loop.

Summary

- Arbitrary topological changes within a dislocation network can be carried out
  using just two elementary operators, *merge* two nodes into one and *split* one
  node into two.

- *Add* and *delete* operations can be regarded as special cases of *split* and *merge*
  when applied to the discretization nodes. Using *add* and *delete* to remesh the
  lines, it is possible to significantly improve the quality and efficiency of the
  nodal representation.

- Dislocation reactions are enabled by detecting unconnected segments
  that come in close contact with each other. New nodes are added
  onto the colliding segments and are subsequently *merged* together. The
  resulting multi-arm node is then allowed to *split* into two new nodes
  following the topological switch that maximizes the rate of energy
  dissipation.

Problems

10.4.1. Run DDLab simulations to study the dynamics of binary dislocation collisions. Create two straight dislocations with Burgers vec-
tors $\mathbf{b}_1 = 1/2[011]$, $\mathbf{b}_1 = 1/2[1\bar{1}0]$, line directions $\boldsymbol{\xi}_1 = 1/\sqrt{6}[21\bar{1}]$,
$\boldsymbol{\xi}_2 = 1/\sqrt{6}[11\bar{2}]$, and lengths $l_1 = l_2 = 1000$ nm. Position the lines so
that they intersect each other at their mid-points. For simplicity, ignore
the glide constraints in the mobility function by assuming the mobility
is the same in both glide and climb directions. Run DDLab simulation
with the generic FCC mobility function (eq. 10.37) and observe junction
formation. Compute the length and orientation of the resulting junction
segment.

10.4.2. Create a set of configurations similar to the one described in Problem
10.4.1, by varying the initial directions of the two dislocations. Observe
how the behavior of the two dislocations changes from attractive interaction, resulting in zipping long junctions, to repulsion. Compare your
observations with an equation derived by Kroupa [7] that gives the force
between two infinite straight dislocations, as a function of their Burgers
vectors $\mathbf{b}_1$, $\mathbf{b}_2$ and line directions $\boldsymbol{\xi}_1$, $\boldsymbol{\xi}_2$:

$$\mathbf{F}_{12} = \frac{\mu}{\|\boldsymbol{\xi}_1 \times \boldsymbol{\xi}_2\|} \frac{\mathbf{R}_{12}}{R_{12}} \left\{ \frac{1}{2}(\mathbf{b}_1 \cdot \boldsymbol{\xi}_1)(\mathbf{b}_2 \cdot \boldsymbol{\xi}_2) - (\mathbf{b}_1 \times \mathbf{b}_2) \cdot (\boldsymbol{\xi}_1 \times \boldsymbol{\xi}_2) \right.$$

$$\left. + \frac{1}{1-\nu} \left[(\mathbf{b}_1 \times \boldsymbol{\xi}_1) \cdot \frac{\mathbf{R}_{12}}{R_{12}} \right] \left[(\mathbf{b}_2 \times \boldsymbol{\xi}_2) \cdot \frac{\mathbf{R}_{12}}{R_{12}} \right] \right\}, \qquad (10.57)$$

where $\mathbf{R}_{12}$ is the shortest vector from dislocation 1 to dislocation 2. It is
reasonable to expect the two dislocations to zip a junction for orientations
where the Kroupa force is attractive. Create a dislocation configuration
that, according to the above equation, is neutral (zero interaction force).
Find out if the transition from attraction (junction zipping) to repulsion
indeed takes place at the orientation predicted by the Kroupa equation.
Some atomistic and line DD simulations of dislocation junction reactions
can be found in [114, 115].

10.4.3. Repeat the series of calculations described in Problem 10.4.2, but this
time use the glide constraints to restrict dislocation segments to move
only in their glide planes, $(1\bar{1}1)$ and (111) respectively. For the junction
segment, allow only two end nodes (with three arms) to move and only
along the $[10\bar{1}]$ intersection line of two glide planes. When two dislocations form a crossed configuration, allow the four-arm node to move
along $[10\bar{1}]$ as well. All these constraints are realistic for dislocations

in FCC metals at moderate temperatures. Depending on the initial dislocation directions, the two dislocations may continue to stay crossed at a point. Find out if the dislocations attract or repel each other in the crossed states. If they do not form a junction, why?

10.4.4. Use one of the junction configurations in Problem 10.4.3 to examine the effect of applied stress on junction behavior. Run a DDLab simulation under uniaxial tensile stress of 100 MPa applied along [101] direction. Raise the stress in 50 MPa increments, each time followed by a series of DD steps until the forces on all nodes become close to zero. Record the level of stress at which the junction fully unzips. Change the direction of tensile stress axis to [0$\bar{1}$1] and repeat the calculations starting from 100 MPa. Compare the minimum stress required to unzip the junction to the previous case. Double the length of both lines and zip the junction again. Calculate the minimum stress required to unzip this junction for the [0$\bar{1}$1] direction of the tensile axis. What is the ratio of the unzipping stresses computed for the two different values of the dislocation length?

10.4.5. Similar to Problem 10.4.1, prepare a configuration of two crossing dislocations by laying them along $\boldsymbol{\xi}_1 = 1/\sqrt{6}\,[21\bar{1}]$ and $\boldsymbol{\xi}_2 = 1/\sqrt{6}\,[112]$. This time, however, assign the same Burgers vector to both dislocations $\mathbf{b}_1 = \mathbf{b}_2 = 1/2[110]$. Let the dislocations move in response to the interaction forces. First perform a simulation with no glide constraints and then perform a similar simulation in which two lines are constrained to move in ($1\bar{1}1$) and ($1\bar{1}\bar{1}$) planes, respectively. Observe dislocation recombination taking place in both cases and the formation of two angular nodes (in the constrained case only).

10.5 Parallel Simulations

Probably the most important application of the line DD method is to the simulations of crystal plasticity. In principle, using line DD simulations one can directly calculate the stress produced in a crystal under a variety of straining conditions. All that is required is to be able to simulate the simultaneous motion of a large number of interacting dislocations over a sufficiently long time. Through a series of such direct simulations it should be possible to compute the material's *constitutive equation* relating the flow stress to the deformation conditions. Such an equation is an all important material input for simulations of macroscopically large solid structures with complex shapes used in various engineering applications. The material's constitutive equation is usually constructed phenomenologically, by fitting to experimental data. Obtaining it directly from a DD simulation instead will finally connect the engineering crystal plasticity model to its physical roots in the underlying physics of dislocations. Furthermore, a DD simulation is not only a means for computing the plastic strength, but also a perfect virtual microscope for observing, *in situ*, why the material responds to the applied stress the way it does.

While the idea of computing the *constitutive equation* directly by line DD simulations sounds simple, it has remained a challenge for materials physicists for a long time. The origin of the difficulties can be traced to dislocation behavior. In a plastically deformed crystal, dislocations tend to bunch together into dense bundles forming spatially heterogeneous patterns over the scales of multiple microns (see Problem 10.5.1). At the same time, some of the smaller details of dislocation patterns consist of short dislocation segments that are only nanometers long. Therefore, a statistically representative simulation volume should contain a large number ($\sim 10^6$) of dislocation segments, whose detailed dynamics needs to be followed for a sufficiently long time. This requires a lot of number crunching, especially because dislocation interactions are long range and expensive to compute. At the moment, the only means to acquire the required computational muscle is through massively parallel computing, which involves running a single simulation simultaneously on a large number ($\sim 10^3$) of processors. How to take advantage of massively parallel computers for line DD simulations is not a trivial issue. This section describes the methods that make it possible.

10.5.1 Scalability

The key parameter that quantifies how efficiently a code runs on a parallel computer is the code's *scalability*. A perfect *strong scalability* means that the code can solve the same computational problem N times faster than before when running on N times more processors. A perfect *weak scalability* means that the code solves a problem N times larger than before in the same time when running on N times more processors. In practice, a code seldom shows perfect scalability of either type but is still considered *scalable* if it can solve the same problem in less time, or a larger problem in the same time, when it runs on more processors. Often, the strong scalability is more difficult to achieve than the weak scalability.

Scalability can be limited for several reasons. Here we will discuss two of them. The most common source of non-scalable computation is the time spent on communications between the processors. Since no such communication is required on a single-processor computer, this is an extra cost of parallel computing. Should the communication time increase with the number of processors,[16] it could overtake the time spent on computation (problem solving) itself. In particularly bad cases, the code may even run slower when more processors are used, meaning that the

---

[16] This is usually the case, especially when global communications are needed such that every processor has to talk to every other processor.

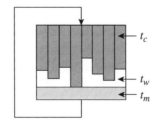

FIG. 10.10. Distribution of time in a parallel simulation. The dark gray bars show
the amounts of time (t_c) it takes each processor to finish its respective calculations required to advance the simulation by a single time step. After all
processors finish their calculations, they have to communicate with each other
before continuing on to the next cycle. The white areas show the amounts of
time (t_w) wasted by each processor waiting for the last one of them to finish its
calculations. The light gray bar represent the communication time (t_m), i.e. time
spent on messaging.

code has lost scalability completely. To enhance scalability, it is desirable to reduce
communication time as much as possible.

An imbalance of computational load among the processors can also lead to a loss
of scalability. In most parallel algorithms, the processors must communicate with
each other after finishing one step of their respective computations, before going
on to the next cycle, as illustrated in Fig. 10.10. If the time it takes to finish the
computation varies among the processors, then a certain amount of time is wasted
while other processors are waiting for the last one to finish its calculations. Let t_c
be the total amount of time spent on computation summed over all the processors,
t_w be the total time wasted by the processors waiting for each other, and t_m be
the total communication (message passing) time. Then the parallel efficiency of
an algorithm is $\eta = t_c/(t_c + t_w + t_m)$. Obviously, to achieve high efficiency it is
desirable to minimize both t_m and t_w. Below we describe methods for reducing
communication cost and load imbalance in line DD simulations.

10.5.2 Spatial Domain Decomposition

Exactly how to break the entire simulation problem into sub-problems and assign
them to different processors is usually the first and most important decision to
make for the developer of a parallel code. A partitioning scheme that minimizes
the need for communications between the processors is more likely to result in a
parallel code with good scalability. For simulations that deal only with short-range
interactions, spatial domain decomposition is a good partitioning scheme.

For the time being, let us assume that the interactions between the nodes in a
line DD simulation are short range, so the nodes do not interact beyond a relatively
short cut-off distance R_c, as is the case for short-range interatomic interactions

(Chapter 2). In this case, one can partition the entire simulation volume into a set of sub-volumes or domains, each assigned to its own processor. Provided the dimensions of the individual domains are larger than R_c, the nodes inside one domain will interact only with the nodes in the same domain and in the nearest-neighbor domains, but no further. Consequently, each domain will have to communicate only with its neighbor domains. When only local communications are required, weak scalability is often readily achievable.

In a line DD simulation, the interaction among dislocation segments is long range. Consequently, every processor will have to communicate with every other processor, no matter which partitioning scheme is used. Such global communications can become a severe communication bottleneck and impair scalability. In some cases, however, decent scalability can still be achieved even in the presence of long-range interactions. As illustrated in Fig. 10.11, the basic idea of the fast multipole method (FMM) is that interactions between dislocations that are sufficiently far away from each other do not have to be accounted for individually. Instead, the dislocations can be lumped into groups whose collective contribution

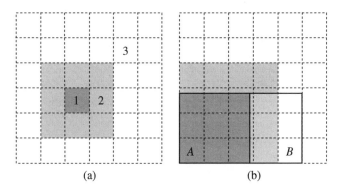

(a)   (b)

FIG. 10.11. (a) The problem space is divided into equal-sized cells (dashed lines) and the task is to compute stress at a point in cell 1 (dark gray). The stress field produced by the segments in cell 1 and its neighboring cells (such as cell 2 and other cells shown in light gray) is computed explicitly, segment by segment. At the same time, the stress produced by the segments in remote cells (such as cell 3 and other cells shown in white), is computed from a multipole expansion. (b) In a parallel simulation, each processor is assigned a rectangular domain (solid lines). At every step, processor A operates on the nodal data in the dark gray region. To calculate the stress inside the domain, processor A has to receive the information about the nodes in the light gray region from its neighbors, such as processor B, by local communications. Furthermore, processor A has to obtain the remote stress produced by all other cells (white area) by global communications. The latter stress is computed using the FMM method.

to the long-range interaction is represented by a few moments of the multipole expansion [116]. This approach has been successfully applied in the context of line DD simulations [117].

10.5.3 Dynamic Load Balance

A simple way to divide the computational burden is to assign equally-sized domains to all processors. Unfortunately, such an approach may lead to serious load imbalance and poor scalability in the line DD simulations. The reason is that, in the course of a simulation, the dislocations tend to bundle together and produce spatially heterogeneous microstructures (see Problem 10.5.1). Because of the large variations in the local dislocation density, the computational load for each processor would vary significantly if equal space partitioning were to be used. Assume that it takes processor i time $\{t_i\}$ to finish its calculations before the next round of communications. Let us define a measure of load imbalance as $I = t_{max}/t_{avg} - 1$, where t_{max} is the maximum time among all processors and t_{avg} is the average over all processors. Essentially, I measures the time wasted on average by a processor waiting for the most overloaded processor to finish its calculations, as a fraction of the total time between two consequtive rounds of communications. The load imbalance can easily reach 100 per cent or higher when a uniform decomposition is used, meaning that more time can be spent idling than computing. The result is low parallel efficiency. To reduce I, it is necessary to use a different partitioning model.

An obvious way to improve load balance is to let the domains have different sizes. Assuming (as a rough approximation) that the time it takes each processor to complete the calculations is proportional to the number of nodes in its domain, let us distribute the nodes equally among all processors. This can be accomplished by partitioning the simulation space recursively along x, y, z directions. First, the entire volume is divided into N_x slabs along the x direction so that each slab contains an equal number of nodes. Then, every slab is divided into N_y columns along the y direction with an approximately equal number of nodes in each column. Lastly, every column is divided into N_z boxes along the z direction. In the end, every box should contain roughly the same number of nodes. The following algorithm describes this recursive procedure in more detail. The resulting domains are shown in Fig. 10.12.

Algorithm 10.9

1. Assuming that the simulation space is a rectangular box, the coordinates of all nodes satisfy the following conditions: $x \in [x_0, x_m]$, $y \in [y_0, y_m]$, $z \in [z_0, z_m]$.

2. Sort the nodes in the simulation space by their x coordinates. Divide the simulation space along the x direction into N_x slabs, each containing the

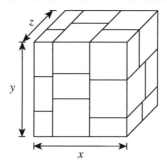

FIG. 10.12. Partitioning of the simulation space into $3 \times 3 \times 2$ domains along x, y, z axes. Each domain is assigned to a separate processor. As a result of recursive space partitioning, the domains will not necessarily have the same volume.

same number (within ± 1) of nodes. The slabs correspond to intervals $x \in [x_{i-1}, x_i]$, $i = 1, 2, \ldots, N_x$ ($x_{N_x} \equiv x_m$).

3. Within each slab $x \in [x_{i-1}, x_i]$, sort the nodes by their y coordinates. Divide the slabs along the y direction into N_y columns each containing the same number of nodes. The columns correspond to $x \in [x_{i-1}, x_i]$, $y \in [y_{j-1}, y_j]$, $j = 1, 2, \ldots, N_y$ ($y_{N_y} \equiv y_m$).

4. Within each column $x \in [x_{i-1}, x_i]$, $y \in [y_{j-1}, y_j]$, sort the nodes by their z coordinates. Divide each column along the z direction into N_z domains each containing the same number of nodes. The domains correspond to $x \in [x_{i-1}, x_i]$, $y \in [y_{j-1}, y_j]$, $z \in [z_{k-1}, z_k]$, $k = 1, 2, \ldots, N_z$ ($z_{N_z} \equiv z_m$).

Such partitioning is certain to improve load balance compared to the uniform domain decomposition. In practice, significant imbalance could still remain, even though each domain contains approximately the same number of nodes. This is because our assumption that each node commands the same number of floating-point operations is inaccurate. Indeed, a node in a high-density region would have more neighbors to interact with and would require more time to complete the calculations than a node in a low-density region. Furthermore, the distribution of nodes in space may change during the simulation, leading to deterioration of the load balance.

To further improve our load balancing procedure, it is desirable to adjust the domain boundaries during the simulation: this is called *dynamic load balancing*. The adjustment can be based on the actual time T_{ijk} it takes for processor (ijk) to finish its computation, where $i = 1, \ldots, N_x$, $j = 1, \ldots, N_y$, $k = 1, \ldots N_z$. Let us try to adjust the boundaries so as to reduce the maximum-to-average ratio in the array of $\{T_{ijk}\}$ for the next cycle. This can be done by shifting the domain boundaries towards domains with higher values of $\{T_{ijk}\}$. Furthermore, in our recursive (hierarchical) partitioning, the boundaries between the "super-domains" (columns and slabs) can be similarly adjusted. The following algorithm performs the required adjustments during the simulation.

Algorithm 10.10

1. Given compute times T_{ijk}, sum them over for each column $t_{ij} = \sum_{k=1}^{N_z} T_{ijk}$ and then for each slab $t_i = \sum_{j=1}^{N_y} t_{ij}$.

2. Adjust positions of the partition planes between the slabs $\{x_i\}$, $i = 1, \ldots, N_x - 1$ by $x_i := x_i + \lambda(t_{i+1} - t_i)$.

3. For each slab i, adjust the partition planes between columns $\{y_j\}$, $j = 1, \ldots, N_y - 1$ by $y_j := y_j + \lambda(t_{i,j+1} - t_{ij})$.

4. For each column $x \in [x_{i-1}, x_i]$, $y \in [y_{j-1}, y_j]$, adjust the partition planes between domains $\{z_k\}$, $k = 1, \ldots, N_z - 1$ by $z_k := z_k + \lambda(T_{ij,k+1} - T_{ijk})$.

Parameter λ can be chosen empirically to ensure stability (λ should not be too large) and efficiency (λ should not be too small). In a typical ParaDiS simulation, e.g. 50 000 dislocation nodes on 512 processors, the above algorithm maintains the imbalance ratio below $I \approx 16$ per cent.

10.5.4 Single-processor Performance

Although this section focuses on the efficient usage of many processors in parallel, performance of each individual processor is just as important. Even seemingly minute differences in the implementation of the same algorithm may lead to dramatic differences in single-processor efficiency (by an order of magnitude) in the actual simulation. This points to the importance of fine tuning the single-processor performance after the algorithm has been designed and implemented [118].

The usual measure of single-processor efficiency is the flop (floating point operation) rate in Mflops/s or Gflops/s (Mega-flops or Giga-flops per second) expressed as a fraction of the peak flop rate quoted for the processor. The flop rate itself measures how many floating point operations are executed per second by the processor. Although it may sound embarrassingly low, codes running as is, without any tuning, typically reach only about 5 per cent of the theoretical peak rate. At the same time, performance of the subroutines in some of the highly optimized numerical libraries may reach over 50 per cent of the theoretical peak. Thus, optimization of the single-processor performance can potentially yield an order-of-magnitude improvement. In a parallel code the effective gain can be enormous, amplified by the number of processors running in parallel. The reader may be interested to test the single-processor performance of the line DD code ParaDiS to be discussed in more detail in the next section.

Summary

- To make efficient use of parallel computing, the line DD algorithm should minimize communications between the processors and balance their computational loads. This can achieved by using recursive space partitioning and dynamic load balancing.

Problem

10.5.1. Let us estimate the computational complexity of line DD simulations of crystal plasticity. For a simulation to be statistically representative, the dynamics of a *large* number of dislocations needs to be followed over a *long* time interval. To size up the challenge of the length scale, consider a typical dislocation microstructure that develops spontaneously in molybdenum during plastic deformation (see Fig. 10.13). This structure is highly heterogeneous, containing features ranging in scales from a few nanometers to several microns. To be able to reproduce this behavior, a simulation box of about $L = 10\,\mu m$ would be necessary. The experimental estimates of dislocation density in such conditions range between $\rho = 10^{13} m^{-2}$ and $\rho = 10^{14} m^{-2}$. Estimate the total length Λ of dislocation lines in the simulation box. Then, assuming that the average segment length in the line DD simulation is $d = 10\,nm$, estimate the total number N of segments in such a simulation. How does N scale with the size of the simulation volume?

10.6 A Virtual Straining Test

In the preceding sections, we discussed the main ingredients of the line DD method. To see how they all work together, let us now consider a specific simulation of a

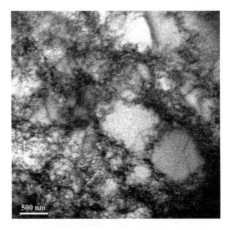

FIG. 10.13. Dislocation microstructure formed in single crystal molybdenum under compressive straining at 500 K (courtesy of L. L. Hsiung).

single crystal deformed at a constant strain rate. Similar to the laboratory experiments, our simulation produces the stress–strain response of the virtual specimen. Also available from the same simulation are all details of the evolution of the dislocation network in the same deformation test.

The simulation described below was performed using code ParaDiS (Parallel Dislocation Simulator) developed at the Lawrence Livermore National Laboratory. The code runs on thousands of processors in parallel, providing the computational power needed for direct simulations of crystal plasticity in bulk crystals [119, 120]. The procedure for acquiring a version of ParaDiS is described at the book's web site.

10.6.1 Mobility Function

In this simulations we employ the generic mobility model for BCC metals described in Section 10.2. The drag coefficient for screw dislocations is set at $B_s = 1\,\mathrm{Pa\,s} \cdot b^{-1}$ and the drag coefficient for the glide motion of edge dislocations is $B_{eg} = 0.1\,\mathrm{Pa\,s} \cdot b^{-1}$. The drag coefficient for climb is $B_{eg} = 10^6\,\mathrm{Pa\,s} \cdot b^{-1}$. The elastic constants and magnitude of the Burgers vector are set to values corresponding to BCC molybdenum: $\mu = 130\,\mathrm{GPa}$ for shear modulus, $\nu = 0.309$ for Poisson's ratio and $b = 2.7\,\text{Å}$ for the Burgers vector.

10.6.2 Boundary and Initial Conditions

The simulation volume is a $10 \times 10 \times 10\,\mu\mathrm{m}^3$ cube with edges aligned along $x = [100]$, $y = [010]$, $z = [001]$ directions. Periodic boundary conditions (PBC) are applied in all three directions. The initial dislocation structure is chosen to mimic that of single crystals of pure molybdenum, consisting of long dislocation lines aligned preferentially along the screw orientations, as shown in Fig. 10.14(a). To enforce PBC, each line that exits the periodic box is made to connect to its own tail on the opposite side of the box. In the initial structure, the Burgers vectors are assigned so that an approximately equal length of dislocations of each of the four Burgers vectors is present.

10.6.3 Loading Condition

The most common deformation test consists of straining the specimen at a fixed rate. In our simulation, the virtual specimen is uniaxially elongated along the x axis at a rate of $1\,\mathrm{s}^{-1}$. Both normal stress components in the transverse directions y and z and all shear stress components remain zero. Implementation of this loading condition in the line DD requires that the average stress is recalculated at every time step, as in the procedure described below.

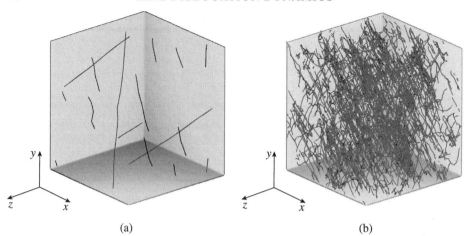

(a)   (b)

FIG. 10.14. (a) The initial dislocation configuration for the line DD simulation
    of BCC Molybdenum under uniaxial straining (see text). (b) The dislocation
    microstructure attained at strain 0.3 per cent.

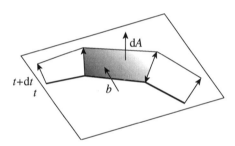

FIG. 10.15. Plastic strain increment produced by the motion of a single dislocation
    segment (see text).

As shown in Fig. 10.15, every time a dislocation segment moves, the following
increment of plastic strain is produced (see also eq. (1.18))

$$\mathrm{d}\varepsilon_{ij}^{\mathrm{p}} = \frac{b_i n_j + b_j n_i}{2\,\Omega}\,\mathrm{d}A, \tag{10.58}$$

where b_i is the Burgers vector, $\mathrm{d}A$ is the incremental area swept out by the segment,
n_i is the normal vector of the glide plane, and Ω is the simulation volume. Total
plastic strain produced at each time step is computed by summing contributions
(10.58) over all dislocation segments. The result is then added to the accumulated
plastic strain tensor $\boldsymbol{\varepsilon}^{\mathrm{p}}(t)$. The xx component of this tensor, $\varepsilon_{xx}^{\mathrm{p}}(t)$ is the time-
dependent plastic elongation along the x axis.

Our straining condition specifies that the total strain along the x axis should be
$\varepsilon_{xx}^{\mathrm{tot}}(t) = 1 \cdot t$. Any mismatch between the total strain and the plastic strain has to

be accommodated by the elastic strain $\varepsilon_{xx}^{el}(t) = \varepsilon_{xx}^{tot}(t) - \varepsilon_{xx}^{p}(t)$. In turn, the elastic strain defines the external stress through the elastic constants. In the considered case of uniaxial elongation, the time-dependent external stress is

$$\sigma_{xx}^{ext}(t) = Y \varepsilon_{xx}^{el}(t) = Y [\varepsilon_{xx}^{tot}(t) - \varepsilon_{xx}^{p}(t)], \qquad (10.59)$$

where $Y = 2\mu(1 + v)$ is the Young's modulus.

10.6.4   Results

Output from this line DD simulation is plotted in Fig. 10.16. During the simulation, the external stress and the dislocation density are recorded as functions of the total strain. Initially, the stress increases steeply along a straight line defined by the Young's modulus—this is because the initial dislocation density is so low that the applied strain is accommodated mostly by the elastic strain. This continues until the stress reaches $\sim$130 MPa, at which point the crystal begins to yield. Following a sharp drop, the stress levels off at about 100 MPa and then continues to grow at a slower rate. This growth is a manifestation of *strain hardening*. The slope of the stress–strain curve, $\theta_1 = d\sigma/d\varepsilon$, is the strain hardening rate, which is about 6 GPa here. Along with hardening, the dislocations multiply at a rapid rate, as shown in Fig. 10.16(b). Line DD simulations like this provide interesting insights into the mechanisms of strain hardening and dislocation multiplication.

That the dislocation density increases during the simulation is a challenge for line DD because the number of nodes has to increase accordingly. What started as a small simulation with just a few lines (Fig. 10.14(a)), grew into a dense bundle of dislocations (Fig. 10.14(b)). One possible strategy for handling the incessant dislocation multiplication is to increase the number of processors during the simulation. Thus, the simulation shown in Fig. 10.14 was run first on a Linux cluster with 16

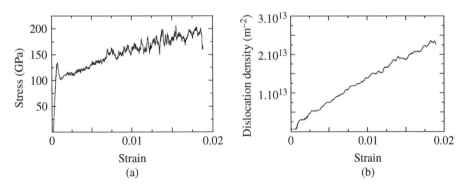

FIG. 10.16. A line DD simulation of single-crystal molybdenum strained at constant rate $\dot{\varepsilon}_{xx}^{ext} = 1 \, \text{s}^{-1}$: (a) the stress–strain curve and (b) the dislocation density as a function of strain.

processors until a 0.3 per cent strain was reached. It was then continued on a 200 processor Linux cluster and, subsequently, on a 512 processor IBM/SP machine.

Summary

- Line DD simulations are used to predict the stress–strain response of a single crystal during a deformation test.

- The plastic strain is computed as the sum of incremental contributions from the dislocation segments. To maintain a constant rate of straining, the instantaneous value of the average stress is obtained from the mismatch between the imposed total strain and the accumulated plastic strain.

11

PHASE FIELD METHOD

The phase field method (PFM) can be used as an approach to dislocation dynamics simulations alternative to the line DD method discussed in Chapter 10. The degrees of freedom in PFM are continuous smooth fields occupying the entire simulation volume, and dislocations are identified with locations where the field values change rapidly. As we will see later, as an approach to dislocation dynamics simulations PFM holds several advantages. First, it is easier to implement into a computer code than a line DD model. In particular, the complex procedures for making topological changes (Section 10.4) are no longer necessary. Second, the implementation of PFM can take advantage of well-developed and efficient numerical methods for solving partial differential equations (PDEs). Another important merit of PFM is its applicability in a wide range of seemingly different situations. For example, it is possible to simulate the interaction and co-evolution of several types of material microstructures, such as dislocations and alloying impurities, within a unified model.

PFM has become popular among physicists and materials scientists over the last 20 years, but as a numerical method it is not new. After all, it is all about solving PDEs on a grid. Numerical integration of PDEs is a vast and mature area of computational mathematics. A number of efficient methods have already been developed, such as the finite difference method [121], the finite element method [122], and spectral methods [123], all of which have been used in PFM simulations. The relatively new aspects of PFM are associated with the method's formulation and applications, which are partly driven by the growing interest in understanding material microstructures.

In Section 11.1, we begin with the general aspects of PFM demonstrated by two simple applications of the method not related to dislocations. Section 11.2 describes the elements required to adapt PFM to dislocation simulations. There we will briefly venture into the field of micromechanics and consider the concept of eigenstrain. The elastic energy of an arbitrary eigenstrain field is derived in Section 11.3. Section 11.4 discusses an example in which the PFM equations for dislocations are solved using the fast Fourier transform method. Section 11.5 continues with a more complex example that demonstrates the versatility of PFM. The final section compares the relative advantages and disadvantages of PFM and line DD methods for modelling dislocation dynamics.

11.1 General Phase Field Approach

The very term *phase fields* probably derives from one of the method's initial
applications to the evolution of phases in melts and alloys. Instead of using particles (as in MD) or mathematical lines (as in line DD), PFM represents material
microstructure by one or several continuum fields $\phi(\mathbf{x})$. The specific meaning of
the fields varies depending on the situation of interest. For example, in a crystal
growing from the melt, a phase field can be used to describe how different phases
(or states) are distributed in the material volume. For instance, it is possible to
associate the solid phase with $\phi \approx 1$ and the liquid phase with $\phi \approx -1$. Alternatively, $\phi(\mathbf{x})$ can describe the concentration of a chemical species at location $\mathbf{x}$ in a
binary alloy. As another example, different values of ϕ can correspond to different
martensite variants in a shape memory alloy. Likewise, dislocations, voids, cracks,
and other structural defects in a crystal can be represented by eigenstrain fields to
be discussed in the next section.

Consider a system specified by a set of phase fields $\{\phi_\alpha(\mathbf{x})\}$. The main assumption in PFM is that the fields should evolve in a way that reduces the total free
energy F, which is a functional of $\{\phi_\alpha(\mathbf{x})\}$. The equations of motion for the fields
are derived by taking the variational derivatives of the free energy F with respect
to $\{\phi_\alpha(\mathbf{x})\}$. The resulting partial differential equations are then solved numerically,
typically on a regular grid. Hence setting up a PFM simulation generally involves
the following four steps.

1. Select phase fields $\{\phi_\alpha(\mathbf{x})\}$ and define their variation ranges.

2. Write down the free energy functional $F[\phi_\alpha(\mathbf{x})]$.

3. Obtain evolution equations for $\{\phi_\alpha(\mathbf{x})\}$.

4. Select numerical methods to solve the evolution equations.

The choice for the free-energy functional reflects the nature of the problem but
leaves much room to exercise one's intuition. For simplicity, let us consider a case
of only one phase field $\phi(\mathbf{x})$. A widely used form for F is the following:

$$F[\phi(\mathbf{x})] = \int [f(\phi(\mathbf{x})) + \epsilon |\nabla\phi(\mathbf{x})|^2]\, \mathrm{d}^3\mathbf{x}, \qquad (11.1)$$

where $f(\phi)$ is the *bulk* contribution to the free energy density. The second term
inside the integrand places a penalty on the phase-field variations, giving rise to an
interfacial energy between domains of different phases. The free-energy expression
in eq. (11.1) is *local*, because the integrand (free energy density at $\mathbf{x}$) is completely
determined by the phase field and its derivatives at point $\mathbf{x}$, independent of the phase
field values at anyother points. Provided there is no need to account for long-range
interactions, the local form for free energy is often sufficient. In other situations it

may be necessary to use a *non-local* form such as

$$F[\phi(\mathbf{x})] = \int [f(\phi(\mathbf{x})) + \epsilon |\nabla \phi(\mathbf{x})|^2] d^3\mathbf{x}$$

$$+ \iint \phi(\mathbf{x}) \Xi(\mathbf{x} - \mathbf{x}')\phi(\mathbf{x}') d^3\mathbf{x} d^3\mathbf{x}'. \qquad (11.2)$$

The second term on the right-hand side includes a *non-local* kernel function, $\Xi(\mathbf{x} - \mathbf{x}')$, which describes the interaction energy between the phase field at two different points, $\mathbf{x}$ and $\mathbf{x}'$. The non-local form for free energy is necessary when long-range elastic interactions have to be accounted for, such as for dislocation simulations, which are discussed in the next section.

In this section we begin our discussion with the general aspects of the PFM approach, assuming first no long-range interactions. We will use eq. (11.1) to derive evolution equations for two simple but representative situations. In the first case, the phase field is not conserved, leading to the Ginzburg–Landau equation. In the second case, the phase field is conserved, resulting in the Cahn–Hilliard equation often used to describe the diffusion of chemical species.

11.1.1   Case A (Relaxational or Ginzburg–Landau)

As an example, consider a model of crystal growth by solidification from the melt. In the simplest formulation, it is sufficient to use a single phase field, $\phi(\mathbf{x})$, such that $\phi(\mathbf{x}) \approx +1$ if the material at position $\mathbf{x}$ is solid, and $\phi(\mathbf{x}) \approx -1$ if it is liquid. To account for the system's preference to be in either one of two states, $+1$ or -1, its bulk free-energy density can be written as

$$f(\phi) = U(\phi - 1)^2(\phi + 1)^2. \qquad (11.3)$$

As shown in Fig. 11.1(a), $f(\phi)$ has two minima of equal depth at $\phi = \pm 1$ and can describe the coexistence of solid and liquid phases. The global minimum of the free-energy functional $F[\phi(\mathbf{x})]$ is reached when $\phi(\mathbf{x})$ is either $+1$ or -1 everywhere, i.e. when the system is either solid or liquid in the entire volume.

The negative variational derivative of F with respect to $\phi(\mathbf{x})$, i.e. $-\delta F[\phi]/\delta\phi$, can be regarded as a generalized force on the phase field $\phi(\mathbf{x})$. Changing $\phi(\mathbf{x})$ in the direction of $-\delta F[\phi]/\delta\phi$ by a small amount is guaranteed to reduce F, as long as $-\delta F[\phi]/\delta\phi$ is not zero. Because $\phi(\mathbf{x})$ describes whether the system is solid or liquid at location $\mathbf{x}$, it does not have to be conserved. A simple model for the system's evolution towards its minimum free-energy state is to make the time derivative of $\phi(\mathbf{x})$ proportional to $-\delta F[\phi]/\delta\phi$, i.e. following the steepest-descent

PHASE FIELD METHOD

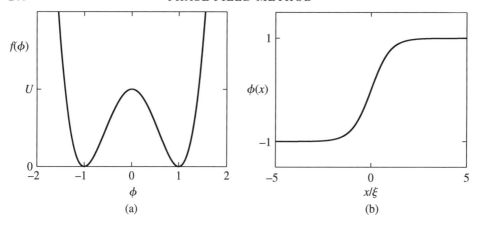

FIG. 11.1. (a) The bulk free-energy density function given by eq. (11.3). (b) The
   solution for solid–liquid co-existence.

direction (Section 2.3),

$$\frac{\partial \phi}{\partial t} = -k \frac{\delta F[\phi]}{\delta \phi} \tag{11.4}$$

$$= -k \left[\frac{\partial f}{\partial \phi} - 2\epsilon \nabla^2 \phi \right], \tag{11.5}$$

where k is a relaxation rate constant. This equation is often called the Ginzburg–
Landau or Allen–Cahn equation. A more detailed derivation is given below. In the
literature, the right-hand side of this equation sometimes includes an extra "noise"
term to account for thermal fluctuations. For simplicity, no such noise is considered
here.

Variational derivative of F* Notice that the argument in $F[\phi(\mathbf{x})]$ is function $\phi(\mathbf{x})$,
whereas F itself is a number. Such functions of functions are called *functionals*.
The variational derivative of functional F with respect to its argument function
$\phi(\mathbf{x})$ is a function denoted $\delta F[\phi]/\delta \phi$ and defined as follows.
   Consider a small change in function $\phi(\mathbf{x})$, i.e. $\phi(\mathbf{x}) \rightarrow \phi(\mathbf{x}) + \lambda(\mathbf{x})$. If the
corresponding change of F can be written as

$$\delta F = \int A(\mathbf{x})\lambda(\mathbf{x}) \, \mathrm{d}^3\mathbf{x}, \tag{11.6}$$

then

$$\frac{\delta F[\phi]}{\delta \phi} \equiv A(\mathbf{x}) \tag{11.7}$$

is defined as the variational derivative of F with respect to $\phi(\mathbf{x})$.

For example, when $F[\phi(\mathbf{x})]$ takes the form of eq. (11.1), then the variation of F is

$$\delta F = \int \left[\frac{\partial f}{\partial \phi} \lambda(\mathbf{x}) + 2\epsilon \nabla \phi(\mathbf{x}) \cdot \nabla \lambda(\mathbf{x}) \right] d^3 \mathbf{x}. \tag{11.8}$$

Using integration by parts and assuming that $\phi(\mathbf{x})$ satisfies either periodic or fixed boundary conditions, we obtain

$$\delta F = \int \left[\frac{\partial f}{\partial \phi} - 2\epsilon \nabla^2 \phi(\mathbf{x}) \right] \lambda(\mathbf{x}) \, d^3 \mathbf{x}. \tag{11.9}$$

Hence

$$\frac{\delta F[\phi]}{\delta \phi} = \frac{\partial f}{\partial \phi} - 2\epsilon \nabla^2 \phi. \tag{11.10}$$

Let us now consider a steady-state solution in which the solid and the liquid phases coexist. Because $\partial \phi / \partial t = 0$ for a steady state, $\phi(\mathbf{x})$ must satisfy the following equation:

$$\frac{\partial f}{\partial \phi} = 2\epsilon \nabla^2 \phi. \tag{11.11}$$

Assuming that $\phi(\mathbf{x})$ only varies along the x axis, and that the bulk free-energy density takes the form of eq. (11.3), we arrive at a one-dimensional equation:

$$\epsilon \frac{d^2}{dx^2} \phi(x) = 2U\phi(x)[\phi^2(x) - 1]. \tag{11.12}$$

The analytic solution of this equation with boundary conditions $\phi(-\infty) = -1$ (liquid) and $\phi(+\infty) = 1$ (solid) is

$$\phi(x) = \tanh(x/\xi), \tag{11.13}$$

where $\xi = \sqrt{\epsilon/U}$ characterizes the width of the solid–liquid interface (see Problem 11.1.1). This is quite similar to the arctan solution of the Peierls–Nabarro model (Chapter 8).

Substituting this solution into the volume integral (11.1), the free energy becomes $\gamma = \frac{8}{3}\sqrt{U\epsilon}$. This quantity can be regarded as the excess free energy of the solid–liquid interface. Together with the expression for the interface width $\xi = \sqrt{\epsilon/U}$, it is possible to adjust two parameters ϵ and U to reproduce any desired combination of interface energy γ and width ξ. In particular, γ can be matched directly to the interface energy from experiments or atomistic simulations. However, the value of ξ usually does not match the width of real material interfaces. This is because the latter can be very small, down to a single atomic spacing, whereas the typical resolution of PFM models is much coarser. In practice ξ is usually selected to be a few grid spacings of PFM, to enable stable numerical integration of the corresponding PDEs. It is for this reason that PFM is sometimes referred to as the *diffuse interface* method. As we will discuss in the subsequent sections, compared with the line DD model in the previous chapter, dislocations in the PFM can have very large core width.

To model the evolution of ϕ under more complex initial and boundary conditions, eq. (11.5) will have to be solved numerically. The simple and widely used method of *finite differences* is employed here. Another method based on the fast Fourier transform algorithm will be discussed later, after PFM is extended to deal with dislocations.

The finite difference method begins with replacing all (spatial and temporal) differential operators with finite-difference operators. Applications of this method to ordinary differential equations (ODEs) were demonstrated in Chapter 2 (for molecular dynamics) and Chapter 10 (for line DD). Similarly, let us consider a set of grid points on the x axis with spacing h and define $x_+ \equiv x + h$, $x_- \equiv x - h$. The first and second spatial derivatives of $\phi(x)$ with respect to x can be replaced with the following centered differences:

$$\frac{\partial \phi}{\partial x} \to \frac{\phi(x_+) - \phi(x_-)}{2h} \tag{11.14}$$

$$\frac{\partial^2 \phi}{\partial x^2} \to \frac{\phi(x_+) - 2\phi(x) + \phi(x_-)}{h^2}. \tag{11.15}$$

Likewise, on a square grid in two dimensions

$$\nabla^2 \phi \to \frac{\phi(x_+, y) + \phi(x, y_+) + \phi(x_-, y) + \phi(x, y_-) - 4\phi(x, y)}{h^2}. \tag{11.16}$$

It can be shown that the errors associated with the use of these finite differences are order $\mathcal{O}(h^2)$. In the Euler forward method, the time derivative is replaced by

$$\frac{\partial \phi}{\partial t} \to \frac{\phi(t + \Delta t) - \phi(t)}{\Delta t}, \tag{11.17}$$

as before. Thus, the discretized version of the evolution eq. (11.5) is,

$$\phi(x, y; t + \Delta t) = \phi(x, y; t) - k\Delta t \left(\frac{\mathrm{d}f(\phi)}{\mathrm{d}\phi}\right)_{\phi=\phi(x,y;t)}$$
$$+ \frac{2k\epsilon \Delta t}{h^2} \cdot [\phi(x_+, y; t) + \phi(x, y_+; t)$$
$$+ \phi(x_-, y; t) + \phi(x, y_-; t) - 4\phi(x, y; t)]. \tag{11.18}$$

Given the values of $\phi(x, y; t)$ at time t, this equation returns $\phi(x, y; t + \Delta t)$ at the next time step $t + \Delta t$.

As an example, consider a two-dimensional phase field $\phi(x, y)$ on a 64×64 square grid with grid size $h = 1$ whose free-energy functional is given by eq. (11.1) and eq. (11.3). Let us use parameters $U = 1$ and $\epsilon = 4$ so that the resulting interface width is $\xi = \sqrt{\epsilon/U} = 2$, which is larger than h. Periodic boundary conditions are applied in both directions to eliminate possible edge effects, i.e.

$$\phi(x, y) = \phi(x + L, y) = \phi(x, y + L), \qquad (11.19)$$

where $L = Nh$, $N = 64$. To ensure numerical stability, time step Δt must be sufficiently small compared to the relaxation time. Setting the relaxation rate constant $k = 1$, $\Delta t = 10^{-3}$ is a reasonable choice. With these choices, everything is now ready to follow the evolution of $\phi(x, y; t)$, as given by eq. (11.18), from an arbitrary initial condition $\phi(x, y; 0)$.

First, let us consider a case when the initial field is constant, $\phi(x, y; 0) = \phi_0$. In this case, field evolution is quite trivial: $\phi(x, y)$ quickly relaxes to $+1$ (if $\phi_0 > 0$) or to -1 (if $\phi_0 < 0$) remaining spatially uniform everywhere during the relaxation. A more interesting scenario develops when the field values on the grid points are initialized using random numbers uniformly distributed in $[-a/2, a/2]$ (here a is a small number, say $a = 0.01$). From this initial "noisy" state, the system gradually evolves into two or several large domains in which $\phi \approx 1$ or $\phi \approx -1$. The time evolution of the system's free energy is plotted in Fig. 11.2(a). Between $t = 0$ and $t = 2$, most of the ϕ values remain close to zero, giving rise to a high free energy. This unphysical state results from the rather peculiar choice of the initial condition.

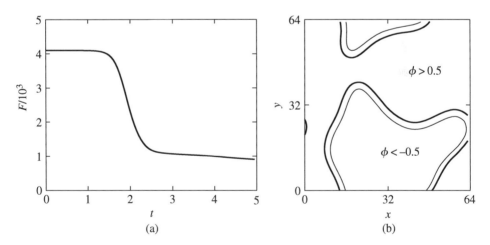

FIG. 11.2. (a) The evolution of free energy F as a function of time t during the relaxation following eq. (11.5). (b) A snapshot of the phase field $\phi(x, y)$ at $t = 5$. The thick line is the contour $\phi = 0.5$ and the thin line is the contour $\phi = -0.5$. The lines delineate the areas where $\phi > 0.5$ (solid) and $\phi < -0.5$ (liquid). The narrow strip between the two lines is the interface.

However, by about $t = 2.5$, $\phi(x, y)$ evolves into a more physically meaningful state consisting of two well-formed solid ($\phi \approx 1$) and liquid ($\phi \approx -1$) domains. After $t = 2.5$, the relaxation slows down considerably and proceeds by "sluggish" adjustments of the interfaces between two domains. A snapshot of $\phi(x, y)$ at $t = 5$ is plotted in Fig. 11.2(b), showing two large domains of liquid and solid.

11.1.2 Case B (Diffusional or Cahn–Hilliard)

In the second example, let us assume that field ϕ is the concentration of a chemical species in a substitutional binary alloy AB. In the absence of chemical transformations, the total amount of either species must remain constant, providing an example of a conserved field. Let us identify $n_A = (\phi + 1)/2$ with the concentration of species A. Then $\phi = 1$ corresponds to the state of pure A and $\phi = -1$ corresponds to the state of pure B. With the field so defined, it is now possible to use the same free energy density function (11.3) to describe this seemingly different situation.

The statement that $\phi(\mathbf{x})$ is a conserved field means that $\phi(\mathbf{x})$ cannot vary arbitrarily, but any increase of ϕ at $\mathbf{x}$ must be accompanied by a depletion of ϕ in the neighborhood of $\mathbf{x}$ and vice versa. This conservation statement can be expressed through the continuity equation:

$$\frac{\partial \phi(\mathbf{x})}{\partial t} + \nabla \cdot \mathbf{J}(\mathbf{x}) = 0, \tag{11.20}$$

where $\mathbf{J} = (J_x, J_y, J_z)$ is the flux vector of field ϕ. For example, J_x is the amount of ϕ flowing in the x direction per unit area per unit time. Equation (11.20) establishes that the local accumulation rate of $\phi(\mathbf{x})$ must be equal to the difference between *in-flux* and *out-flux* of ϕ at position $\mathbf{x}$.

Similar to the previous example, reduction of the total free energy provides the driving force for field evolution. However, in this case, the free energy can only decrease by *redistribution* of field ϕ from one place to another. It is reasonable to expect that the field would flow in such a way as to make the driving force, $-\delta F/\delta \phi(\mathbf{x})$, more evenly distributed for all $\mathbf{x}$. Therefore, the simplest evolution model in this case is to let the flux J be proportional to the spatial gradient of the driving force, i.e.

$$\mathbf{J} = -M \nabla \mu \tag{11.21}$$

$$\mu \equiv \frac{\delta F}{\delta \phi}, \tag{11.22}$$

where μ is sometimes called the *generalized chemical potential* and M is a *diffusion coefficient*. Therefore, the equation of motion is

$$\frac{\partial \phi}{\partial t} = \nabla \cdot \left[M \nabla \left(\frac{\delta F}{\delta \phi} \right) \right]. \tag{11.23}$$

Again, assuming that the free-energy functional has the form of eq. (11.1), we obtain

$$\frac{\partial \phi}{\partial t} = M\nabla^2 \left[\frac{\partial f}{\partial \phi} - 2\epsilon\nabla^2\phi \right]$$

$$= M\left[\frac{\partial^2 f}{\partial \phi^2}\nabla^2\phi + \frac{\partial^3 f}{\partial \phi^3}(\nabla\phi)^2 - 2\epsilon\nabla^4\phi \right] \equiv g[\phi], \qquad (11.24)$$

which can also be solved numerically using the finite difference method. Similar to the previous example, the spatial differential operators ∇, ∇^2, ∇^4 are replaced by appropriate centered differences. However, the previously used explicit Euler forward method, eq. (11.17), becomes rather unstable when applied to eq. (11.24), unless a very small time step Δt is used. This is related to the fact that eq. (11.24) involves spatial derivatives of higher order (up to order 4). It is better to employ a more stable method, such as the implicit trapezoidal integrator introduced in Section 10.3. Given eq. (11.24), the implicit method requires solving the following equation:

$$\frac{\phi(t + \Delta t) - \phi(t)}{\Delta t} = \frac{g[\phi(t)] + g[\phi(t + \Delta t)]}{2} \qquad (11.25)$$

or

$$\phi(t + \Delta t) = \phi(t) + \frac{\Delta t}{2} \cdot (g[\phi(t)] + g[\phi(t + \Delta t)]). \qquad (11.26)$$

Equation (11.26) is called implicit because its right-hand side contains the unknown $\phi(t + \Delta t)$. This equation can be solved iteratively to a prespecified accuracy level δ. The following algorithm evolves the field forward by a single step based on eq. (11.26).

Algorithm 11.1

1. Compute the predictor $\tilde{\phi}_0 := \phi(t) + g(\phi(t)) \cdot \Delta t$.
2. Compute the corrector $\tilde{\phi}_1 := \phi(t) + (g[\phi(t)] + g[\tilde{\phi}_0]) \cdot \Delta t/2$.
3. If max $|\tilde{\phi}_1 - \tilde{\phi}_0| < \delta$, return with $\phi(t + \Delta t) := \tilde{\phi}_1$.
4. Otherwise, $\tilde{\phi}_0 := \tilde{\phi}_1$, go to 2.

As a specific example, consider a simulation with the following set of parameters, $U = 1$, $\epsilon = 4$, $M = 1$, $N = 64$, $h = 1$, $\Delta t = 10^{-3}$, and $\delta = 10^{-8}$. To compare the behaviors, let us use exactly the same initial values for $\phi(x, y)$ as those in the previous case of the non-conserved fields. The predicted evolution of the total free energy is shown in Fig. 11.3(a). The general behavior is similar to that in

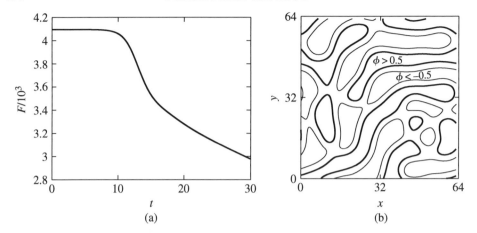

FIG. 11.3. (a) The evolution of free energy F as a function of time obtained by
numerical solution of eq. (11.24). (b) A contour plot of $\phi(x, y)$ at $t = 30$. The
thick line is the contour $\phi = 0.5$ and the thin line is the contour $\phi = -0.5$. The
contours lines delineate domains $\phi > 0.5$ (species A) from domains $\phi < -0.5$
(species B).

Fig. 11.2(b), but the relaxation kinetics is considerably slower. After $t = 15$, $\phi(x, y)$
reaches a state in which ϕ is close to $+1$ or -1 in most of the material volume.
A snapshot of the microstructure at $t = 30$ is shown in Fig. 11.3(b). Domains with
enhanced concentration of element A ($\phi \approx 1$) are elongated and interlaced with
domains of enhanced concentration of element B ($\phi \approx -1$). This is a result of
conservation of field ϕ: to create an A-rich domain, the excess of species B must
move away to nearby locations, resulting in the development of alternating A-rich
and B-rich domains. This behavior is widely known to physicists and materials
scientists as *spinodal decomposition*.

Summary

- The phase field method relies on continuous fields to provide a coarse-grained
  description of material microstructure. In PFM, the phase fields evolve in
  ways that reduce the system's free energy.

- Finite differences are commonly used to solve partial differential equations
  appearing in PFM. The spatial and temporal grids and numerical integrators
  have to be chosen carefully to ensure numerical stability.

- Whether or not the phase field is conserved results in different evolution
  equations that predict different microstructures, even though the free energy
  functional may be exactly the same.

Problems

11.1.1. Show that $\phi(x) = \tanh(x/\xi)$, $\xi = \sqrt{\epsilon/U}$ is the solution of eq. (11.12). Verify that the excess free energy of this solution is $F = \frac{8}{3}\sqrt{U\epsilon} \equiv \gamma$. (Hint: $\tanh'(x) = 1 - \tanh^2(x)$.)

11.1.2. It is possible to use a gradient relaxation method to minimize eq. (11.1) numerically. Consider a 1D case where $f(\phi)$ is given by eq. (11.3). Find an optimal solution for $\phi(x)$ subject to the boundary condition $\phi(-\infty) = -1$, $\phi(\infty) = 1$. For this purpose, write the trial solution in the form of $\phi(x) = \tanh(x/c)$, where c is the optimization parameter to be determined. The free-energy integral can be approximated by the following sum: $\tilde{F} = \sum_{n=-N}^{N} [f(\phi(n\,h)) + \epsilon(\phi'(n\,h))^2] \cdot h$, where h is a grid size. Write down the expression for $d\tilde{F}/dc$. Assuming $U = 1$ and $\epsilon = 1$, use a steepest-descent or a conjugate-gradient algorithm to minimize $\tilde{F}$ with respect to c. Obtain a converged value for c and $\tilde{F}$ and compare the result to the analytical solution.

11.2 Dislocations as Phase-field Objects

Given that a dislocation is the boundary line between slipped and unslipped areas on a crystal plane (Section 1.2), the amount of slip across the plane makes a natural candidate for the phase field associated with the dislocations. This slip field is a specific example of a more general concept of *eigenstrain* introduced by John Eshelby nearly fifty years ago [124].

To see what an eigenstrain is, let us consider an elastic body Ω that is initially free from any external and internal stress. Now imagine carving out of Ω a platelet-shaped volume A, as shown in Fig. 11.4.[1] Next, let A undergo a permanent shape change such that its stress-free shape becomes A'. For example, spontaneous shape changes like this can be induced by a martensitic phase transformation. Positions $\mathbf{x}'$ and $\mathbf{x}$ of the same point in A' and A are related to each other by a linear transformation $x_i' = (\delta_{ij} + \eta_{ij})x_j$, where η_{ij} is a rank-2 tensor.[2] In this example, η_{ij} is a constant tensor inside volume A. Assuming that the transformation is uniform over the platelet volume, the eigenstrain is defined as a symmetrized part of tensor η_{ij}, i.e. $e_{ij}^* = (\eta_{ij} + \eta_{ji})/2$.[3] In the definition of the eigenstrain, the part of η_{ij} that

---

[1] To introduce the notion of eigenstrain, volume A can be of any shape. The platelet shape is used here because it is most appropriate for representing dislocations.

[2] Here and in the following we use the Einstein convention—any pair of repeating Latin indices is summed over from 1 to 3, unless explicitly stated otherwise. δ_{ij} is the Kronecker delta that equals 1 when $i = j$ and 0 when $i \neq j$.

[3] This is similar to the infinitesimal strain tensor in the linear elasticity theory.

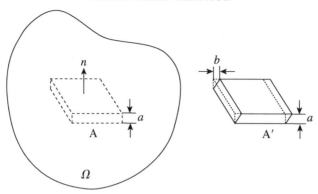

F IG . 11.4. Shear transformation of a platelet inclusion A of thickness a. Starting
    from the stress-free state, the upper surface of the platelet shifts by **b** with respect
    to the lower surface.

accounts for the rigid body rotation of A is excluded. Volume A is often referred
to as an *inclusion* [124].

In reality, inclusion A remains embedded inside Ω and, thus, cannot freely
attain its new stress-free shape (as quantified by eigenstrain e_{ij}^*). Instead, as a result
of accommodation of the transformed inclusion A' inside Ω, stress both inside
and outside the inclusion is produced. The problem of finding the field of internal
stress induced by an arbitrary eigenstrain is commonly referred to as the Eshelby's
inclusion problem. Its solution has found numerous applications in engineering and
materials sciences. What matters to us here is that the same solution can be used to
derive the stress field produced by an arbitrary dislocation network.

Consider a transformation that results in a relative shift of the upper and lower
surfaces of the platelet by **b**, as shown in Fig. 11.4. The eigenstrain for this
transformation is

$$e_{ij}^* = \frac{b_i n_j + b_j n_i}{2a}, \tag{11.27}$$

where **n** is the unit vector normal to the platelet plane and a is the platelet thickness.
When a is equal to the spacing between two adjacent planes in a crystal lattice, the
effect of such eigenstrain on the stress field is equivalent to that of a dislocation loop
of Burgers vector **b**. In the limit $a \to 0$, the perimeter line of platelet A' becomes a
Volterra dislocation (Section 4.1). In PFM, a is set to be equal to the grid spacing,
which is usually much larger than the atomic dimensions.

This equivalence between the dislocation loops and the inclusions makes it
possible to reproduce arbitrary dislocation arrangements by superposition of eigenstrain fields. Because dislocations usually move on a finite set of crystallographic
planes, it is sensible to define a separate phase field for each of the two orthogonal slip directions on every slip plane. For example, in FCC crystals the dominant

slip planes are (111), $(\bar{1}11)$, $(1\bar{1}1)$, and $(11\bar{1})$. Therefore, eight phase fields are necessary to describe an arbitrary dislocation network in FCC crystals. Here, each of the eight fields ϕ_α represents a particular combination of the glide plane normal vector $\mathbf{n}^{(\alpha)}$ and the slip vector $\mathbf{b}^{(\alpha)}$ on this plane.[4] The total eigenstrain is then the superposition of all phase fields, i.e.

$$e_{ij}^*(\mathbf{x}) = \sum_\alpha e_{ij}^{*(\alpha)} \phi_\alpha(\mathbf{x}), \qquad (11.28)$$

where each eigenstrain $e_{ij}^{*(\alpha)}$ is defined by $\mathbf{b}^{(\alpha)}$ and $\mathbf{n}^{(\alpha)}$ through eq. (11.27).

Following the procedure outlined in the previous section, the next step in constructing a phase-field model of dislocations is to express the system's free energy as a functional of $\{\phi_\alpha(\mathbf{x})\}$. The total free energy is written as a sum of three contributions, i.e.

$$F[\phi_\alpha(\mathbf{x})] = E_{\text{el}}[\phi_\alpha(\mathbf{x})] + E_{\text{latt}}[\phi_\alpha(\mathbf{x})] + E_{\text{grad}}[\phi_\alpha(\mathbf{x})]. \qquad (11.29)$$

The first term is the elastic energy; this is usually the largest contribution to the free energy. The second term accounts for the crystal's natural preference to have the slip vector constrained to integer multiples of the lattice vectors. This term is very similar to the generalized stacking fault energy in the Peierls–Nabarro model (Chapter 8).[5] The third term is a gradient energy added here to artificially increase the dislocation core width to provide for better numerical behavior. Because the first and the second terms in the above free energy find counterparts in the Peierls–Nabarro model, the PFM description of dislocations can be viewed as a three-dimensional, coarse-grained version of the PN model.

Since $\phi_\alpha(\mathbf{x})$ represents local shape transformation at point $\mathbf{x}$, it does not need to be conserved. Thus its evolution equation is of the Ginzburg–Landau type, i.e.

$$\frac{\partial \phi_\alpha}{\partial t} = -k \frac{\delta F[\phi_\alpha]}{\delta \phi_\alpha}, \qquad (11.30)$$

where k is a rate constant. In the following we discuss all three contributions to the free energy in more detail.

---

[4] This is a bit different from the usual definition of dislocation phase fields in FCC crystals, which are considered to have 12 slip systems, namely three different $\langle 110 \rangle$ Burgers vectors on each of the four $\{111\}$ planes. However, two phase fields—one for each component of the in-plane slip displacement—are minimally sufficient to describe all three slip systems on a given slip plane.

[5] Similar to the case of solid–liquid interfaces discussed in the preceding section, it is possible to adjust the magnitudes of the second and third terms to make the dislocations spread over several grid spacing (for integrability) and, simultaneously, to ensure that the dislocation core energy matches its actual value (for physical realism). Given this difference, we are reluctant to refer to the second term in eq. (11.29) by its common name, the Peierls term. Here we will call it the *lattice* term instead.

11.2.1 The Elastic Energy

The general expression for the energy of infinite, homogenous, linear elastic medium containing eigenstrain phase fields ϕ_α is

$$E_{el} = \frac{1}{2} \sum_{\alpha,\beta} \iint \Xi^{(\alpha\beta)}(\mathbf{x} - \mathbf{x}') \phi_\alpha(\mathbf{x}) \phi_\beta(\mathbf{x}') \, d^3\mathbf{x} \, d^3\mathbf{x}', \tag{11.31}$$

where Ξ is an elastic *kernel* that subsumes the tensor of elastic constants C_{ijkl} and the eigenstrains $e_{ij}^{*(\alpha)}$. The elastic energy is a double integral with a non-local kernel corresponding to the long-range interactions between the eigenstrains representing dislocations. This is similar to the elastic energy term in the Peierls–Nabarro model, eq. (8.6). Analytic expressions for the elastic kernel will be given in the next section. The computation of the elastic energy and its derivatives is the most time-consuming part of the numerical PFM models of dislocations, mainly because of this non-locality.

11.2.2 The Lattice Energy

In a crystal, the slip vectors on each slip plane are close to integer multiples of the lattice Burgers vectors. A simple way to enforce this behavior in the phase-field model is to add an energy penalty function for deviations of ϕ_α from integer values [125], e.g.

$$E_{latt}[\phi_\alpha] = U \sum_\alpha \int \sin^2[\pi \phi_\alpha(\mathbf{x})] \, d^3\mathbf{x}. \tag{11.32}$$

However, the above expression implies that two Burgers vectors on each glide plane $\mathbf{n}^{(\alpha)}$ are orthogonal,[6] which is usually not the case, except for a simple cubic crystal.

Here, we opt to use the two-dimensional generalized stacking fault energy (γ-surface) already familiar to us from the formulation of the Peierls–Nabarro model. For example, a γ-function that accounts for the symmetry of the {111} planes in FCC crystals is given in eq. (8.30). The corresponding lattice energy in the phase-field model becomes

$$E_{latt}[\phi_\alpha] = \zeta \sum_{p=1}^{N/2} \int \gamma \left(b\phi_\alpha(\mathbf{x}), b\phi_\beta(\mathbf{x}) \right) \, d^3\mathbf{x}, \tag{11.33}$$

where $\alpha = 2p - 1$ and $\beta = 2p$ are the indices of two phase fields sharing the same slip plane, i.e. $\mathbf{n}^{(\alpha)} = \mathbf{n}^{(\beta)}$. Here, N is the total number of phase fields, b is the magnitude of the slip vector and parameter ζ is introduced to scale the amplitude of the γ-function as needed for stable numerical integration.

---

[6] This is because we have defined two phase fields on plane α such that their slip vectors $\mathbf{b}^{(\alpha)}$ are orthogonal to each other.

11.2.3 The Gradient Energy

In the case of an infinite straight dislocation, the first two energy terms in the PFM free energy are exactly the same as in the Peierls–Nabarro model. The width of a PN dislocation, over which $\phi_\alpha(\mathbf{x})$ changes from one integer to the next, is of the order of few atomic spacings. On the other hand, the PFM evolution equations are typically solved on a grid with a spacing that is considerably greater than the width of a PN dislocation. For the sake of numerical stability, it is then useful to add an artificial gradient energy term, to help regulate the dislocation width and make it spread over a few grid spacings. The gradient penalty term to be used in PFM simulations of dislocations must be a bit more complicated than that in eq. (11.1). In particular, it should not penalize the gradient across the area enclosed by a dislocation loop, such as surface S_1 in Fig. 11.5. Otherwise, there will appear an unphysical interface with an extra energy proportional to the area (instead of the length) of the dislocation loop. To avoid this problem, the following form can be used, which only penalizes the in-plane components of the field gradients,

$$E_{\text{grad}}[\phi_\alpha] = \epsilon \sum_\alpha \int [(\mathbf{n}^{(\alpha)} \times \nabla)\, \phi_\alpha(\mathbf{x})]^2 \, \mathrm{d}^3\mathbf{x}. \qquad (11.34)$$

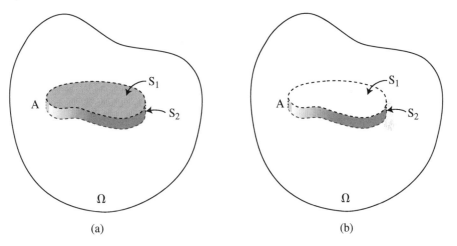

(a)   (b)

FIG. 11.5. (a) A dislocation loop is defined as a platelet inclusion A inserted in an elastic solid body Ω. Because the value of the phase field (eigenstrain) is constant inside A and is zero outside, the phase field varies rapidly from a constant value inside the platelet (gray) to zero outside (white). (b) The extra energy associated with the field gradient component in the direction normal to plane S_1 should be set to zero to avoid creating an unphysical interface (white). At the same time, the in-plane components of the field gradient are non-zero around the perimeter S_2 (gray) and represent the dislocation core.

The latter gradient energy is non-zero only around the loop perimeter and can be thought of as an additional contribution to the dislocation core energy. By selecting a numerical value of ϵ, the width of the interface (dislocation line) can be adjusted. In practice, the choice of ϵ is dictated by numerical requirements: narrow interfaces offer better resolution but require small meshes for stable integration.

Summary

- In PFM, a dislocation can be represented by an appropriate distribution of eigenstrains.

- The energy functional of a PFM model of dislocations includes elastic, lattice, and gradient energy contributions. The gradient term is introduced to control the width of the dislocation core as required for stable numerical integration.

11.3   Elastic Energy of Eigenstrain Fields

In Fig. 11.4, a dislocation loop is represented by a platelet-shaped distribution of eigenstrain field $e_{ij}^*(\mathbf{x})$ that is constant inside the inclusion and zero outside. To be able to treat arbitrary dislocation configurations within the phase-field model, it is necessary to consider continuous fields of eigenstrain $e_{ij}^*(\mathbf{x})$. In this section, we obtain the elastic energy of an arbitrary eigenstrain field $e_{ij}^*(\mathbf{x})$ and use the result to define the elastic kernel function Ξ in eq. (11.31). Given that several excellent texts exist on the subject of micromechanics, e.g. [60, 126], we will only sketch the derivation and highlight results necessary for numerical implementation of the phase-field model of dislocations.

The first important result needed for subsequent discussion is that, in the absence of external stress, the elastic energy E_{el} can be expressed in terms of the internal stress field $\sigma_{ij}(\mathbf{x})$ induced by the eigenstrain field $e_{ij}^*(\mathbf{x})$:

$$E_{\mathrm{el}} = -\frac{1}{2} \int_\Omega \sigma_{ij}(\mathbf{x}) e_{ij}^*(\mathbf{x}) \, \mathrm{d}^3 x. \qquad (11.35)$$

Elastic energy of an eigenstrain field Equation (11.35) can be rationalized as follows. Consider an eigenstrain field $\lambda e_{ij}^*(\mathbf{x})$. Because the solid Ω is linear elastic, the internal stress field induced by this eigenstrain is $\lambda \sigma_{ij}(\mathbf{x})$. Let us gradually turn the eigenstrain field on by switching λ from 0 to 1, so that the eigenstrain changes from 0 to $e_{ij}^*(\mathbf{x})$. As the eigenstrain is incremented from $\lambda e_{ij}^*(\mathbf{x})$ to $(\lambda + \mathrm{d}\lambda) e_{ij}^*(\mathbf{x})$, the amount of work done against the internal stress is

$$\mathrm{d}W = \int_\Omega -[\lambda \sigma_{ij}(\mathbf{x})][\mathrm{d}\lambda e_{ij}^*(\mathbf{x})] \, \mathrm{d}^3 x. \qquad (11.36)$$

The elastic energy of the final state is equal to the total work done against the internal stress, i.e.

$$E_{\mathrm{el}} = \int_0^1 \frac{\mathrm{d}W}{\mathrm{d}\lambda}\, \mathrm{d}\lambda = -\frac{1}{2}\int_\Omega \sigma_{ij}(\mathbf{x})e_{ij}^*(\mathbf{x})\, \mathrm{d}^3\mathbf{x}. \tag{11.37}$$

This and eq. (11.35) should not be confused with the usual expression for the elastic strain energy density applicable in the absence of eigenstrain:

$$E_{\mathrm{el}} = \frac{1}{2}\int_\Omega \sigma_{ij}(\mathbf{x})\varepsilon_{ij}(\mathbf{x})\, \mathrm{d}^3\mathbf{x}, \tag{11.38}$$

where $\varepsilon_{ij}(\mathbf{x})$ is the elastic strain field induced by external loads. The major difference is the minus sign in eq. (11.35), reflecting that the eigenstrain energy is the work done *against* the internal stress while creating the inclusion. A further difference is that elastic stress and elastic strain induced by external loads are proportional to each other at every point in Ω, i.e. $\sigma_{ij}(\mathbf{x}) = C_{ijkl}\varepsilon_{kl}(\mathbf{x})$, where C_{ijkl} is the tensor of elastic constants. In contrast, the relationship between internal stress $\sigma_{ij}(\mathbf{x})$ and the eigenstrain is non-local because eigenstrain at point $\mathbf{x}'$ induces stress at every other point $\mathbf{x}$ in Ω. Solving for the stress field induced by an eigenstrain field is known as Eshelby's problem.

Given eq. (11.35), our task is reduced to finding the solution for the stress field $\sigma_{ij}(\mathbf{x})$ induced by the eigenstrain field $e_{ij}^*(\mathbf{x})$. In an infinite linear elastic continuum, the most general expression relating these two fields is

$$\sigma_{ij}(\mathbf{x}) = -\int_\Omega \Theta_{ijkl}(\mathbf{x}-\mathbf{x}')e_{kl}^*(\mathbf{x})\, \mathrm{d}^3\mathbf{x}\, \mathrm{d}^3\mathbf{x}'. \tag{11.39}$$

(An explicit expression for $\Theta_{ijkl}(\mathbf{x}-\mathbf{x}')$ will be given shortly). With this, the elastic energy becomes

$$E_{\mathrm{el}} = \frac{1}{2}\int_\Omega\int_\Omega \Theta_{ijkl}(\mathbf{x}-\mathbf{x}')e_{ij}^*(\mathbf{x})e_{kl}^*(\mathbf{x})\, \mathrm{d}^3\mathbf{x}\, \mathrm{d}^3\mathbf{x}'. \tag{11.40}$$

Notice that the variational derivative of E_{el} with respect to eigenstrain $e_{ij}^*(\mathbf{x})$ is simply the stress field (with a minus sign):

$$\frac{\delta E_{\mathrm{el}}}{\delta e_{ij}^*(\mathbf{x})} = \int \Theta_{ijkl}(\mathbf{x}-\mathbf{x}')e_{kl}^*(\mathbf{x}')\, \mathrm{d}^3\mathbf{x}' = -\sigma_{ij}(\mathbf{x}). \tag{11.41}$$

When the eigenstrain field is a superposition of several phase fields, as in eq. (11.28), then the non-local kernel in the elastic energy term in eq. (11.31) can be expressed as follows:

$$\Xi^{(\alpha\beta)}(\mathbf{x} - \mathbf{x}') = \Theta_{ijkl}(\mathbf{x} - \mathbf{x}')e_{ij}^{*(\alpha)}e_{kl}^{*(\beta)}. \tag{11.42}$$

The variational derivative of E_{el} with respect to phase field ϕ_α is then

$$\frac{\delta E_{\text{el}}}{\delta \phi_\alpha(\mathbf{x})} = -\sigma_{ij}(\mathbf{x})\, e_{ij}^{*(\alpha)}. \tag{11.43}$$

Thus, the elastic component of the driving force on the phase fields is defined by the local stress field. Because of the non-locality of expression (11.39), evaluation of the local stress field is time consuming.

The solution for $\Theta_{ijkl}(\mathbf{x} - \mathbf{x}')$ was first obtained by Eshelby in terms of the elastic Green's function $G_{ij}(\mathbf{x})$ of the solid (see below):

$$\Theta_{ijkl}(\mathbf{x}) = C_{ijrs}[C_{pmkl}G_{rm,sp}(\mathbf{x}) + \delta_{kr}\delta_{sl}\delta(\mathbf{x})]. \tag{11.44}$$

Here, C_{ijkl} is the tensor of elastic constants. In an isotropic solid

$$C_{ijkl} = \lambda\delta_{ij}\delta_{kl} + \mu(\delta_{ik}\delta_{jl} + \delta_{il}\delta_{jk}) \tag{11.45}$$

and the Green's function is

$$G_{ij}(\mathbf{x}) = \frac{1}{16\pi\mu(1-v)|\mathbf{x}|}\left[(3-4v)\delta_{ij} + \frac{x_i x_j}{|\mathbf{x}|^2}\right], \tag{11.46}$$

where $\lambda = 2\mu v/(1-2v)$, μ is the shear modulus and v is Poisson's ratio.

Elastic Green's function The elastic Green's function is the tensor $G_{ij}(\mathbf{x}, \mathbf{x}')$ that relates the i component of the equilibrium displacement induced at point $\mathbf{x}$ to the j component of a point force applied at point $\mathbf{x}'$. If a linear elastic solid is subjected to a distributed body force $f_j(\mathbf{x})$, the corresponding displacement field is

$$u_i(\mathbf{x}) = \int G_{ij}(\mathbf{x}, \mathbf{x}')f_j(\mathbf{x}')d\mathbf{x}'. \tag{11.47}$$

In an infinite and elastically homogeneous solid, the Green's function depends only on the relative position of the source and field points, i.e.

$$G_{ij}(\mathbf{x}, \mathbf{x}') = G_{ij}(\mathbf{x} - \mathbf{x}'). \tag{11.48}$$

This function satisfies the following equation of mechanical equilibrium:

$$C_{kpim}G_{ij,mp}(\mathbf{x}) + \delta_{jk}\delta(\mathbf{x}) = 0, \tag{11.49}$$

where $\delta(\mathbf{x})$ is the three-dimensional Dirac delta function representing a unit point force at the origin, while the indices m and p after the comma denote partial derivatives of the Green's function with respect to x_m and x_p.

For a generally anisotropic elastic solid, no analytic expressions for the Green's function exist. At the same time, the above equilibrium equation for the Green's function can be solved in Fourier (reciprocal) space. Let us define the Fourier transform of $G_{ij}(\mathbf{x})$ as

$$g_{ij}(\mathbf{k}) = \int \exp(-i\mathbf{k} \cdot \mathbf{x})G_{ij}(\mathbf{x})\,d^3\mathbf{x}. \tag{11.50}$$

Rewritten in Fourier space, eq. (11.49) becomes

$$C_{kpim}k_m k_p\, g_{ij}(\mathbf{k}) = \delta_{jk}. \tag{11.51}$$

The solution of this equation is

$$g_{ij}(\mathbf{k}) = \frac{(zz)_{ij}^{-1}}{k^2}, \tag{11.52}$$

where $k \equiv \|\mathbf{k}\|$, $\mathbf{z} \equiv \mathbf{k}/k$, $(zz)_{ki} \equiv C_{pkim}z_p z_m$ and $(zz)_{ij}^{-1}$ is the inverse of $(zz)_{ij}$ defined by $(zz)_{ij}^{-1}(zz)_{jk} = \delta_{ik}$. In the elastically isotropic case, the Green's function is particularly simple:

$$g_{ij}(\mathbf{k}) = \int \exp(-i\mathbf{k} \cdot \mathbf{x})G_{ij}(\mathbf{x})\,d^3\mathbf{x} = \frac{1}{\mu k^2}\left[\delta_{ij} - \frac{k_i k_j}{2(1 - \nu)|\mathbf{k}|^2}\right]. \tag{11.53}$$

The Green's function in real space can be written as the inverse Fourier transform:

$$G_{ij}(\mathbf{x}) = \frac{1}{(2\pi)^3}\int \exp(i\mathbf{k} \cdot \mathbf{x})\frac{(zz)_{ij}^{-1}}{k^2}\,d^3\mathbf{k}. \tag{11.54}$$

For the isotropic case this integral can be evaluated analytically to give eq. (11.46). More details on the elastic Green's function can be found in [60].

11.3.1 Useful Expressions in the Fourier Space

The stress and the non-local kernel are conveniently expressed through their Fourier components. For example,

$$\hat{\sigma}_{ij}(\mathbf{k}) = C_{ijkl}[C_{pmrs}g_{km}(\mathbf{k})\hat{e}_{rs}(\mathbf{k})k_l k_p - \hat{e}_{kl}(\mathbf{k})], \tag{11.55}$$

where $\hat{\sigma}_{ij}(\mathbf{k})$ and $\hat{e}_{ij}(\mathbf{k})$ are the Fourier transforms of $\sigma_{ij}(\mathbf{x})$ and $e_{ij}^*(\mathbf{x})$, respectively. Assume that $\hat{\phi}_\alpha(\mathbf{k})$ is the Fourier transform of phase field $\phi_\alpha(\mathbf{x})$. Then

$$\hat{e}_{ij}(\mathbf{k}) = \sum_\alpha e_{ij}^{(\alpha)}\hat{\phi}_\alpha(\mathbf{k}). \tag{11.56}$$

Defining

$$\hat{\sigma}_{ij}^{(\alpha)}(\mathbf{k}) \equiv C_{ijkl}[C_{pmrs}g_{km}(\mathbf{k})e_{rs}^{(\alpha)}k_l k_p - e_{kl}^{(\alpha)}], \tag{11.57}$$

the internal stress produced by the phase fields becomes

$$\hat{\sigma}_{ij}(\mathbf{k}) = \sum_\alpha \hat{\sigma}_{ij}^{(\alpha)}(\mathbf{k})\hat{\phi}_\alpha(\mathbf{k}) \tag{11.58}$$

and the Fourier transform of the elastic kernel $\Xi^{(\alpha\beta)}(\mathbf{x})$ is

$$\hat{\Xi}^{(\alpha\beta)}(\mathbf{k}) = -\hat{\sigma}_{ij}^{(\alpha)}(\mathbf{k})e_{ij}^{(\beta)}. \tag{11.59}$$

This suggests an efficient procedure for computing the internal stress in a PFM simulation. First, compute coefficients $\hat{\sigma}_{ij}(\mathbf{k})$ defined by eq. (11.57). For the elastically anisotropic cases, it may be beneficial to precompute and store these coefficients in tables. On every cycle of stress update, it is necessary to compute the Fourier transforms of phase fields $\hat{\phi}_\alpha(\mathbf{k})$. The latter should then be plugged into eq. (11.58) to obtain the internal stress in Fourier space. To complete the stress update, the inverse Fourier transform should be used to obtain the new stress field in the real space.

As an example, consider a case with two phase fields ϕ_α and ϕ_β. Assume that $\mathbf{n}^{(\alpha)}$ and $\mathbf{n}^{(\beta)}$ are along the y axis, whereas $\mathbf{b}^{(\alpha)}$ and $\mathbf{b}^{(\beta)}$ are along the x and z axes, respectively. Also assume that $e_{12}^{(\alpha)} = e_{21}^{(\alpha)} = b/2$ and $e_{23}^{(\beta)} = e_{32}^{(\beta)} = b/2$. In this case the coefficients in eq. (11.57) can be evaluated explicitly (in isotropic elasticity)

$$\hat{\sigma}_{12}^{(\alpha)}(\mathbf{k}) = \mu b \left[-\frac{k_z^2}{k^2} - \frac{2k_x^2 k_y^2}{(1-v)k^4} \right], \tag{11.60}$$

$$\hat{\sigma}_{23}^{(\alpha)}(\mathbf{k}) = \mu b \left[\frac{k_x k_z}{k^2} - \frac{2k_x k_y^2 k_z}{(1-v)k^4} \right], \tag{11.61}$$

$$\hat{\sigma}_{23}^{(\beta)}(\mathbf{k}) = \mu b \left[-\frac{k_x^2}{k^2} - \frac{2k_y^2 k_z^2}{(1-v)k^4} \right]. \tag{11.62}$$

11.3.2   Stress Expressions in Two Dimensions

Consider now a two-dimensional problem in which the same two phase fields $\phi_\alpha(\mathbf{x})$ and $\phi_\beta(\mathbf{x})$ as defined above are non-zero only in the narrow slab around $y = 0$ (as in Fig. 11.4). That is

$$\phi_\alpha(\mathbf{x}) = \begin{cases} \frac{1}{a}\phi_\alpha(x, z), & y \in \left[-\frac{a}{2}, \frac{a}{2}\right] \\ 0, & \text{otherwise} \end{cases} \tag{11.63}$$

and similarly for ϕ_β. In the limit $a \to 0$, phase fields $\phi_\alpha(\mathbf{x})$ remain non-zero only on the $y = 0$ plane because $\phi_\alpha(\mathbf{x}) \to \delta(y)\phi_\alpha(x, z)$.

Let us define $\hat{\sigma}_{ij}(k_x, k_z)$, $\hat{\phi}_\alpha(k_x, k_z)$ and $\hat{\phi}_\beta(k_x, k_z)$ as the two-dimensional Fourier transforms of $\sigma_{ij}(x, z)$, $\phi_\alpha(x, z)$ and $\phi^{(\beta)}(x, z)$, respectively. Then from eq. (11.58) it follows that

$$\hat{\sigma}_{ij}(k_x, k_z) = \hat{\sigma}_{ij}^{(\alpha)}(k_x, k_z)\hat{\phi}^{(\alpha)}(k_x, k_z) + \hat{\sigma}_{ij}^{(\beta)}(k_x, k_z)\hat{\phi}^{(\beta)}(k_x, k_z), \qquad (11.64)$$

where

$$\hat{\sigma}_{ij}^{(\alpha)}(k_x, k_z) = \frac{1}{2\pi}\int_{-\infty}^{\infty}\sigma_{ij}^{(\alpha)}(\mathbf{k})\,dk_y. \qquad (11.65)$$

In the isotropic case, the stress expressions reduce to

$$\hat{\sigma}_{12}^{(\alpha)}(k_x, k_z) = \frac{\mu b}{2\kappa}\left(-k_z^2 - \frac{k_x^2}{1-\nu}\right), \qquad (11.66)$$

$$\hat{\sigma}_{23}^{(\alpha)}(k_x, k_z) = -\frac{\mu b\nu}{2(1-\nu)}\frac{k_x k_z}{\kappa}, \qquad (11.67)$$

$$\hat{\sigma}_{23}^{(\beta)}(k_x, k_z) = \frac{\mu b}{2\kappa}\left(-k_x^2 - \frac{k_z^2}{1-\nu}\right), \qquad (11.68)$$

where $\kappa = \sqrt{k_x^2 + k_z^2}$.

Summary

- The interaction kernel in the double-integral expression for the elastic part of the free-energy function is expressed in terms of the elastic Green's function.

- The variational derivative of the elastic energy with respect to the phase field is proportional to the local stress.

- A solution for the stress field induced by the phase fields is conveniently obtained in the Fourier space.

Problem

11.3.1. Consider an infinite straight dislocation parallel to the z axis. By symmetry, solutions for the fields produced by this dislocation should not depend on z. For the same reason, the only non-zero Fourier components of stress are those with $k_z = 0$. Obtain the following components of the elastic kernel in the Fourier space: $\Xi^{(\alpha\alpha)}(k_x)$, $\Xi^{(\alpha\beta)}(k_x)$,

$\Xi^{(\beta\beta)}(k_x)$. Assuming isotropic elasticity, evaluate the real-space components $\Xi^{(\alpha\alpha)}(x)$, $\Xi^{(\alpha\beta)}(x)$, $\Xi^{(\beta\beta)}(x)$. Show that the elastic energy computed with this kernel is equivalent to the elastic energy of a Peierls–Nabarro dislocation (first two terms of eq. (8.31)).

11.4 A Two-dimensional Example

Having discussed the general formulation of the PFM approach, let us now consider a simple example of its application to dislocation simulations. For simplicity, assume that there is only one slip plane $\mathbf{n}^{(\alpha)}$ and one Burgers vector $\mathbf{b}^{(\alpha)}$ in the system represented by a single phase field $\phi_\alpha(\mathbf{x})$. As before, assume that $\mathbf{b}^{(\alpha)}$ and $\mathbf{n}^{(\alpha)}$ are parallel to the x axis and y axis respectively so that the phase field is confined to the $y = 0$ plane. Thus, the eigenstrain can be written as,

$$e_{ij}^*(\mathbf{x}) = e_{ij}^{(\alpha)}\phi_\alpha(\mathbf{x}) = e_{ij}^{*(\alpha)}\delta(y)\phi_\alpha(x, z), \tag{11.69}$$

where the non-zero components of $e_{ij}^{(\alpha)}$ are

$$e_{12}^{(\alpha)} = e_{21}^{(\alpha)} = \frac{b}{2}. \tag{11.70}$$

Under these assumptions, the free-energy functional becomes

$$F[\phi_\alpha(x, z)] = \frac{1}{2} \iint \Xi^{(\alpha\alpha)}(x - x', z - z')\phi_\alpha(x, z)\phi_\alpha(x', z')\,dx\,dz\,dx'\,dz'$$

$$+ U \iint \sin^2[\pi\phi_\alpha(x, z)]\,dx\,dz$$

$$+ \epsilon \iint \{[\partial_x\phi_\alpha(x, z)]^2 + [\partial_z\phi_\alpha(x, z)]^2\}\,dx\,dz. \tag{11.71}$$

The first term in the above functional is the elastic energy. The second term is the lattice energy that penalizes the deviations of the field from the integer values. The third term is a gradient energy introduced here for numerical purposes, namely to artificially increase the width of the dislocation cores. The first variational derivative of this functional with respect to field ϕ_α is

$$\frac{\partial F[\phi_\alpha]}{\partial\phi_\alpha} = -\sigma_{ij}(x, z)e_{ij}^{*(\alpha)} + U\pi \sin[2\pi\phi_\alpha(x, z)] - \epsilon(\partial_x^2 + \partial_z^2)\phi_\alpha(x, z), \tag{11.72}$$

where $\sigma_{ij}(x, z)$ is the stress field in plane $y = 0$.

To avoid possible edge effects in the numerical implementation, let us apply periodic boundary conditions (PBC) along the x and z directions

$$\phi_\alpha(x, z) = \phi_\alpha(x + L, z) = \phi_\alpha(x, z + L), \tag{11.73}$$

where L is the linear dimension of the square-shaped simulation cell. The two-dimensional Fourier transform of the stress field is

$$\hat{\sigma}_{ij}(k_x, k_z) = \hat{\sigma}_{ij}^{(\alpha)}(k_x, k_z)\hat{\phi}_\alpha(k_x, k_z). \tag{11.74}$$

Because field $\phi_\alpha(x, z)$ is periodic, its Fourier transform is non-zero only when k_x and k_z are integer multiples of $2\pi/L$. In such a case, it is appropriate to use the following discrete form of the Fourier transform:

$$\hat{\phi}_\alpha(k_x, k_z) = \frac{1}{L^2} \int_0^L \int_0^L e^{-i(k_x x + k_z z)} \phi_\alpha(x, z) \, dx \, dz, \tag{11.75}$$

$$\phi_\alpha(x, z) = \sum_{k_x k_z} e^{i(k_x x + k_z z)} \hat{\phi}_\alpha(k_x, k_z) \, dx \, dz. \tag{11.76}$$

If necessary, the elastic energy can be conveniently computed in the Fourier space,

$$E_{el} = \frac{L^2}{2} \sum_{k_x k_z} \Xi^{(\alpha\alpha)}(k_x, k_z) \, \hat{\phi}_\alpha(k_x, k_z)\hat{\phi}_\alpha^*(k_x, k_z), \tag{11.77}$$

where $\hat{\phi}_\alpha^*(k_x, k_z)$ is the complex conjugate of $\hat{\phi}_\alpha(k_x, k_z)$.

The transforms in eqs. (11.75) and (11.76) are computed using the fast Fourier transform (FFT) algorithm. FFT is a very efficient algorithm: given function $\phi_\alpha(x, z)$ represented by an array of M numbers, the FFT algorithm computes its Fourier transform $\hat{\phi}_\alpha(k_x, k_z)$ in just $\mathcal{O}(M \log M)$ operations. Furthermore, once the solution for stress $\hat{\sigma}_{ij}(k_x, k_z)$ is obtained in the Fourier space from eq. (11.74), the real-space solution for stress $\sigma_{ij}(x, z)$ can be obtained by the inverse FFT, and again in just $\mathcal{O}(M \log M)$ operations. The lattice and gradient energies, as well as their variations, are local and present no difficulty: they are computed in real space using finite differences, in much the same way as in Section 11.1. We now have a complete procedure for computing the right-hand side of the field evolution equations,

$$\frac{\partial \phi_\alpha}{\partial t} = g[\phi_\alpha] \equiv -k \frac{\delta F[\phi_\alpha]}{\delta \phi_\alpha}$$

$$= k\{\sigma_{ij}(x, z)e_{ij}^{*(\alpha)} - U\pi \sin[2\pi\phi_\alpha(x, z)] + \epsilon(\partial_x^2 + \partial_z^2)\phi_\alpha(x, z)\}.$$

As a numerical example, consider a phase field on a 128×128 square grid with spacing $h = 1$ and $L = 128$. Let us fix the amplitude of the lattice potential $U = 1$ and the gradient penalty parameter $\epsilon = 4$ and solve the above evolution equation numerically using the implicit trapezoid integrator described in Algorithm 11.1, with time step $\Delta t = 10^{-2}$ and convergence tolerance $\delta = 10^{-4}$.

To specify the initial condition, the values of the phase field are set to 1 within a 32×32 square at the center of the simulation cell, while all the remaining field values are set to zero. This initial condition corresponds to a square-shaped dislocation loop with core width of the order of h. Numerical integration of the equation of motion shows that the phase field relaxes over time and eventually becomes zero everywhere: the dislocation loop shrinks under zero applied stress. A more interesting behavior is observed if we let the dislocation loop relax to its equilibrium shape while keeping its total area fixed. This requires that we solve a different evolution equation:

$$\frac{\partial \phi_\alpha}{\partial t} = g[\phi_\alpha] - \lambda \int_0^L \int_0^L \phi_\alpha(x, z) \, dx \, dz, \qquad (11.78)$$

where λ is a Lagrange multiplier introduced here to enforce the constraint[7]

$$\int_0^L \int_0^L \phi_\alpha(x, z) \, dx \, dz = \text{constant}.$$

A possible physical analogy for the effect of the Lagrange multiplier is with the action of external stress (see Problem 11.4.1), whose magnitude adjusts instantly to cancel the net driving force $g[\phi]$ on the loop. The evolution of the free energy computed under this constraint is shown in Fig. 11.6(a). The initial rapid decrease of the free energy over the first few time steps coincides with a rapid spreading of the interface (dislocation core). Subsequently, the free energy continues to decrease more gradually while the dislocation loop changes its shape from the initial square to a more rounded shape. By $t = 20$, the relaxation is mostly completed resulting in the shape shown in Fig. 11.6(b). The relaxed loop has an elliptical shape elongated along the x direction. This result is consistent with linear elasticity theory, predicting that edge dislocation segments have higher elastic energy than screws (by a factor of $1/(1 - \nu)$). Under the constant-area constraint, the loop elongates in the direction of its Burgers vector, making the screw segments longer at the expense of the edge segments.

Summary

- In phase-field models of dislocations under periodic boundary conditions, the lattice energy and the gradient energy terms are computed in real space,

---

[7] This is quite analogous to the constrained minimization discussed in Section 7.3.

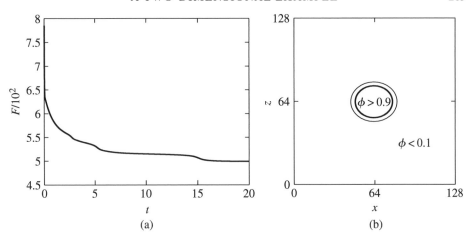

FIG. 11.6. (a) Time evolution of free energy F computed for the constrained case in which the total area of the loop remains constant (see text). (b) Contour plot of $\phi_\alpha(x, z)$ at $t = 20$. The thick line is contour $\phi_\alpha = 0.9$ and the thin line is contour $\phi_\alpha = 0.1$. The two contours enclose the dislocation core.

while the elastic energy is evaluated in the Fourier space using the fast Fourier transform algorithm.

Problems

11.4.1. To account for the effect of external stress σ_{ij}^{ext}, the following term has to be added to the elastic energy:

$$-\sigma_{ij}^{\text{ext}} e_{ij}^{(\alpha)} \iint \phi_\alpha(x, z)\, dx\, dz. \qquad (11.79)$$

Derive the additional contribution to the driving force $g[\phi_\alpha]$ due to this term. Depending on the magnitude and sign of the applied stress, a dislocation loop will either expand or shrink (unless its area is artificially constrained). Starting from the configuration obtained in the end of the simulation shown in Fig. 11.6(b), remove the constraint on the total area of the loop and apply a constant stress. Assume that σ_{xy} is the only non-zero component of the applied stress and simulate the loop's evolution under $\sigma_{xy} = 3, 4$, and 5. Which of these three stress magnitudes is sufficient to make the loop expand? Find the magnitude of stress σ_{xy} that is minimally sufficient to prevent the loop from shrinking.

11.4.2. Consider the $\{111\}$ slip plane in an FCC crystal containing three $\frac{1}{2}\langle 110 \rangle$ Burgers vectors. Use the coordinate system in which the slip plane is the $y = 0$ plane. An arbitrary dislocation on this plane can be modeled by two phase fields ϕ_α and ϕ_β that quantify the amount of slip in the x and z directions, respectively. Use the following lattice energy functional:

$$E_{\text{latt}}[\phi_\alpha, \phi_\beta] = \iint \gamma(b\,\phi_\alpha(x, z), b\,\phi_\beta(x, z))\,dx\,dz, \qquad (11.80)$$

where potential γ is the same as in eq. (8.30). For ϕ_α use the same initial values as in the example discussed in this section, but for ϕ_β set the initial values at $\phi_\beta(x, z) = \phi_\alpha(x - x_0, z - z_0)$, where $x_0 = 20$, $z_0 = 20$. Simulate the evolution of the phase fields and observe the types of dislocations formed.

11.5   Dislocation–Alloy Interaction

The ability to deal with various types of material microstructure in one model is a valuable attribute of the PFM approach. This versatility is illustrated in this section, which discusses a simple extension of the PFM model presented in the preceding sections. The particular extension is introduced below to account for the effect of heterogeneous particles on dislocation behavior under stress.

In general, the interaction of dislocations and alloy microstructure is one of the most important and least understood issues in physical metallurgy. This interaction comes in many forms. For example, variations in the local alloy composition result in variations of the elastic constants or the lattice parameter, which produce additional internal stress that is likely to affect dislocation motion. Dislocation core properties can also vary significantly following dislocation motion from one location to another. Furthermore, the interaction is mutual, meaning that dislocations can also affect the behavior of the alloy microstructure. Some aspects of dislocation–alloy interactions are relatively easy to incorporate in the phase-field models, while others (such as changes in the dislocation core) can be subtle and require more detailed atomistic analyses. In this section we consider a relatively simple effect, while a few other, more complex, situations are considered in the problem set.

Consider an AB binary alloy represented by a phase field ϕ_{AB}. As shown in Fig. 11.3(b), the material can develop a microstructure with alternating A-rich and B-rich domains. Let us assume that the only effect of the alloy composition ϕ_{AB} is on the local magnitude of the lattice energy, which provides resistance to dislocation motion. One way to take this into account is to make the amplitude U of the lattice energy in eq. (11.71) depend on the local value of field ϕ_{AB}, such as

$$E_{\text{latt}}[\phi_\alpha] = \iint U(x, z)\sin^2[\pi\,\phi_\alpha(x, z)]\,dx\,dz, \qquad (11.81)$$

where

$$U(x, z) = U_0[1 + (A - 1)(\phi_{AB}(x, z) + 1)/2] \qquad (11.82)$$

and A is a constant. With this modification, the magnitude of the lattice potential in the pure-A phase ($\phi_{AB} = 1$) becomes AU_0 while in the pure-B phase ($\phi_{AB} = -1$) it is U_0. It is certainly possible to define a model in which both alloy and dislocation phase fields evolve simultaneously. All that has to be done is to construct an appropriate free-energy functional and write down the evolution equation for the fields (see Problem 11.5.5). Here, for simplicity, we limit our discussion to a situation in which the alloy field ϕ_{AB} is fixed and only the dislocation field is allowed to evolve. To be specific, let us assume that the alloy field ϕ_{AB} has a fixed distribution such that the amplitude of the lattice energy has the following Gaussian form:

$$U(x, z) = U_0 + (A - 1)U_0 \exp\left[-\frac{(x - x_0)^2 + (z - z_0)^2}{2w^2}\right], \qquad (11.83)$$

where w is a constant describing the size of the A-rich region (alloy particle). In this case, the variational derivative of the total free energy with respect to ϕ_α is

$$\frac{\partial F[\phi_\alpha]}{\partial \phi_\alpha} = -\sigma_{ij}(x, z)\, e_{ij}^{(\alpha)} - \sigma_{ij}^{\text{ext}}\, e_{ij}^{(\alpha)}$$

$$+ \pi U(x, z) \sin[2\pi \phi_\alpha(x, z)]$$

$$- \epsilon(\partial_x^2 + \partial_z^2)\phi_\alpha(x, z), \qquad (11.84)$$

where σ_{ij}^{ext} is the externally applied stress (see Problem 11.4.1).

The parameter values used in the numerical simulation are $U_0 = 1$, $A = 3.5$, $x_0 = 32$, $z_0 = 32$, $w = 4$, while all other parameters are the same as those in Section 11.4. The initial values for ϕ_α are the same as those at the end of the simulation illustrated in Fig. 11.6(b). The position and size of the alloy particle are shown in Fig. 11.7(a) using the contour plot of $U(x, y) = 2.8$.

When external stress $\sigma_{xy}^{\text{ext}} = 0.45$ is applied, the dislocation loop expands. By $t = 40$, its motion is visibly hindered by the alloy particle, as shown in Fig. 11.7(b). However, the loop continues its expansion in other directions away from the particle. The expansion is notably faster along the x direction. Over time, the angle θ between the dislocation arms on two sides of the particle continues to decrease. θ reaches its minimum value θ_c at $t = 57$, which is shown in Fig. 11.7(c). At the very next time step, the dislocation breaks away from the particle, which is now sheared by one Burgers vector along the x direction, as shown in Fig. 11.7(d). Subsequently, the dislocation recombines across the periodic boundary and the phase field becomes close to 1 everywhere, meaning that the entire plane is now sheared by one Burgers vector.

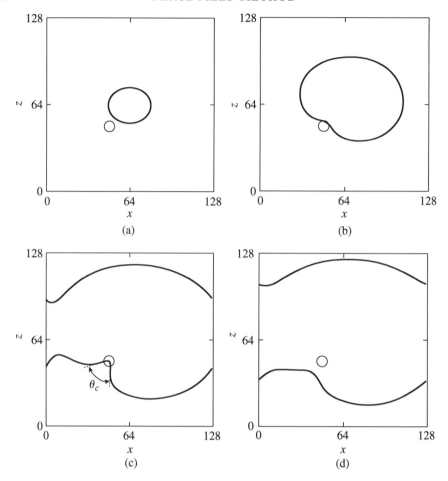

FIG. 11.7. A series of simulation snapshots showing the dislocation (thick contour of $\phi_\alpha(x, z) = 0.5$) cutting through a particle (thin circular contour of
$U(x, z) = 2.8$). (a) Initial values of $\phi_\alpha(x, z)$ at $t = 0$ are the same as those in
Fig. 11.6(b). (b) At $t = 40$ the dislocation encounters the particle. (c) At $t = 57$
the dislocation is bowing around the particle and the angle between its arms
on two sides of the particle reaches minimum θ_c. Part of the expanding loop
has recombined with its periodic images. (d) At $t = 60$ the dislocation passes
through and shears the particle.

With only one particle in the plane, the dislocation loop is found always to
sweep through the plane, although exactly how this happens depends on the obstacle
strength. When the obstacle is strong (large A), a small dislocation loop (sometimes
called the *Orowan loop*) may be left behind encircling the particle, while the rest of
the dislocation continues to march forward sweeping the plane (see Problem 11.5.2).
The value of the slip field within the loop remains close to 0 even if it becomes

close to 1 everywhere else, meaning that a strong alloy particle is not sheared by the dislocation. If, on the other hand, multiple particles are introduced in the plane, the dislocations may become completely immobilized by these particles (see Problem 11.5.3). For a given obstacle strength (A), the critical stress to break away from the particles increases with decreasing interparticle spacing.

Summary

- External stress contributes a constant term to the driving force on the strain-inducing phase fields.

- With appropriate modifications in the free-energy functional, the PFM approach can describe the interaction between different types of phase fields, e.g. between alloy composition fields and dislocations.

- Local variations of the lattice energy mimic the resistance of alloy particles to dislocation motion.

Problems

11.5.1. Repeat the simulation described in this section with $A = 3, 4, 5$. For each A, find the critical angle θ_c between dislocation arms on two sides of the particle, just before the dislocation breaks away.

11.5.2. In the case when $A = 5$, as described in Problem 11.5.1, the dislocation by-passes the obstacle and leaves behind an Orowan loop. If, after that, another dislocation with the same Burgers vector passes around the obstacle, it may either create a second Orowan loop concentric with the first or force the first loop to shear through the particle. To simulate this behavior, let us create a source that can generate dislocation loops with the same Burgers vector. This can be done by imposing a time-dependent boundary condition, $\phi_\alpha(x_s, z_s) = St$, where x_s, z_s are the coordinates of the dislocation source and S is a parameter controlling the rate of dislocation generation. Simulate the phase-field evolution with $A = 5, 6, 7$, with $x_s = z_s = 64$ and $S = 0.01$. How many Orowan loops can the obstacle sustain for each value of A?

11.5.3. Repeat the simulations described in this section with four identical alloy particles instead of one. Position the particles at $(48, 48)$, $(48, 80)$, $(80, 48)$, and $(80, 80)$. For $A = 3, 4, 5$, compute the minimum stress required for a single dislocation to break through and loop around the obstacles.

11.5.4. In addition to offering increased resistance to dislocation motion, alloy particles can induce extra stress, which can also affect dislocation behavior. For example, an alloy particle may possess its own eigenstrain. Consider an eigenstrain distribution $e_{ij}^{(\beta)}(\mathbf{x}) = B\delta_{ij}$ that is uniform inside a sphere $(x - x_0)^2 + (y - y_0)^2 + (z - z_0)^2 \leq R$. Derive the stress field generated by this eigenstrain. Assume that $U(x, z) = 1$, $x_0 = 10$, $y_0 = 0$, $z_0 = 10$ and $R = 5$ and, starting from the configuration corresponding to Fig. 11.7(b), run a series of simulations with $B = 0.1, 0.2, 0.3$ under zero applied stress. Raise the stress in increments to observe at which stress levels the dislocation cuts through or by-passes the particle.

11.5.5. To model co-evolution of alloy composition fields and dislocation microstructures, let us construct a PFM model that includes both phase fields, ϕ_{AB} and ϕ_α. As an example, consider the following free-energy functional,

$$F[\phi_{AB}, \phi_\alpha]$$

$$= U_1 \iint [\phi_{AB}(x, z) - 1]^2 [\phi_{AB}(x, z) + 1]^2 \, dx \, dz$$

$$+ \epsilon_1 \iint [\partial_x \phi_{AB}(x, z)]^2 + [\partial_z \phi_{AB}(x, z)]^2 \, dx \, dz$$

$$+ \frac{1}{2} \iint \Xi^{(\alpha\alpha)}(x - x', z - z')\phi_\alpha(x, z)\phi_\alpha(x', z') \, dx \, dz \, dx' \, dz'$$

$$- \sigma_{ij}^{ext} e_{ij}^{(\alpha)} \iint_L \phi_\alpha(x, z) \, dx \, dz$$

$$+ U_2 \int_L \int_L dx \, dz \, (\phi_{AB}(x, z) + p) \sin^2[\pi \phi_\alpha(x, z)]$$

$$+ \epsilon_2 \int_L \int_L dx \, dz[\partial_x \phi_\alpha(x, z)]^2 + [\partial_z \phi_\alpha(x, z)]^2, \qquad (11.85)$$

where U_1, ϵ_1, U_2, ϵ_2 and p are constants. The form of the lattice energy (second-to-last term) implies that the ratio between the amplitude of the lattice potential in the pure-A state ($\phi_{AB} = 1$) to that in the pure-B state ($\phi_{AB} = -1$), is $(p + 1)/(p - 1)$. Notice that ϕ_{AB} is conserved and should follow the Cahn–Hilliard equation of motion (11.23), while ϕ_α is not conserved and should evolve according to the Ginzburg–Landau equation (11.4). Use $U_1 = U_2 = 1$, $\epsilon_1 = \epsilon_2 = 4$ and $p = 1.8$, and set the initial values of ϕ_{AB} and ϕ_α to those corresponding to Fig. 11.3(b) and Fig. 11.6(b), respectively. Simulate the co-evolution of ϕ_{AB} and ϕ_α as functions of time and plot a series of snapshots. Use contour plots to visualize the dislocations and the domain boundaries.

11.6   PFM or Line DD?

The phase-field method discussed in this chapter and the line DD method described in Chapter 10 are two different approaches to modeling dislocations, both based on linear elasticity theory. In a wider context, these two methods reflect two different styles of dealing with the material interfaces. So, which of the two representations should one use? The answer depends on the particular problem at hand.

An advantageous feature of PFM is that there is no need to track the interfaces explicitly. PFM evolves field values on the grid points and the interfaces can be visualized as contour lines (in two dimensions) or iso-surfaces (in three dimensions) on which the fields take some specified values. In contrast, in front-tracking methods, which explicitly resolve and evolve the interfaces, bookkeeping of the evolving topology of lines or surfaces becomes a major challenge in the numerical implementation. For example, the line DD model discussed in Chapter 10 needs to handle topological changes due to dislocation multiplication, recombination or junction formation by splitting and merging nodes, while maintaining consistency of the entire data structure. Complexity of such bookkeeping increases manyfold in a parallel line-tracking code, where handling of topological rearrangements involving data owned by different processors can be cumbersome.

Phase-field dislocation dynamics has several other advantages as well. One of them is that it handles anisotropic elasticity with the same ease and efficiency as the isotropic case. In contrast, most line DD codes presently in use are limited to isotropic elasticity, and bringing in elastic anisotropy can increase the computational load by orders of magnitude. This is because the elastic Green's function does not have a closed-form solution in real space for a general anisotropic medium. This is a serious argument for the use of the phase-field method when elastic anisotropy is important.

As mentioned in the previous section, it is relatively straightforward to model the co-evolution of dislocations and various other elements of material microstructure in PFM. Since the phase-field approach was originally developed and used in the context of phase transformations and alloy microstructure evolution, there is much more experience in these application areas than in dislocation simulations. This versatility in coupling dislocation simulations with other well developed PFM models is certainly very attractive. Finally, the computational efficiency of the existing PFM implementations of dislocation dynamics rides on the FFT algorithm, which has been perfected over decades of hard work by applied mathematicians and computer scientists.

Given all these impressive pros, what are the cons? If PFM can deliver more for much less effort, why do we even bother with the front-tracking methods such as line DD? Below, we point out a serious limitation of the phase-field dislocation dynamics where the line DD method appears to be a more reasonable choice [127].

The spatial resolution required in some important situations cannot be achieved in the phase-field method. A useful measure of resolution is the ratio of the size L of the simulation volume to the dimension d of the smallest feature resolved in the model. For example, for dislocation dynamics to accurately simulate plastic deformation and strain hardening, a simulation cell with $L \geq 10 \, \mu m$ is needed for the model to be statistically representative. At the same time, to resolve areas of dense dislocation tangles, a spacial resolution of $d \leq 1 \, nm$ is required. This leads to a resolution ratio exceeding $L/d = 10^4$.

On the other hand, the maximum resolution of the phase-field method is determined by the size of data arrays that FFT can handle. For three-dimensional FFT with N^3 grid points, the resolution is $L/d = N$. Unfortunately, the size of arrays that FFT can comfortably deal with on present-day workstations is quite limited. For example, parallel computing is required to achieve a reasonable speed in PFM simulations even for $N = 128$. Because in three dimensions the computational load scales as $N^3 \log N$, FFT with larger N becomes exceedingly more expensive. The largest three-dimensional FFT tested to date is $N = 2048$ on 4096 processors of the Earth Simulator supercomputer [128]. It stands to reason that even this heroic calculation does not come close to the resolution of $L/d = 10^4$ required for realistic strain-hardening simulations.

Fortunately, the line DD method can meet this resolution demand even with the presently available computer resources. Consider a cubic simulation box with $L = 10 \, \mu m$ and an average dislocation density $\rho = 10^{13} \, m^{-2}$ resulting in total dislocation length $\rho L^3 = 10^{-2} \, m$. Dividing the lines into segments 1 nm long would produce 10^7 segments, which is within the reach of the `ParaDiS` code (Chapter 10). This should be compared with the 10^{12} grid points required for a PFM simulation with a similar resolution. Given that only 10^{-5} of the total volume is occupied by dislocations, PFM appears to be astonishingly wasteful in attempting to evolve the field everywhere in the material volume. In comparison, the line-tracking method is more intelligent because it only allocates degrees of freedom at important locations, i.e. where the dislocation lines actually are. Of course, the price one pays for these savings is the need to trace the topological changes of nodal connectivity. But this is a price well worth paying, given the huge reduction in the total number of degrees of freedom.

To see how the lack of resolution can adversely affect simulation results, consider a PFM dislocation dynamics simulation on a cubic grid with 256 grid points in a cube-shaped $10 \times 10 \times 10 \, \mu m^3$ box. Because the smallest feature that can be resolved in this simulation is no larger than one grid spacing ($d = 40 \, nm$), dislocations are not quite the thin lines with the atomistic core diameter that we are used to thinking about. Rather, dislocations in such a PFM model are thick "worms" 40 nm in diameter[8]. This is a severe problem because microscopic processes taking

---

[8] The actual resolution is even lower because the dislocation core is typically spread over a few grid points (typically $3d$ to $5d$). A recipe to partially remedy this problem was recently suggested [129].

place on length scales smaller than 40 nm do influence the dislocation behavior. For example, dislocation dipoles with width smaller than d can contribute significantly to the yield and hardening behavior in real crystals (see Fig. 11.8). The inability of PFM simulations to resolve this and other relevant details of dislocation interactions may lead to a significant loss of realism, and to incorrect predictions of plastic yield stress, flow stress, and strain hardening rates.

Conceivably, it should be possible to make the PFM methods "smarter" by using multi-resolution methods such as adaptive mesh refinement (AMR) techniques [130]. The idea behind AMR is to use fine grids only where the field gradient is high, while using coarser grids everywhere else, resulting in a considerable reduction in the total number of degrees of freedom. This is a valid idea, except that, by going along this path, the phase-field approach is likely to gain efficiency while losing one of its most appealing features—simplicity.

Given the method's simplicity and ease of implementation, PFM is well suited for situations where the required resolution ratio does not exceed $L/d = 256$. However, in the most demanding situations when both high resolution and a large simulation volume are simultaneously required, line DD is likely to retain its leading role. Although considerably more difficult to implement, the line DD method is more computationally efficient given the huge reduction in the number of degrees of freedom it affords. It seems likely that dislocation dynamics simulations can benefit from the development of hybrid algorithms that will combine some of the better features of the two approaches.

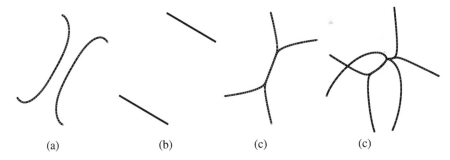

(a)   (b)   (c)   (c)

FIG. 11.8. A line DD simulation of two Frank–Read sources (segments with fixed ends) with opposite Burgers vectors. (a) The two sources attract each other even under zero stress. (b) When the sources are on the same slip plane, recombination occurs in the middle, leading to the formation of two new Frank–Read sources. (c) When the sources are on two parallel slip planes separated by 10 nm, they lock each other by forming a dislocation dipole. (d) The dipole provides significant resistance to other mobile dislocations. A typical PFM DD simulation uses grid sizes larger than 10 nm. Hence, it will fail to account for this and other similar behaviors that otherwise affect dislocation motion and contribute to strain hardening.

Summary

- Compared to line DD, PFM dislocation dynamics is much easier to implement in a computer code.

- Line DD is much more powerful when high-resolution ratio is required. The extra effort in implementation of line DD is more than offset by a great reduction in the total number of degrees of freedom.

BIBLIOGRAPHY

[1] V. Volterra. Sur l'équilibre des corps élastiques multiplement connexes. *Annales Scientifiques de l'École Normale Supérieure*, Sér. 3(24): 401–517, 1907.

[2] G. I. Taylor. Plastic deformation of crystals. *Proceedings of the Royal Society of London*, 145: 362–404, 1934.

[3] M. Polanyi. Lattice distortion which originates plastic flow. *Zeitschrift fur Physik*, 89(9–10): 660–662, 1934.

[4] E. Orowan. Plasticity of crystals. *Zeitschrift fur Physik*, 89(9–10): 605–659, 1934.

[5] P. B. Hirsch, R. W. Horne, and M. J. Whelan. Direct observations of the arrangement and motion of dislocations in aluminium. *Philosophical Magazine*, 1(7): 677–684, 1956.

[6] D. Hull and D. J. Bacon. *Introduction to Dislocations*. Butterworth-Heinemann, Oxford, 4th edition, 2001.

[7] J. P. Hirth and J. Lothe. *Theory of Dislocations*. Wiley, New York, 1982.

[8] A. Kelly, G. W. Groves, and P. Kidd. *Crystallography and Crystal Defects*. Wiley, Chichester, New York, revised edition, 2000.

[9] C. Kittel. *Introduction to Solid State Physics*. Wiley, New York, 7th edition, 1996.

[10] N. W. Ashcroft and N. D. Mermin. *Solid State Physics*. Saunders College, Philadelphia, 1976.

[11] MD++ simulation package is available from the book web site. http://micro.stanford.edu.

[12] W. H. Beamer and C. R. Maxwell. Physical properties of polonium. ii. X-ray studies and crystal structure. *Journal of Chemical Physics*, 17: 1293–1298, 1949.

[13] F. Prinz and A. S. Argon. Dislocation cell formation during plastic deformation of copper single crystals. *Physica Status Solidi A*, 57(2): 741–753, 1980.

[14] W. Cai, V. V. Bulatov, J. Chang, J. Li, and S. Yip. Dislocation core effects on mobility. In F. R. N. Nabarro and Hirth. J. P., editors, *Dislocations in Solids*, volume 12, pages 1–80. Elsevier, Amsterdam, 2004.

[15] A. P. Sutton. *Electronic Structure of Solids*. Oxford University Press, Oxford, 1996.

[16] R. M. Martin. *Electronic Structure: Basic Theory and Practical Methods*. Cambridge University Press, Cambridge, 2003.

[17] M. W. Finnis. *Interatomic Forces in Condensed Matter*. Oxford University Press, Oxford, 2003.

[18] F. H. Stillinger and T. A. Weber. Computer simulation of local order in condensed phases of silicon. *Physical Review B*, 31(8): 5262–5271, 1985.

[19] M. W. Finnis and J. E. Sinclair. A simple empirical n-body potential for transition metals. *Philosophical Magazine A*, 50(1): 45–55, 1984.

[20] M. P. Allen and D. J. Tildesley. *Computer Simulation of Liquids*. Oxford University Press, Oxford, 1989.

[21] M. C. Payne, M. P. Teter, D. C. Allan, T. A. Arias, and J. D. Joannopoulos. Iterative minimization techniques for ab initio total-energy calculations: molecular dynamics and conjugate gradients. *Reviews of Modern Physics*, 64(4): 1045–1097, 1992.

[22] Vienna Ab initio Simulation Package (VASP). http://cms.mpi.univie.ac.at/vasp.

[23] C. Z. Wang, C. T. Chan, and K. M. Ho. Empirical tight-binding force model for molecular-dynamics simulation of Si. *Physical Review B*, 39(12): 8586–8592, 1989.

[24] F. F. Abraham, R. Walkup, H. J. Gao, M. Duchaineau, T. D. De la Rubia, and M. Seager. Simulating materials failure by using up to one billion atoms and the world's fastest computer: Work-hardening. *Proceedings of the National Academy of Sciences of the United States of America*, 99(9): 5783–5787, 2002.

[25] J. Q. Broughton, F. F. Abraham, N. Bernstein, and E. Kaxiras. Concurrent coupling of length scales: Methodology and application. *Physical Review B*, 60: 2391–2403, 1999.

[26] D. Chandler. *Introduction to Modern Statistical Mechanics*. Oxford University Press, Oxford, 1987.

[27] T. L. Hill. *An Introduction to Statistical Thermodynamics*. Dover, New York, 1986.

[28] W. H. Press, B. P. Flannery, S. A. Teukolsky, and W. T. Vetterling. *Numerical Recipies in C: The Art of Scientific Computing*. Cambridge University Press, Cambridge, 1992.

[29] J. H. Holland. *Adaptation in Natural and Artificial Systems: An Introductory Analysis with Applications to Biology, Control, and Artificial Intelligence*. MIT Press, Cambridge, Mass., 1992.

[30] M. Mitchell. *Introduction to Genetic Algorithms*. MIT Press, Cambridge, Mass., 1998.

[31] R. L. Haupt and S. E. Haupt. *Practical Genetic Algorithms*. Wiley, New York, 2nd edition, 2004.

[32] S. Kirkpatrick, C. D. Gelatt, and M. P. Vecchi. Optimization by simulated annealing. *Science*, 220(4598): 671–680, 1983.

[33] S. Kirkpatrick. Optimization by simulated annealing: quantitative studies. *Journal of Statistical Physics*, 34(5/6): 975–986, 1984.

[34] L. D. Landau and E. M. Lifshitz. *Mechanics*. Pergamon Press, New York, 3rd edition, 1976.

[35] C. W. Gear. *Numerical Initial Value Problems in Ordinary Differential Equations*. Prentice-Hall, Englewood Cliffs, NJ, 1971.

[36] D. Frenkel and B. Smit. *Understanding Molecular Simulation: From Algorithms to Applications*. Academic Press, San Diego, CA, 2002.

[37] Ephemeris generator. http://ssd.jpl.nasa.gov/cgi/eph.

[38] J. R. Barber. *Elasticity*. Kluwer, Dordrecht, 2nd edition, 2002.

[39] J. E. Sinclair, P. C. Gehlen, R. G. Hoagland, and J. P. Hirth. Flexible boundary-conditions and non-linear geometric effects in atomic dislocation modeling. *Journal of Applied Physics*, 49(7): 3890–3897, 1978.

[40] W. Cai, M. de Koning, V. V. Bulatov, and S. Yip, Minimizing Boundary Reflections in Coupled-Domain Simulations, Physical Review Letters, 85, 3213–3216 (2000).

[41] M. de Koning, W. Cai, and V. V. Bulatov. Anomalous dislocation multiplication in fcc metals. *Physical Review Letters*, 91(2): 025503–4, 2003.

[42] S. M. Foiles, M. I. Baskes, and M. S. Daw. Embedded-atom-method functions for the fcc metals Cu, Ag, Au, Ni, Pd, Pt, and their alloys. *Physical Review B*, 33(12): 7983–7991, 1986.

[43] F. H. Stillinger and T. A. Weber. Hidden structure in liquids. *Physical Review A*, 25(2): 978–989, 1982.

[44] C. L. Kelchner, S. J. Plimpton, and J. C. Hamilton. Dislocation nucleation and defect structure during surface indentation. *Physical Review B*, 58(17): 11085–11088, 1998.

[45] M. H. Kalos and P. A. Whitlock. *Monte Carlo methods*. Wiley, New York, 1986.

[46] J. A. Zimmerman, E. B. Webb, J. J. Hoyt, R. E. Jones, P. A. Klein, and D. J. Bammann. Calculation of stress in atomistic simulation. *Modelling and Simulation in Materials Science and Engineering*, 12(4): S319–332, 2004.

[47] A. M. Pendas. Stress, virial, and pressure in the theory of atoms in molecules. *Journal of Chemical Physics*, 117(3): 965–979, 2002.

[48] J. Gao and J. H. Weiner. Excluded-volume effects in rubber elasticity. 1. Virial stress formulation. *Macromolecules*, 20(10): 2520–2525, 1987.

[49] G. Marc and W. G. McMillan. The virial-theorem. *Advances in Chemical Physics*, 58: 209–361, 1985.

[50] M. Parrinello and A. Rahman. Polymorphic transitions in single crystals: a new molecular dynamics method. *Journal of Applied Physics*, 52(12): 7182–7190, 1981.

[51] G. C. Lynch and B. M. Pettitt. Grand canonical ensemble molecular dynamics simulations: reformulation of extended system dynamics approaches. *Journal of Chemical Physics*, 107(20): 8594–8610, 1997.

[52] S. Nosé. A molecular dynamics method for simulations in the canonical ensemble. *Molecular Physics*, 52(2): 255–268, 1984.

[53] W. G. Hoover. Canonical dynamics: equilibrium phase-space distributions. *Physical Review A*, 31(3): 1695–1697, 1985.

[54] B. L. Holian, A. J. De Groot, W. G. Hoover, and C. G. Hoover. Time-reversible equilibrium and nonequilibrium isothermal–isobaric simulations with centered-difference Stoermer algorithms. *Physical Review A (Statistical Physics, Plasmas, Fluids, and Related Interdisciplinary Topics)*, 41(8): 4552–4553, 1990.

[55] J. P. Chang, W. Cai, V. V. Bulatov, and S. Yip. Dislocation motion in bcc metals by molecular dynamics. *Materials Science and Engineering A*, 309: 160–163, 2001.

[56] J. P. Chang, W. Cai, V. V. Bulatov, and S. Yip. Molecular dynamics simulations of motion of edge and screw dislocations in a metal. *Computational Materials Science*, 23(1–4): 111–115, 2002.

[57] H. Gao, Y. Huang, P. Gumbsch, and A. J. Rosakis. On radiation-free transonic motion of cracks and dislocations. *Journal of the Mechanics and Physics of Solids*, 47(9): 1941–1961, 1999.

[58] P. Gumbsch and H. Gao. Dislocations faster than the speed of sound. *Science*, 283(5404): 965–968, 1999.

[59] W. Cai, V. V. Bulatov, J. Chang, J. Li, and S. Yip. Periodic image effects in dislocation modelling. *Philosophical Magazine*, 83(5): 539–567, 2003.

[60] D. J. Bacon, D. M. Barnett, and R. O. Scattergood. Anisotropic continuum theory of lattice defects. *Progress in Materials Science*, 23(2/4): 53–262, 1978.

[61] V. V. Bulatov and W. Cai. Nodal effects in dislocation mobility. *Physical Review Letters*, 89(11): 115501–115504, 2002.

[62] J. F. Justo, M. de Koning, W. Cai, and V. V. Bulatov. Vacancy interaction with dislocations in silicon: the shuffle–glide competition. *Physical Review Letters*, 84(10): 2172–2175, 2000.

[63] M. de Koning, W. Cai, A. Antonelli, and S. Yip. Efficient free-energy calculations by the simulation of nonequilibrium processes. *Computing in Science and Engineering*, 2(3): 88–96, 2000.

[64] G. J. Martyna, M. L. Klein, and M. Tuckerman. Nosé–Hoover chains: the canonical ensemble via continuous dynamics. *Journal of Chemical Physics*, 97(4): 2635–2643, 1992.

[65] P. Hänggi, P. Talkner, and M. Borkovec. Reaction-rate theory: fifty years after Kramers. *Reviews of Modern Physics*, 62(2): 251–341, 1990.

[66] D. Chandler. Statistical mechanics of isomerisation dynamics in liquids and the transition state approximation. *Journal of Chemical Physics*, 68(6): 2959–2970, 1978.

[67] C. Dellago, P. G. Bolhuis, F. S. Csajka, and D. Chandler. Transition path sampling and the calculation of rate constants. *Journal of Chemical Physics*, 108(5): 1964–1977, 1998.

[68] G. Henkelman and H. Jonsson. A dimer method for finding saddle points on high dimensional potential surfaces using only first derivatives. *Journal of Chemical Physics*, 111(15): 7010–7022, 1999.

[69] G. Henkelman, B. P. Uberuaga, and H. Jonsson. A climbing image nudged elastic band method for finding saddle points and minimum energy paths. *Journal of Chemical Physics*, 113(22): 9901–9904, 2000.

[70] G. Henkelman and H. Jonsson. Improved tangent estimate in the nudged elastic band method for finding minimum energy paths and saddle points. *Journal of Chemical Physics*, 113(22): 9978–9985, 2000.

[71] D. J. Wales, J. P. K. Doye, M. A. Miller, P. N. Mortenson, and T. R. Walsh. Energy landscapes: from clusters to biomolecules. *Advances in Chemical Physics*, 115: 1–111, 2000.

[72] W. Cai, M. H. Kalos, M. De Koning, and V. V. Bulatov. Importance sampling of rare transition events in Markov processes. *Physical Review E*, 66(4): 1–10, 2002.

[73] W. E, W. Q. Ren, and E. Vanden-Eijnden. String method for the study of rare events. *Physical Review B*, 66(5): 052301–052304, 2002.

[74] W. E, W. Q. Ren, and E. Vanden-Eijnden. Finite temperature string method for the study of rare events. *Journal of Physical Chemistry B*, 109(14): 6688–6693, 2005.

[75] W. E, W. Q. Ren, and E. Vanden-Eijnden. Transition pathways in complex systems: Reaction coordinates, isocommittor surfaces, and transition tubes. *Chemical Physics Letters*, 413(1–3): 242–247, 2005.

[76] W. Ren, E. Vanden-Eijnden, P. Maragakis, and W. E. Transition pathways in complex systems: Application of the finite-temperature string method to the alanine dipeptide. *Journal of Chemical Physics*, 123(13): 134109, 2005.

[77] V. V. Bulatov, S. Yip, and A. S. Argon. Atomic modes of dislocation mobility in silicon. *Philosophical Magazine A*, 72(2): 453–496, 1995.

[78] R. Elber and M. Karplus. A method for determining reaction paths in large molecules: application to myoglobin. *Chemical Physics Letters*, 139(5): 375–380, 1987.

[79] M. Nastar, V. V. Bulatov, and S. Yip. Saddle-point configurations for self-interstitial migration in silicon. *Physical Review B*, 53(20): 13521–13527, 1996.

[80] V. Bulatov, M. Nastar, J. Justo, and S. Yip. Atomistic modeling of crystal-defect mobility and interactions. *Nuclear Instruments and Methods in Physics Research, Section B*, 121(1/4): 251–256, 1997.

[81] C. Dellago, P. G. Bolhuis, and P. L. Geissler. Transition path sampling. *Advances in Chemical Physics*, 123: 1–78, 2002.

[82] J. E. Sinclair and R. Fletcher. A new method of saddle-point location for the calculation of defect migration energies. *Journal of Physics C (Solid State Physics)*, 7(5): 864–870, 1974.

[83] M. R. Sorensen and A. F. Voter. Temperature-accelerated dynamics for simulation of infrequent events. *Journal of Chemical Physics*, 112(21): 9599–9606, 2000.

[84] A. F. Voter, F. Montalenti, and T. C. Germann. Extending the time scale in atomistic simulation of materials. *Annual Review of Materials Research*, 32: 321–346, 2002.

[85] F. Montalenti and A. F. Voter. Applying accelerated molecular dynamics to crystal growth. *Physica Status Solidi B*, 226(1): 21–27, 2001.

[86] R. E Peierls. The size of a dislocation. *Proc. Phys. Soc.*, 52: 34–37, 1940.

[87] F. R. N. Nabarro. Dislocations in a simple cubic lattice. *Proc. Phys. Soc.*, 59: 256–272, 1947.

[88] R. Miller, R. Phillips, G. Beltz, and M. Ortiz. A non-local formulation of the Peierls dislocation model. *Journal of the Mechanics and Physics of Solids*, 46(10): 1845–1867, 1998.

[89] V. V. Bulatov and E. Kaxiras. Semidiscrete variational Peierls framework for dislocation core properties. *Physical Review Letters*, 78(22): 4221–4224, 1997.

[90] G. Xu. A variational boundary integral method for the analysis of three-dimensional cracks of arbitrary geometry in anisotropic elastic solids. *Transactions of the ASME. Journal of Applied Mechanics*, 67(2): 403–408, 2000.

[91] G. Xu, A. S. Argon, and M. Oritz. Critical configurations for dislocation nucleation from crack tips. *Philosophical Magazine A*, 75(2): 341–367, 1997.

[92] G. Xu, A. S. Argon, and M. Ortiz. Nucleation of dislocations from crack tips under mixed modes of loading: implications for brittle against ductile behaviour of crystals. *Philosophical Magazine A*, 72(2): 415–451, 1995.

[93] A. H. W. Ngan. A new model for dislocation kink-pair activation at low temperatures based on the Peierls–Nabarro concept. *Philosophical Magazine A (Physics of Condensed Matter: Structure, Defects and Mechanical Properties)*, 79(7): 1697–1720, 1999.

[94] P. Rosakis. Supersonic dislocation kinetics from an augmented Peierls model. *Physical Review Letters*, 86(1): 95–98, 2001.

[95] H. C. Huang, G. H. Gilmer, and T. D. de la Rubia. An atomistic simulator for thin film deposition in three dimensions. *Journal of Applied Physics*, 84(7): 3636–3649, 1998.

[96] M. Rak, M. Izdebski, and A. Brozi. Kinetic Monte Carlo study of crystal growth from solution. *Computer Physics Communications*, 138(3): 250–263, 2001.

[97] A. C. Levi and M. Kotrla. Theory and simulation of crystal growth. *Journal of Physics: Condensed Matter*, 9(2): 299–344, 1997.

[98] W. Cai, V. V. Bulatov, S. Yip, and A. S. Argon. Kinetic Monte Carlo modeling of dislocation motion in bcc metals. *Materials Science and Engineering A-Structural Materials Properties Microstructure and Processing*, 309: 270–273, 2001.

[99] A. B. Bortz, M. H. Kalos, and J. L. Lebowitz. A new algorithm for Monte Carlo simulation of ising spin systems. *Journal of Computational Physics*, 17(1): 10–18, 1975.

[100] V. V. Bulatov, J. F. Justo, W. Cai, S. Yip, A. S. Argon, T. Lenosky, M. de Koning, and T. D. de la Rubia. Parameter-free modelling of dislocation motion: the case of silicon. *Philosophical Magazine A*, 81(5): 1257–1281, 2001.

[101] W. Cai, A. Arsenlis, C. R. Weinberger, and V. V. Bulatov. A non-singular continuum theory of dislocations. *Journal of the Mechanics and Physics of Solids*, 54(3): 561–587, 2006.

[102] J. Spence and C. Koch. Experimental evidence for dislocation core structures in silicon. *Scripta Materialia*, 45(11): 1273–1278, 2001.

[103] S. Redner. *A Guide to First-Passage Processes*. Cambridge University Press, Cambridge, 2001.

[104] W. Cai, V. V. Bulatov, and S. Yip. Kinetic Monte Carlo method for dislocation glide in silicon. *Journal of Computer-Aided Materials Design*, 6(2): 175–183, 1999.

[105] C. S. Deo and D. J. Srolovitz. First passage time Markov chain analysis of rare events for kinetic Monte Carlo: double kink nucleation during dislocation glide. *Modelling and Simulation in Materials Science and Engineering*, 10(5): 581–596, 2002.

[106] D. T. Gillespie. *Markov Processes: An Introduction for Physical Scientists*. Academic Press, San Diego, CA, 1992.

[107] M. C. Miguel, A. Vespignani, S. Zapperi, J. Weiss, and J. R. Grasso. Intermittent dislocation flow in viscoplastic deformation. *Nature*, 410(6829): 667–671, 2001.

[108] B. Devincre and L. P. Kubin. Mesoscopic simulations of dislocations and plasticity. *Materials Science and Engineering A*, 234/236: 8–14, 1997.

[109] K. W. Schwarz. Simulation of dislocations on the mesoscopic scale. I. Methods and examples. *Journal of Applied Physics*, 85(1): 108–119, 1999.

[110] K. W. Schwarz. Simulation of dislocations on the mesoscopic scale. II. Application to strained-layer relaxation. *Journal of Applied Physics*, 85(1): 120–129, 1999.

[111] N. M. Ghoniem and L. Z. Sun. Fast-sum method for the elastic field of three-dimensional dislocation ensembles. *Physical Review B*, 60(1): 128–140, 1999.

[112] R. V. Kukta. *Observations on the Kinetics of Relaxation in Epitaxial Films Grown on Conventional and Compliant Substrates: A Continuum Simulation of Dislocation Glide Near an Interface*. Ph.D. Thesis, Brown University, 1998.

[113] W. Cai and V. V. Bulatov. Mobility laws in dislocation dynamics simulations. *Materials Science and Engineering A*, 387–389(1–2 Spec. Iss.): 277–281, 2004.

[114] D. Rodney and R. Phillips. Structure and strength of dislocation junctions: an atomic level analysis. *Physical Review Letters*, 82(8): 1704–1707, 1999.

[115] V. B. Shenoy, R. V. Kukta, and R. Phillips. Mesoscopic analysis of structure and strength of dislocation junctions in fcc metals. *Physical Review Letters*, 84(7): 1491–1494, 2000.

[116] L. Greengard. *The Rapid Evaluation of Potential Fields in Particle Systems*. MIT Press, Cambridge, Mass., 1987.

[117] Z. Q. Wang, N. Ghoniem, and R. LeSar. Multipole representation of the elastic field of dislocation ensembles. *Physical Review B*, 69(17): 174102, 2004.

[118] S. Goedecker. *Performance Optimization of Numerically Intensive Codes*. SIAM, Philadelphia, PA, 2001.

[119] W. Cai, V. V. Bulatov, T. Pierce, M. Hiratani, M. Rhee, M. Bartelt, and M. Tang. Massively-parallel dislocation dynamics simulations. In *Solid Mechanics and Its Applications*, volume 115, page 1. Kluwer, Dordrecht, 2004.

[120] V. Bulatov, W. Cai, J. Fier, M. Hiratani, G. Hommes, T. Pierce, M. Tang, M. Rhee, K. Yates, and A. Arsenlis. Scalable line dynamics in Para Dis. *IEEE/ACM SC2004 Conference, Proceedings*, http://www.sc-conference.org/sc2004/schedule/pdfs/ pap206.pdf, pages 367–378, 2004.

[121] K. W. Morton and D. F. Mayers. *Numerical Solution of Partial Differential Equations*. Cambridge University Press, Cambridge, 1994.

[122] T. J. R. Hughes. *The Finite Element Method: Linear Static and Dynamic Finite Element Analysis*. Dover, Mieola, NY, 2000.

[123] E. Brigham. *Fast Fourier Transform and Its Applications*. Prentice-Hall, Englewood Cliff, NJ, 1988.

[124] J. D. Eshelby. Elastic inclusions and inhomogeneities. *Progress in Solid Mechanics*, 2: 89–140, 1961.

[125] Y. U. Wang, Y. M. Jin, A. M. Cuitino, and A. G. Khachaturyan. Nanoscale phase field microelasticity theory of dislocations: model and 3d simulations. *Acta Materialia*, 49(10): 1847–1857, 2001.

[126] T. Mura. *Micromechanics of Defects in Solids*. Kluwer, Dordrecht, 1987.

[127] V. V. Bulatov. Current developments and trends in dislocation dynamics. *Journal of Computer-Aided Materials Design*, 9(2): 133–144, 2002.

[128] K. Itakura, A. Uno, M. Yokokawa, M. Saito, T. Ishihara, and Y. Kaneda. Performance tuning of a CFD code on the Earth Simulator. *NEC Research and Development*, 44(1): 115–120, 2003.

[129] D. Rodney, Y. Le Bouar, and A. Finel. Phase field methods and dislocations. *Acta Materialia*, 51(1): 17–30, 2003.

[130] N. Provatas, N. Goldenfeld, and J. Dantzig. Adaptive mesh refinement computation of solidification microstructures using dynamic data structures. *Journal of Computational Physics*, 148(1): 265–290, 1999.

[131] S. D. Bond, B. J. Leimkuhler, and B. B. Laird. The Nosé-Poincaré method for constant temperature molecular dynamics. *J. Comp. Phys.* 151: 114–134, 1999.

[132] F. Ercolessi and J. B. Adams interatomic potentials from first-principles calculations: the force-matching method. *Europhysics Letters*, 26: 583–588, 1994.

[133] M. Challacombe, C. White, and M. Head-Gordon. Periodic boundary conditions and the fast multipole method. *Journal of Chemical Physics*. 107: 10131–10140, 1997.

[134] W. Cai, V. V. Bulatov, J. F. Justo, S. Jip, and A. S. Argon. Intrinsic mobility of a dissociated dislocation in silicon. *Physical Review Letters*, 84: 3346–3349, 2000.

SUBJECT INDEX